T0189297

Lecture Notes of the Institute for Computer Sciences, Social Informatics and Telecommunications Engineering 553

The LNICST series publishes ICST's conferences, symposia and workshops.
LNICST reports state-of-the-art results in areas related to the scope of the Institute.
The type of material published includes

- Proceedings (published in time for the respective event)
- Other edited monographs (such as project reports or invited volumes)

LNICST topics span the following areas:

- General Computer Science
- E-Economy
- E-Medicine
- Knowledge Management
- Multimedia
- Operations, Management and Policy
- Social Informatics
- Systems

Jingchao Li · Bin Zhang · Yulong Ying

Editors

6GN for Future Wireless Networks

6th EAI International Conference, 6GN 2023
Shanghai, China, October 7–8, 2023
Proceedings, Part I

 Springer

Editors
Jingchao Li ⓘ
Shanghai Dianji University
Shanghai, China

Bin Zhang
Kanagawa University
Yokohama, Japan

Yulong Ying ⓘ
Shanghai University of Electric Power
Yangpu, China

ISSN 1867-8211 ISSN 1867-822X (electronic)
Lecture Notes of the Institute for Computer Sciences, Social Informatics
and Telecommunications Engineering
ISBN 978-3-031-53400-3 ISBN 978-3-031-53401-0 (eBook)
https://doi.org/10.1007/978-3-031-53401-0

This Springer imprint is published by the registered company Springer Nature Switzerland AG
The registered company address is: Gewerbestrasse 11, 6330 Cham, Switzerland

Paper in this product is recyclable.

Preface

We are delighted to introduce the proceedings of the sixth edition of the European Alliance for Innovation (EAI) International Conference on 6G for Future Wireless Networks (6GN 2023). This conference brought together researchers, developers and practitioners around the world who are leveraging and developing smart grid technology for a smarter and more resilient grid. The theme of EAI 6GN 2023 was "Smart Grid Inspired Future Technologies: A smarter and more resilient grid".

The technical program of EAI 6GN 2023 consisted of 60 full papers, including 3 invited papers in oral presentation sessions at the main conference tracks. The conference tracks were: Track 1 - 6G Communications for UAVs; Track 2 - Signal Gene Characteristics; Track 3 - Signal Gene Characteristics. Aside from the high-quality technical paper presentations, the technical program also featured three invited keynote speeches. The first invited talk was presented by Cesar Briso from Technical University of Madrid, Spain. The title of the talk was A Connected Sky: 6G Communications for UAVs. The second invited talk was presented by Peng Chen from Southeast University, China. The title of the talk was Efficient DOA Estimation Method for Reconfigurable Intelligent Surfaces Aided UAV Swarm. The third invited talk was presented by Jingchao Li from Shanghai Dianji University, China. The title of the talk was Physical layer authentication method for the Internet of Things based on radio frequency signal gene.

Coordination with the steering chairs, Imrich Chlamtac, Victor C.M. Leung and Kun Yang, was essential for the success of the conference. We sincerely appreciate their constant support and guidance. It was also a great pleasure to work with such an excellent organizing committee team for their hard work in organizing and supporting the conference. In particular, the Technical Program Committee completed the peer-review process of technical papers and made a high-quality technical program. We are also grateful to all the authors who submitted their papers to the EAI 6GN 2023 conference and workshops.

We strongly believe that EAI 6GN 2023 provided a good forum for all researchers, developers and practitioners to discuss all science and technology aspects that are relevant to wireless networks. We also expect that 6GN conferences will be as successful and stimulating as indicated by the contributions presented in this volume.

Yulong Ying

Organization

Organizing Committee

General Chairs

Junjie Yang	Shanghai Dianji University, China
Yulong Ying	Shanghai University of Electric Power

General Co-chairs

Cheng Cai	Shanghai Dianji University, China
Yulong Ying	Shanghai University of Electric Power, China

TPC Chair and Co-chairs

Yudong Zhang	University of Leicester, UK
Peng Chen	Southeast University, China
Jingchao Li	Shanghai Dianji University, China
Wanying Shi	Portland State University, USA
Pengpeng Zhang	Shanghai Dianji University, China
Na Wu	Nanjing University of Posts and Telecommunications, China

Sponsorship and Exhibit Chair

Siyuan Hao	Qingdao University of Technology, China

Local Chair

Ming Li	Shanghai Dianji University, China

Workshops Chairs

Haijun Wang	Shanghai Dianji University, China
Zhimin Chen	Shanghai Dianji University, China

Publicity and Social Media Chair

Tingting Sui   Shanghai Dianji University, China

Publications Chairs

Bin Zhang   Kanagawa University, Japan
Ao Li   Harbin University of Science and Technology,
                                      China

Web Chair

Xiaoyong Song   Shanghai Dianji University, China

Posters and PhD Track Chair

Yang Xu   Guizhou University, China

Panels Chair

Shuihua Wang   University of Leicester, UK

Demos Chair

Yue Zeng   Jinling Institute of Technology, China

Tutorials Chair

Pengyi Jia   Western University, Canada

Technical Program Committee

Jingchao Li   Shanghai Dianji University, China
Bin Zhang   Kanagawa University, Japan
Peng Chen   Southeast University, China

Contents – Part I

Artificial Intelligent Techniques for 6G Networks

Power and Energy Systems

Contents – Part II

**Communications Systems and Networking & Control and Automation
Systems**

Computer Systems and Applications

Intelligent Systems

3D Spatial Information Reflected 2D Mapping for a Guide Dog Robot

Toya Aoki$^{(\boxtimes)}$, Bin Zhang, and Hun-ok Lim

Kanagawa University, Yokohama 2218686, Japan
`r202370095nt@jindai.jp`

Abstract. This paper proposes a risk map generation method that considers the occupied space of objects and their characteristics. The objective is to guide visually impaired individuals safely to their intended destinations using a guide dog robot that can assist them in walking. The number of guide dogs in active service in Japan has been continuously declining, which has prompted the development of guide dog robots. The robot utilized in this study employs Mecanum wheels, the RoboSense RS-LiDAR-16 sensor, and Intel's RealSense Depth Camera D435 to scan the surrounding environment and measure distance up to 150 m. To prevent visually impaired individuals from entering spaces potentially occupied by objects, the three-dimensional spatial information of the objects is projected onto a two-dimensional map, and object recognition is performed to project the potential risks of objects onto the map. The generated risk map is used to path planning that considers the risk levels established according to object properties. The effectiveness is proven by experiments of guiding visually impaired individuals to destinations while avoiding potential occupied spaces.

Keywords: Guide dog robot · 2D mapping · Risk map

1 Introduction

According to the Japan Council of Social Welfare Facilities for the Blind, a Social Welfare Corporation, the number of guide dogs in active service has been continuously declining for 14 years, dropping from 1,070 dogs in 2009 to 848 dogs in 2022. However, it is estimated that among the 310,000 visually impaired individuals in Japan, around 3,000 people desire to live with guide dogs. To tackle this issue, extensive research and development into guide dog robots, which can assist visually impaired individuals in walking, is currently being conducted. Such as traffic light and moving object detection [1], voice expression system of visual environment [2], flying guide dog robot [3], obstacle avoidance system [4] and robotic navigation guide dog [5]. It is widely known that white canes and guide dogs, which are typically used as walking aids for the visually impaired, struggle to accurately perceive and identify objects and face difficulties in avoiding potential risks [6]. Furthermore, previous guide dog robots have been reported to occasionally invade areas potentially occupied by objects though the places are empty,

© ICST Institute for Computer Sciences, Social Informatics and Telecommunications Engineering 2024
Published by Springer Nature Switzerland AG 2024. All Rights Reserved
J. Li et al. (Eds.): 6GN 2023, LNICST 553, pp. 3–9, 2024.
https://doi.org/10.1007/978-3-031-53401-0_1

such as underneath desks or around doors, due to their inability to consider the characteristics of the objects [7]. Given that the primary objective of the guide dog robot is to ensure the safety of visually impaired individuals, it is essential to incorporate avoidance and guidance mechanisms that are tailored to the properties of objects. Therefore, in this study, we propose a risk map generation method that considers the potential occupied space that reflects the characteristics of identified objects. By applying the generated map to guide visually impaired individuals safely to their intended destinations, we conducted practical machine experiments to validate the effectiveness of our proposed system.

2   Guide Dog Robot

The robotic guide dog utilized in the present study is illustrated in Fig. 1, employing Vstone Corporation's Mecanumrover Ver2.1 as the mobile platform. This mobile robot is endowed with Mecanum wheels that enable it to move omnidirectionally. The upper portion of the robotic guide dog is fitted with RoboSense RS-LiDAR-16 sensor, which continuously scans the surrounding environment utilizing 16 lasers in a 360° pattern, with a maximal range of 150 m. Furthermore, the robot is mounted with an RGB-D camera, namely Intel's RealSense Depth Camera D435, which enables it to measure color image data, as well as distance information up to 10 m.

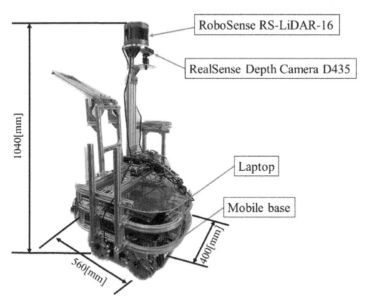

Fig. 1. Guide dog robot.

3   Risk Map Generation Considering Object Characteristics

To avert the ingress of visually impaired individuals into spaces potentially preoccupied by objects, the three-dimensional spatial information of the objects is projected onto a two-dimensional map. Additionally, object recognition is performed to project the associated risks of objects onto the map. The system flow is depicted in Fig. 2.

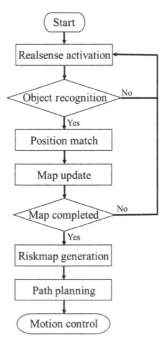

Fig. 2. System flowchart

3.1   Generation of Two-Dimensional Map

To facilitate indoor navigation of the guide dog robot, the RS-LiDAR-16 sensor is used to acquire a three-dimensional point cloud of obstacles based on distance information (refer to Fig. 3(a)). The conversion of the obtained three-dimensional point cloud to a two-dimensional laser within a predetermined height range is shown in Fig. 3(b). Two-dimensional distance information projected onto the floor plane is utilized to generate and update the map. By converting the three-dimensional point cloud to a two-dimensional laser, the reflection of three-dimensional spatial information onto a two-dimensional map is realized. The Cartographer method [8] is used, which accomplishes simultaneous localization and mapping (SLAM), to generate the environment map and to estimate the robot's position. This method can generate maps with high accuracy even in intricate indoor environments.

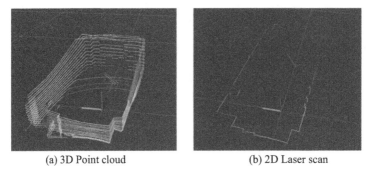

(a) 3D Point cloud   (b) 2D Laser scan

Fig. 3. Converting 3D information to 2D

3.2 Object Recognition Using YOLACT

We capture color images and depth information using the RealSense Depth Camera D435. Utilizing the color images, we engage in deep learning-based object recognition and segment the spatial shape information of the objects. To achieve segmentation, we employ YOLACT [9], which utilizes Instance Segmentation. By partitioning the identified objects' region in the image into pixels, Instance Segmentation is adept at accurately extracting objects' shapes [10] (refer to Fig. 4). We extract the objects' three-dimensional point cloud information from YOLACT. Subsequently, we map the obstacle information from the three-dimensional point cloud of the recognized objects onto the respective region in the map.

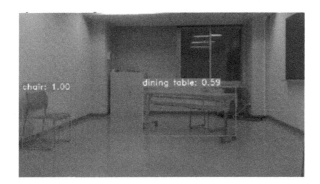

Fig. 4. Object recognition

3.3 Path Planning Based on Risk Maps

A risk map is a cartographic representation of risk levels in an area, reflecting risks in the vicinity of obstacles. Risk levels are established according to object properties. The moving path is generated based on the risk map. The changing risk level in the image and the path planning result are shown in Fig. 5. It is observed that the risks around objects are shown in the map and the generated path avoids these places.

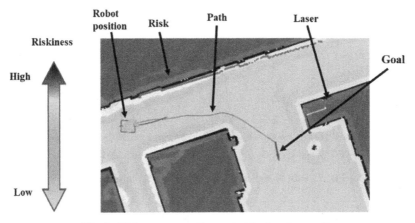

Fig. 5. Risk map generation and path planning result.

4 Experiment

The experimental scene is shown in Fig. 6(a). By applying the proposed method, the robot was capable of moving under the experimental environment and generating a two-dimensional map that considers the three-dimensional spatial information of the objects. Figure 6(b) shows a two-dimensional map that reflects the three-dimensional shapes of objects in the environment.

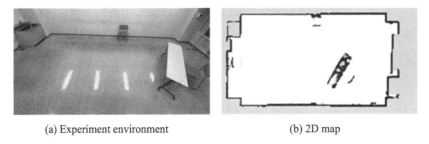

(a) Experiment environment   (b) 2D map

Fig. 6. Environmental scene and generated map.

Potential risks are generated in the map and a path is generated based on the risk map (refer to Fig. 7(a)). The robot was able to avoid risk areas and generate a secure route to the destination by avoiding entering areas with potential risks, which are generated from recognized the 3D objects such as tables and chairs. According to the property of the objects, the risk areas around desks are bigger than that of chairs. The risk areas are not only the areas under the objects but also the places around them, considering the concept of personal spaces for objects. Figure 7(b) shows the generated path without using 3D information, leading to collisions when passing through the empty area under a table.

The experimental scene for generating maps for objects only is shown in Fig. 8(a). Obstacles that cannot be recognized are ignored in the object map, and only the objects

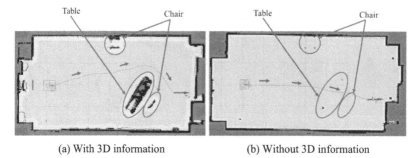

(a) With 3D information   (b) Without 3D information

Fig. 7. Path on risk map with or without 3D information.

are projected onto the map, as shown in Fig. 8(b). The risk map that considers objects combines a two-dimensional map with normal risks and an object map with special risks. Comparing the path generated from the risk map without object recognition (refer to Fig. 9(a)) with the path generated from the risk map utilizing the object map (refer to Fig. 9(b)), it is confirmed that the robot moves along a safe path, avoiding the objects without crossing between them.

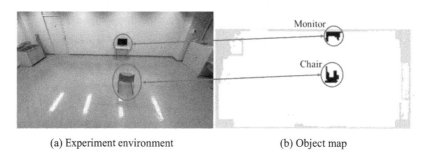

(a) Experiment environment   (b) Object map

Fig. 8. Map projected objects.

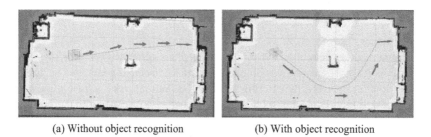

(a) Without object recognition   (b) With object recognition

Fig. 9. Path on risk map.

5   Conclusion

In this study, we proposed a method for generating risk maps that consider the characteristics of recognized objects. By projecting the three-dimensional spatial information of objects and realizing object recognition, the 2D risk map is generated. Real-world experiments demonstrated that path planning with avoidance of risk areas is achievable by generating maps that consider three-dimensional shapes. Additionally, more safe paths are generated to reach the destination. The effectiveness of the proposed method is confirmed.

References

1. Chen, Q., Chen, Y., Zhu, J., De Luca, G., Zhang, M., Guo, Y.: Traffic light and moving object detection for a guide-dog robot. J. Eng. **13**, 675–678 (2020)
2. Ichikawa, R., Zhang, B., Lim, H.O.: Voice expression system of visual environment for a guide dog robot. In: 2022 8th International Symposium on System Security, Safety, and Reliability (ISSSR), pp. 191–192. IEEE (2022)
3. Tan, H., et al.: Flying guide dog: Walkable path discovery for the visually impaired utilizing drones and transformer-based semantic segmentation. In: 2021 IEEE International Conference on Robotics and Biomimetics (ROBIO), pp. 1123–1128. IEEE (2021)
4. Shoval, S., Borenstein, J., Koren, Y.: Mobile robot obstacle avoidance in a computerized travel aid for the blind. In: Proceedings of the 1994 IEEE International Conference on Robotics and Automation, pp. 2023–2028. IEEE (1994)
5. Wang, L., Zhao, J., Zhang, L.: NavDog: robotic navigation guide dog via model predictive control and human-robot modeling. In: Proceedings of the 36th Annual ACM Symposium on Applied Computing, pp. 815–818 (2021)
6. Mitsou, N.C., Tzafestas, C.S.: Temporal occupancy grid for mobile robot dynamic environment mapping. In: 2007 Mediterranean Conference on Control & Automation, pp. 1–8. IEEE (2007)
7. Guerrero, L.A., Vasquez, F., Ochoa, S.F.: An indoor navigation system for the visually impaired. Sensors **12**(6), 8236–8258 (2012)
8. Hess, W., Kohler, D., Rapp, H., Andor, D.: Real-time loop closure in 2D LIDAR SLAM. In: 2016 IEEE International Conference on Robotics and Automation (ICRA), pp. 1271–1278. IEEE (2016)
9. Bolya, D., Zhou, C., Xiao, F., Lee, Y.J.: YOLACT: real-time instance segmentation. In: Proceedings of the IEEE/CVF International Conference on Computer Vision, pp. 9157–9166 (2019)
10. He, K., Gkioxari, G., Dollár, P., Girshick, R.: Mask R-CNN. In: Proceedings of the IEEE International Conference on Computer Vision, pp. 2961–2969 (2017)

Development of an Image Generation System for Expressing Auditory Environment to Hearing-Impaired People

Bin Zhang$^{(\boxtimes)}$, Toya Aoki, Kazuma Matsuura, and Hun-ok Lim

Kanagawa University, Yokohama 2218686, Japan
zhangbin@kanagawa-u.ac.jp

Abstract. This paper presents the development of an assistive system for hearing impaired people, aiming to convey auditory information through visual images for effective communication. The system integrates speech recognition, morphological analysis, and image generation components, implemented on an assistive robot platform. Experiments were conducted to validate the system's effectiveness, including speech recognition, morphological analysis, and image generation. The results demonstrate the system's ability to accurately convey speech content and its surrounding visual information. This research contributes to the development of assistive technologies for individuals with hearing impairments, enhancing their communication abilities and improving their daily lives.

Keywords: Assisting Robot · Auditory Environment Recognition · Image Generation

1 Introduction

In Japan, there are 358,800 individuals with hearing and speech impairments, who rely on sign language and written communication as means of conversation [1]. However, the number of proficient sign language users is limited, making it challenging for individuals with hearing impairments to effectively communicate with others in their daily lives [2, 3]. To address this issue, the development of assistive systems for individuals with hearing impairments, utilizing artificial intelligence technology, is being conducted both domestically and internationally [4–7]. Conventional conversation support systems often rely on presenting the spoken content in text, but they often fail to accurately convey information to individuals with hearing impairments.

Therefore, this study aims to develop an assistive system for individuals with hearing impairments that can effectively present the content of speech and its surrounding visual information in a user-friendly manner. The system is composed of speech recognition, morphological analysis, and image generation components. The proposed system will be integrated into a hearing impairment support robot, and speech experiments will be conducted to validate the effectiveness of the proposed system.

© ICST Institute for Computer Sciences, Social Informatics and Telecommunications Engineering 2024
Published by Springer Nature Switzerland AG 2024. All Rights Reserved
J. Li et al. (Eds.): 6GN 2023, LNICST 553, pp. 10–16, 2024.
https://doi.org/10.1007/978-3-031-53401-0_2

2   Assistive Robot for Individuals with Hearing Impairments

The assistive robot employed in this study to aid individuals with hearing impairments is depicted in Fig. 1. The locomotion mechanism of this robot employs the Vstone Co., Ltd.'s Mecanum Rover Ver2.1. With the integration of four Mecanum wheels, this robotic platform enables seamless omnidirectional mobility. The robot is equipped with the Sanwa Supply USB Microphone MM-MCUSB25, featuring a flexible arm, affixed at an approximate distance of 1060 mm from the lower surface. This microphone, employing a unidirectional configuration, effectively mitigates the intrusion of ambient noise. Moreover, situated at an approximate height of 940 mm from the base, the monitor facilitates the comprehension and visual representation of audio data captured by the microphone, thereby providing auditory impaired individuals with accessible information.

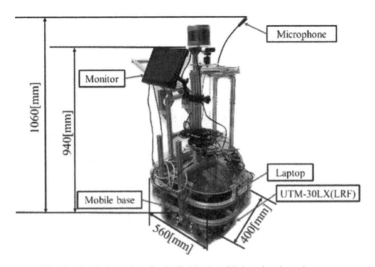

Fig. 1. Assistive robot for individuals with hearing impairments

3   Speech Recognition and Image Generation System

We have devised an assistive system for individuals afflicted by auditory impairments, facilitating seamless communication by conveying the auditory milieu through visual imagery. The schematic diagram outlining the operational flow of this supportive framework is illustrated in Fig. 2. Upon activation via predefined keywords, this system adeptly discerns the utterances of individuals with intact hearing, selectively extracting salient lexicons indicative of the prevailing context. Subsequently, these lexical units are transmuted into visually apprehensible depictions. The resulting synthesized images, in conjunction with accompanying textual annotations and auditory cues, are presented synergistically to foster and streamline dialogues.

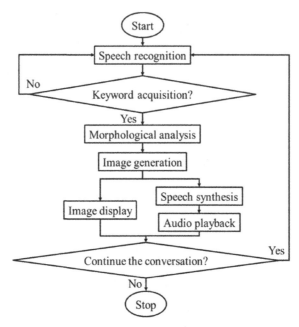

Fig. 2. System flowchart

3.1 Speech Recognition

We conduct speech recognition utilizing the SpeechRecognition module. Through the predefinition of keywords, the system discerns the commencement of communication upon their recognition, followed by the conversion of spoken utterances into textual form. The system remains quiescent until the keywords are identified, cyclically iterating the speech recognition process.

3.2 Morphological Analysis: A Title for an English Research Paper

Utilizing the NLTK morphological analysis library, we engage in morphological analysis of the recognized speech content, aiming to tokenize the text into discrete words and assign part-of-speech tags to each word.

3.3 Word Extraction

When transmitting the auditory environment to the user, it becomes imperative to discern the appropriateness of visualizing the spoken content. For instance, words like "Hello" or "Today" denoting salutations or temporal references pose challenges for visualization. Consequently, in this research, we exclusively extract nouns, such as "Dog," and parts of speech related to verbs, such as "Running dog," for output.

3.4   Image Generation Reflecting Audio Information

To generate images from the extracted text, Diffusion Models (DM) are employed and re-trained by using the open data from the internet. DM serves as a model that progressively eliminates noise from noisy images to produce pristine images. The procedural sequence of this operation is depicted in Fig. 3 [8]. Given that the noisy image is conditioned by the input text, the resulting image manifests attributes harmonious with the text. The encoded feature map undergoes a linear transformation utilizing the weight matrix w of the network. This transformation entails the utilization of the query matrix Q, the key matrix K obtained from the Text Encoder, and the value matrix V, leading to the derivation of a probability map aligned with the text, as delineated in the subsequent equation [9].

$$Attention(Q, K, V) = softmax\left(\frac{QK^T}{\sqrt{d_k}}\right)V \qquad (1)$$

The conditional probability $p_\theta(x_{t-1}|x_t)$ during the process of image generation is estimated and determined by inferring the mean $\mu_\theta(x_t, t)$ and variance $\Sigma_\theta(x_t, t)$ in order to conform to a Gaussian distribution.

$$p_\theta(x_{t-1}|x_t) = N(x_{t-1}; \mu_\theta(x_t, t), \Sigma_\theta(x_t, t)) \qquad (2)$$

Upon simplifying Eq. (2), it can be expressed as follows.

$$p_\theta(x_{t-1}|x_t) = N(x_{t-1}; \mu_\theta(x_t, t), \sigma_t^2 I) \qquad (3)$$

During the process of image generation, noise elimination is executed at each time step t based on Eq. (3). Commencing from the complete noise X_T at time T and traversing backwards from time t to $t - 1$, the Gaussian noise is represented as Z, leading to the formulation presented below.

$$x_{t-1} = \mu_\theta(x_t, t) + \sigma_t Z \qquad (4)$$

By iteratively applying Eq. (4) starting from the noisy image X_T and progressing towards the actual data X_0, the ultimate generated data is eventually attained.

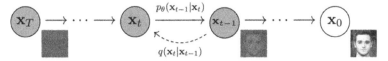

Fig. 3. Process flow [8]

In the event of a sustained conversation, the extracted words are preserved within a list for utilization in the process of image generation.

3.5   Speech Synthesis

Utilizing the outcomes of speech recognition, we employ the "gTTs" speech synthesis library to produce an MP3 audio file derived from the given text. Subsequently, the audio file is played through the "playsound" audio playback library, serving as a means to communicate the recognized text to individuals possessing typical auditory abilities.

4   Experiment

Experiments are conducted in indoor environment, where the robot help the user to communicate with normal people. The robot moves together the hearing-impaired user and would start the translating function when it is called by name (keyword). The robot would cease its movement and initiate the dialogue. The robot's integrated speech recognition system duly acknowledged and comprehended the user's speech, and the resultant dialogue content was employed for image generation via DM, subsequently exhibited on a monitor. To validate the generated content with the speaker, speech synthesis was employed to audibly articulate the content. Throughout the experiment, three distinct conversational exchanges comprising responses were carried out, with the specific details presented in Table.1.

Table 1. Conversation content

Q. (User)	Where have you been?
A. (Interlocutor)	I went to see Mt. Fuji
Q. (User)	What did you see?
A. (Interlocutor)	I could also see the five-storied pagoda
Q. (User)	The view looks good
A. (Interlocutor)	Yes, it is even more beautiful when you see it with cherry blossoms

The illustration depicting the image generated by the auditory-based support system for individuals with hearing impairments is presented in Fig. 4. Firstly, mountain Fuji is recognized and images with the scenes of mountain fuji are generated. When five-storied pagoda is mentioned in the conversation, it is added to the image. At last, cheery blooms are also added to the images to reflect the contents talked by the interlocutor. Here, new items keep being added to the images instead of generating new images with items to show the continuity of the conversation.

Other experiments with different conversation contents are conducted in the same way. The generated images are evaluated based on questionaries. The conversational content and generated images are shown to a group of people with 18 adult individuals, including both males and females. Comprehensive surveys are conducted according to their impressions. The outcomes show that the proposed system is favorable, and the generated images can express the conversations well since more than 94% evaluators give 4 or 5 stars in the five-rated evaluation questionaries. Even the generated images are more impressive than hearing the conversation contents sometimes.

(a) I went to see Mt. Fuji

(b) I could also see the five-storied pagoda.

(c) Yes, it is even more beautiful when you see it with cherry blossoms.

Fig. 4. Image Reflecting Auditory Information

5   Conclusion

In this study, we have developed an image generation system for expressing auditory environment to hearing impaired people, aimed at improving communications between individuals with hearing impairments and normal people. The system achieves this by recognizing the auditory environment by speech recognition, and generating related images by using DM. Within this framework, the robot can understand what people are talking about around the hearing-impaired users and express the important contents to them by showing visual images, which are generated according to the contents. The effectiveness of the proposed system, with the functions of speech recognition, morphological analysis, and image generation, is proven by guiding experiments based on the developed guide mobile robot.

References

1. https://www.japantimes.co.jp/news/2020/07/22/national/social-issues/sign-language-deaf/
2. https://www.accessibility.com/blog/do-all-deaf-people-use-sign-language
3. Toba,Y., et al.: Considering multi-modal speech visualization for deaf and hard of hearing people. In: 2015 10th Asia-Pacific Symposium on Information and Telecommunication Technologies (APSITT), Colombo, Sri Lanka, pp. 1-3 (2015). https://doi.org/10.1109/APSITT.2015.7217102
4. KN, S.K., Sathish, R., Vinayak, S., Pandit, T.P.: Braille assistance system for visually impaired, blind & deaf-mute people in indoor & outdoor application. In: 2019 4th International Conference on Recent Trends on Electronics, Information, Communication & Technology (RTEICT), Bangalore, India, pp. 1505–1509 (2019). https://doi.org/10.1109/RTEICT46194.2019.9016765
5. Yuanming, J., Jiacheng, F., Hongmiao, C., Zhanlu, L., injin, C.: Applied research of speech recognition system-based auxiliary communication device in college PE course for the deaf. In: 2021 International Conference on Information Technology and Contemporary Sports (TCS), Guangzhou, China, pp. 485–488 (2021). https://doi.org/10.1109/TCS52929.2021.00104
6. Jairam, B.G., Ponnappa, D.: gesture based virtual assistant for deaf-mutes using deep learning approach. In: 2023 9th International Conference on Advanced Computing and Communication Systems (ICACCS), Coimbatore, India, pp. 1–7 (2023). https://doi.org/10.1109/ICACCS57279.2023.10112986
7. Furuhashi, M., Nakamura, T., Kanoh, M., Yamada, K.: Touch-based information transfer from a robot modeled on the hearing dog. In: 2015 IEEE International Conference on Fuzzy Systems (FUZZ-IEEE), Istanbul, Turkey, pp. 1–6 (2015). https://doi.org/10.1109/FUZZ-IEEE.2015.7337981
8. Ho, J., Jain, A., Abbeel, P.: Denoising diffusion probabilistic models. In: Advances in Neural Information Processing Systems, vol. 33, pp.1-25 (2020)
9. Rombach, R., Blattmann, A., Lorenz, D., Esser, P., Ommer, B.: High-resolution image synthesis with latent diffusion models.In: 2022 IEEE/CVF Conference on Computer Vision and Pattern Recognition, pp.10674–10685 (2022)

Robust Object Recognition and Command Understanding for a House Tidying-Up Robot

Bin Zhang[✉], Congzhi Ren, Junyan Wang, and Hun-Ok Lim

Kanagawa University, Yokohama-Shi 221-8686, Kanagawa, Japan
`zhangbin@kanagawa-u.ac.jp`

Abstract. In this study, a robust object recognition and command understanding system for a house tidying-up robot is proposed. The robot can understand the user's intentions by using a speech recognition system. When the user in-structs the tidying-up robot to tidy up an object using voice commands, the robot detects and recognizes the object. To detect and recognize multiple objects, we employ an instance segmentation method using deep learning. This method extracts the contour and shape of objects and generates the appropriate grasping posture of the robot arm. An experiment using the tidying-up robot is conducted to verify the effectiveness of the tidying-up system.

Keywords: Tidying-up Robot · Object recognition · Command Understanding

1 Introduction

In recent years, Japan has been facing a serious social problem due to the decreasing population of working-age people caused by a low birth rate and aging population. Therefore, there is a growing expectation for robots that can tidy up in place of hu-mans in homes, offices, convenience stores, and other places [1–3]. Recently developed robots such as HSR [4] are said to be applicable to tidying up work, such as arranging items placed arbitrarily. The YOLO object detection method using deep learning is commonly used to recognize objects [5]. A method has been proposed that uses multi-modal information, including visual, auditory, and tactile information, to infer the category of objects, in addition to category classification [6]. However, it has been said that many robots have difficulty tidying up in response to real-time instructions from users.

In this study, a robust object recognition and command understanding system for a house tidying-up robot is proposed. The robot can understand the user's intentions by using a speech recognition system. When the user instructs the tidying-up robot to tidy up an object using voice commands, the robot detects and recognizes the object. To detect and recognize multiple objects, we employ an instance segmentation meth-od using deep learning. This method extracts the contour and shape of objects and generates the appropriate grasping posture of the robot arm. An experiment using the tidying-up robot is conducted to verify the effectiveness of the tidying-up system.

© ICST Institute for Computer Sciences, Social Informatics and Telecommunications Engineering 2024
Published by Springer Nature Switzerland AG 2024. All Rights Reserved
J. Li et al. (Eds.): 6GN 2023, LNICST 553, pp. 17–22, 2024.
https://doi.org/10.1007/978-3-031-53401-0_3

2 Tidying-Up Robot

The tidying-up robot used in this study is an upper-body humanoid robot called "Sciurus17" developed by RT Corporation. The appearance of Sciurus17 is shown in Fig. 1. Its size is 270 × 393 × 665 (mm), and its weight, including the fixed metal for installation, is about 6 kg. The arms have 7 degrees of freedom, and the range of workable area is within a radius of 0.6 m centered on the robot. The head is equipped with a RealSense D415, which can acquire RGB-D environmental information around the robot.

Fig. 1. Overview of the tidying-up robot.

3 Object Grasping Method Based on Object Recognition and Command Understanding

The control system of the cleanup robot is shown in Fig. 2. First, the RGB-D sensor RealSense D415 acquires environmental information in front of the robot and per-forms category and area detection of objects. Next, the user's voice instructions are recognized, and if the instructed object is detected, it is selected as the cleanup target. Then, the contour information of the object is obtained, and the optimal grasping posture is generated. The grasping function of Sciurus17 uses ROS (Robot Operating System) to sequentially grasp the object with the arm closest to the target object and transport it to a predetermined storage location.

3.1 Speech Recognition

Google's speech-to-text API, Google speech recognition function, is used to recognize the user's voice and convert it into text. The noun in the text is detected and matched with the registered object target words to confirm the grasping target.

3.2 Object Recognition

Real-time object recognition and segmentation are performed using YOLACT [7], which is based on deep learning. YOLACT is a method that integrates the object recognition method YOLO and instance segmentation that can be divided into individual objects. The results of object recognition and segmented areas are shown in Fig. 3, accurately recognizing and segmenting objects commonly seen in daily life. The object target words registered in this study are limited to "bottle," "cup," and "sports ball." Objects that cannot be recognized are classified as unknown categories.

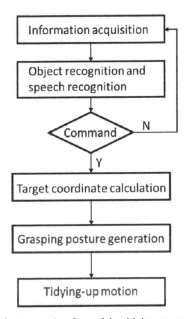

Fig. 2. Processing flow of the tidying-up system.

Fig. 3. Object recognition and region extraction result.

3.3 Grasping Gesture Generation

We obtain 3D point cloud data of the object from the segmented region in the image and calculate the center coordinates of the object (shown in Fig. 4). We then compute the shortest outer diameter that passes through the center point of the object using linear regression, which becomes the grasping posture of the object. We send the position and grasping posture information of the object to the robotic arm to perform the optimal grasping action.

Fig. 4. Object recognition and region extraction result.

4 Experiment

In the experiment, multiple objects, including plastic bottles, paper cups, and balls, were left on a desk as shown in Fig. 5. After the robot Sciurus17 was started, it looked at the desk and waited for user instructions while performing object detection.

Fig. 5. Experimental scene.

Upon understanding the voice command "clean up the plastic bottle," the extracted plastic bottle was matched with the detected target and confirmed as the cleanup target. The center coordinates and shortest diameter of the ball were obtained to determine the grasping posture, and by sending this information to Sciurus17's robot arm, the plastic bottle was cleaned up to the designated location.

The flow of the object grasping operation is shown in Fig. 6. Starting from the initial pose shown in Fig. 6(a), the cleanup target was determined, and the target object was grasped in the optimal grasping posture as shown in Fig. 6(b). Through lifting, transport, and placement operations shown in Fig. 6(c) to (f), the object was cleaned up to the predetermined location. The number, type, and position of the objects were changed, and the experiment was conducted 15 times. The success rate was calculated from the 15 experiment results and summarized in Fig. 7.

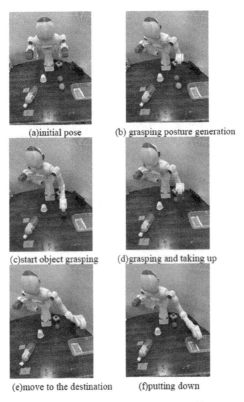

(a)initial pose   (b) grasping posture generation

(c)start object grasping   (d)grasping and taking up

(e)move to the destination   (f)putting down

Fig. 6. Motions when grasping the objects.

The object recognition rate for plastic bottles was over 90%, but the grasping success rate was relatively poor at 86.7%. One reason for this was the introduction of noise when performing linear regression from the 3D point cloud data, leading to errors in calculating the grasping coordinates.

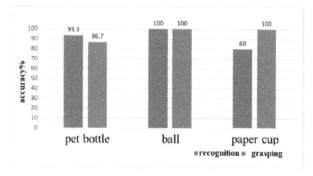

Fig. 7. Experimental results of the experiments.

5 Conclusion

In this study, a control system for a cleaning robot that responds to voice commands was proposed. The grip target was determined by voice recognition, and the object was recognized and categorized using YOLACT. The grip posture was determined by the shortest outer diameter and the center coordinates of the object. Through actual machine experiments, the grip coordinates of the objects were obtained, and the objects were successfully cleaned up in the designated locations.

References

1. Liu, S., Zheng, L., Wang, S., Li, R., Zhao, Y.: Cognitive abilities of indoor cleaning robots. In: 2016 12th World Congress on Intelligent Control and Automation (WCICA), Guilin, China, pp. 1508–1513 (2016)
2. Gopalakrishnan, R., Ramani, U., Maheswari, K.U., Thilagaraj, M.: Design and development of controller based automatic ground cleaning robot. In: 2022 6th International Conference on Computing Methodologies and Communication (ICCMC), Erode, India, pp. 491–494 (2022)
3. Oh, Y.-J., Watanabe, Y.: Development of small robot for home floor cleaning. In: Proceedings of the 41st SICE Annual Conference, Osaka, Japan (2002)
4. Yi, J.-B., Yi, S.-J.: Mobile manipulation for the HSR intelligent home service robot. In: 2019 16th International Conference on Ubiquitous Robots (UR), Jeju, Korea (South), pp. 169–173 (2019)
5. Mahendru, M., Dubey, S.K.: Real time object detection with audio feedback using yolo vs. Yolo_v3. In: 2021 11th International Conference on Cloud Computing, Data Science & Engineering (Confluence), Noida, India, pp. 734–740 (2021)
6. Araki, T., Nakamura, T., Nagai, T., Funakoshi, K., Nakano, M., Iwahashi, N.: Autonomous acquisition of multimodal information for online object concept formation by robots. In: IROS 2011, San Francisco, CA, USA, pp. 1540-1547 (2011)
7. Bolya, D., Zhou, C., Xiao, F., Lee, Y.J.: Yolact: real-time instance segmentation. In: Proceedings of the IEEE/CVF International Conference on Computer Vision, pp. 9157–9166 (2019)

3D Mapping Considering Object Recognition Result Based on 3D LiDAR Information

Congzhi Ren[✉], Bin Zhang, and Hun-Ok Lim

Kanagawa University, Yokohama-shi 221-8686, Kanagawa, Japan
r202270153pj@jindai.jp

Abstract. This paper describes a 3D mapping method considering object recognition result by using 3D LiDAR sensor mounted on a wheeled mobile robot. The object recognition is realized by using a deep learning neural network called Point-Pillars. By comparing the recognition results between two continuous frames, the objects can be matched and for those which are recognized in both frames, and new ID will be given for new detected objects. The recognized objects with determined ID will be reflected to the 3D map.

Keywords: Autonomous mobile robot · SLAM · PointPillars · Object recognition

1 Introduction

In recent years, research on autonomous mobile robots has been actively conducted in various fields [1–3]. Many types of robots, such as cleaning robots [4], service robots [5], autonomous vehicles [6], and delivery robots [7] are being utilized. Recognition of the surrounding environment is essential for robots to move autonomously. Simultaneous Localization and Mapping (SLAM) methods have been proposed to perform simultaneous self-positioning and map generation [8].

Map generation algorithm is mainly classified into 2D and 3D map generation. 2D maps are commonly used for robots with limited vertical movement [9], such as cleaning robots, as they do not consider the height of obstacles. 3D maps are used for robots operating in more complex environments [10], such as autonomous vehicles and delivery robots, as they can reflect 3D environmental information. However, most of the existing map generation methods only reflect the shape of obstacles in the map and did not consider the object type or characteristics, making it impossible to plan avoidance and paths adapted to the properties of the objects.

Therefore, this study proposes a map generation method that can reflect the attributes of recognized objects such as type, size, and position in a 3D map by using 3D point cloud information. We conducted experiments with an actual robot to demonstrate the effectiveness of the proposed method.

© ICST Institute for Computer Sciences, Social Informatics and Telecommunications Engineering 2024
Published by Springer Nature Switzerland AG 2024. All Rights Reserved
J. Li et al. (Eds.): 6GN 2023, LNICST 553, pp. 23–29, 2024.
https://doi.org/10.1007/978-3-031-53401-0_4

2   Design of the Mobile Robot

The mobile robot used in this study is shown in Fig. 1. We used the Mecanum Rover Ver 2.1 from Vstone Corporation as the mobile platform. On top of the robot, we mounted the RS-LiDAR-16 3D laser sensor from RoboSense, which can acquire obstacles distance information up to a maximum distance of 150 m around the robot. The robot also has a notebook PC for control purposes.

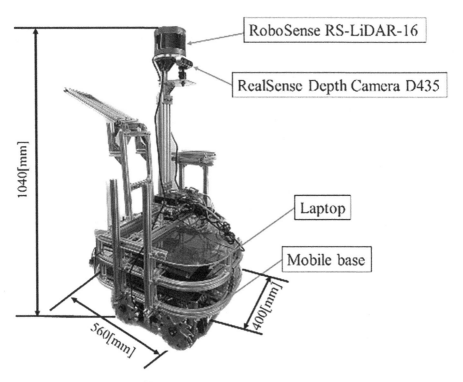

Fig. 1. Overview of the mobile robot.

3   Generating 3D Maps with Object Attributes

To reflect the attribute information such as object type, size, and position in the map, first, we recognize obstacles using a deep learning model called PointPillars [11]. Then, we generate a 3D map using the SLAM method. By matching the type and position of objects between continuous frames, we reflect the confirmed stationary objects in the map. The processing flow is shown in Fig. 2.

3.1   Deep Learning Object Recognition

We acquire 3D point cloud of the environment using RS-LiDAR-16 mounted to the robot. Object recognition is performed using a deep learning model called PointPillars.

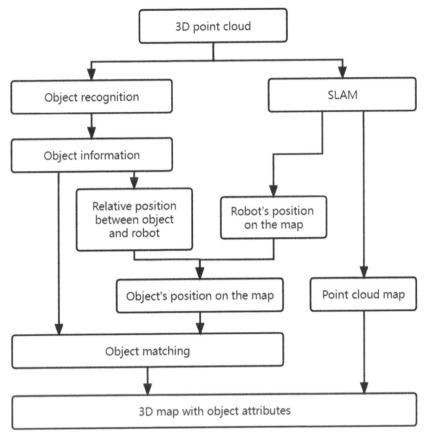

Fig. 2. Processing flow of 3D mapping.

PointPillars is a 3D object detection algorithm widely used in robotics and autonomous driving. It operates on point cloud data collected by lidar sensors, efficiently converting the point clouds into a grid-like representation. By employing a pillar-based approach, PointPillars processes this data efficiently and accurately detects objects in the environment. This algorithm stands out for its speed and accuracy in identifying objects' positions and shapes in 3D space, making it essential for real-time applications where reliable perception of surroundings is crucial.

In this study, we used the OpenPCDet framework for training the object recognition model. The training data utilized a dataset derived from the KITTI public dataset, together with some additional point cloud data and annotated information collected by us. The training of the object detection model incorporated three classes from the KITTI dataset: cars, pedestrians, and bicycles, alongside a new object type introduced from the custom dataset called pillars. This hybrid dataset enriched the training process, allowing the model to learn from diverse scenarios and accommodate a wider range of object classes, ultimately enhancing its ability to detect various objects, including the newly introduced pillar class.

3.2 Self-localization and Mapping

For mapping, we use a method called LeGO-LOAM [12]. LeGO-LOAM is a lightweight ground-optimized lidar localization and mapping algorithm primarily used in the fields of autonomous driving and robotics. By combining data from a lidar sensor and an inertial measurement unit (IMU), it achieves robot motion estimation and environmental mapping. The algorithm comprises two components: localization, which employs lidar data to identify environmental features and estimates robot motion by comparing consecutive scanning information and utilizing IMU data; and mapping, which employs lidar scanning data to construct a three-dimensional map. LeGO-LOAM is characterized by its efficiency and optimized focus on ground features, striking a balance between accuracy and real-time performance, making it suitable for applications like autonomous driving and robotics that demand precise positioning and mapping.

3.3 Object Matching

Recognized objects from the point cloud information acquired by the robot are transformed into the world coordinates. By matching the object information between two consecutive frames, objects of the same type that appear in the same position can be matched. The position, orientation, and size information of the object are updated based on the matched bounding-box. The confidence of the recognized object is accumulated to reflect the confirmed static object in the map. Objects that cannot be matched are considered to be newly appeared which can be matched later, or moving objects which would not be reflected on the map.

4 Experiment

The experimental environment is shown in Fig. 3. The environment contained multiple pillars and several moving people. During the experiment, the robot was manually moved around the environment while recognizing objects and matching the recognition results of consecutive frames to generate a 3D map that reflects the attributes of the recognized objects. Figure 4 shows that the robot accurately detects the bounding boxes of people and pillars in the environment in the robot coordinate system. The tall cuboids (four of them) represent pillars, and the short cuboid (one of them) represents a person. Since the pillars are stationary, they can be recognized in consecutive frames and matched accurately. People are moving around, so they are treated as moving objects and not included in the map.

In addition, Fig. 5 shows a map generated by LeGO-LOAM. 3D information sensed during moving is accumulated, and a global environmental map is generated. The map accurately reflects the walls and pillars in the environment, but treats them all as obstacles, without discrimination. Since pillars are stationary objects and can be recognized continuously, they can be matched accurately between continuous frames. Figure 6 shows that confirmed objects after matching are reflected in the map. The bounding boxes of the recognized object (pillars) are accurately reflected in the 3D map. Pillars are reflected in the map with attributes such as size and position, as well as their type. It can be confirmed that the moving person has also been removed from the map.

Fig. 3. Experimental environment in a building.

Fig. 4. Detected objects with bounding boxes.

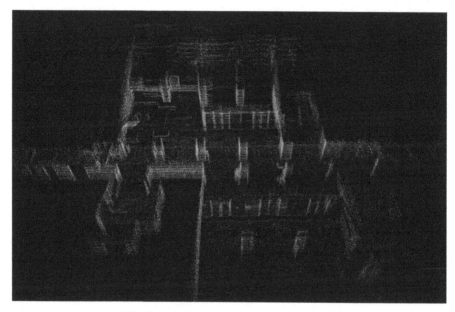

Fig. 5. 3D map generated by LeGO-LOAM.

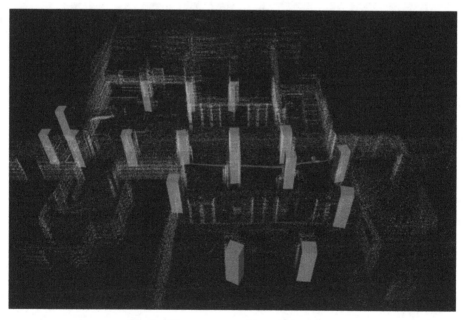

Fig. 6. 3D map generated by our algorithm.

5 Conclusion

In this study, we utilized 3D distance information and performed object recognition using a deep learning model called PointPillars, which was trained to recognize pre-defined objects such as pillars and people. We proposed a map generation method that can reflect the attributes (type, size, position) of recognized static objects in 3D point cloud map. Through actual experiments, we accurately recognized pillars and humans as 3D objects, confirmed the stationary pillars through matching the consecutive frame recognition results, and accurately reflected them as 3D bounding boxes in the 3D map. In the future, we aim to increase the types of objects to be trained, distinguish moving objects, stationary objects, and potentially moving objects on the map, and improve the generality of the generated map. Furthermore, by recognizing the surrounding occupancy state as potential occupancy space from the attributes of the objects and reflecting the occupancy risk on the map, we can plan more natural paths.

References

1. Qin, H., Shao, S., Wang, T., Yu, X., Jiang, Y., Cao, Z.: Review of autonomous path planning algorithms for mobile robots. Drones **7**(3), 211 (2023)
2. Zhang, J., Yang, X., Wang, W., Guan, J., Ding, L., Lee, V.C.: Automated guided vehicles and autonomous mobile robots for recognition and tracking in civil engineering. Autom. Constr. **146**, 104699 (2023)
3. Liu, L., Wang, X., Yang, X., Liu, H., Li, J., Wang, P.: Path planning techniques for mobile robots: review and prospect. Expert Syst. Appl. 120254 (2023)
4. Kim, J., Mishra, A.K., Limosani, R., et al.: Control strategies for cleaning robots in domestic applications: a comprehensive review. Int. J. Adv. Robot. Syst. (2019). https://doi.org/10.1177/1729881419857432
5. Asgharian, P., Panchea, A.M., Ferland, F.: A review on the use of mobile service robots in elderly care. Robotics **11**(6), 127 (2022)
6. Parekh, D., et al.: A review on autonomous vehicles: progress, methods and challenges. Electronics **11**(14), 2162 (2022)
7. Lee, D., Kang, G., Kim, B., Shim, D.H.: Assistive delivery robot application for real-world postal services. IEEE Access **9**, 141981–141998 (2021)
8. Macario Barros, A., Michel, M., Moline, Y., Corre, G., Carrel, F.: A comprehensive survey of visual SLAM algorithms. Robotics **11**(1), 24 (2022)
9. Tee, Y.K., Han, Y.C.: Lidar-based 2D SLAM for mobile robot in an indoor environment: a review. In: 2021 International Conference on Green Energy, Computing and Sustainable Technology (GECOST), pp. 1–7 (2021)
10. Xu, X., et al.: A review of multi-sensor fusion SLAM systems based on 3D LIDAR. Remote Sens. **14**(12), 2835 (2022)
11. Lang, A.H., Vora, S., Caesar, H., Zhou, L., Yang, J., Beijbom, O.: PointPillars: fast encoders for object detection from point clouds. In: 2019 IEEE/CVF Conference on Computer Vision and Pattern Recognition (CVPR), pp.12689–12697 (2019)
12. Tixiao Shan, F., Brendan Englot, S.: LeGO-LOAM: Lightweight and ground-optimized Lidar-Odometry and mapping on variable terrain. In: 2018 IEEE/RSJ International Conference on Intelligent Robots and Systems (IROS), pp. 4758–4765 (2018)

Intelligent Inspection System of Coal Preparation Plant Based on Dynamic QR Code

Zhipeng Wu, Xuefei Xu, Yang Gao, Xin Liang, and Cheng Cai[✉]

Shanghai DianJi University, Shuihua. 300, Shanghai 201306, China
{226003010205,226003010114,226003010102}@st.sdju.edu.cn, {xuxf,
caic}@sdju.edu.cn

Abstract. To reduce the complexity and insecurity of the traditional safety inspection methods of coal preparation plants in the coal industry, an intelligent inspection system of coal preparation plants using dynamic two-dimensional code technology is proposed. The system solution applies dynamic two-dimensional code technology to realize the regular update of the two-dimensional code on the display end of each board. The system server can visually formulate inspection tasks and query inspection records. The system uses cloud development technology to enable users to identify dynamic QR codes through mini programs on mobile Apps, so as to obtain QR code information. Thus, users can fill in the inspection information in the WeChat mini program and upload the inspection information to the server. By comparing with the local QR code to send information, the location positioning of user inspection is realized. Positioning via QR codes is more accurate than traditional GPS positioning. And the use of dynamic QR codes improves system security compared to traditional static QR codes. The system is safe and simple to operate, which can effectively improve the inspection efficiency and reduce the workload.

Keywords: Dynamic QR codes · Intelligent Inspection · WeChat Mini Program

1 Introduction

With the gradual growth of intelligent personnel management demand in the coal preparation industry, a intelligent, simple and safe intelligent inspection system [1] has become an inevitable choice for the equipment management of coal preparation plants. The traditional equipment inspection mode has caused great inconvenience to the staff, and its positioning accuracy and operation convenience cannot be guaranteed. In the face of general intelligent inspection, its positioning accuracy and system security problems urgently need to be improved. Therefore, in the equipment management of coal preparation plants, the intelligent inspection system with convenient operation, accurate positioning and high safety has become an important research content.

J. Li et al. (Eds.): 6GN 2023, LNICST 553, pp. 30–43, 2024.
https://doi.org/10.1007/978-3-031-53401-0_5

2   Research Status

Because the QR code has the characteristics of large capacity, fast recognition speed and good anti-counterfeiting effect, it can be well applied to the authentication of mobile terminals. Most of mobile terminal authentication are based on QR codes to implement equipment inspection, but the process of verifying the location of inspectors [2] is as follows: (1) Store the device number and device GPS location information on the server. (2) Generate a fixed QR code from the device number and paste it on the corresponding device. (3) The user scans the QR code with the mobile phone and sends the obtained device information and the location information obtained by the GPS positioning of the mobile phone to the server for information verification. Then, the verification can prove - the inspection to the current device. Due to different application scenarios, the fixed QR code used in the above scheme is not strong in security and can be scanned after taking photos. GPS positioning requires the mobile phone to be in a location with high GPS signal powers, and the accuracy of positioning is not high, if it needs to be indoors or needs to have high positioning accuracy requirements, it is not applicable. In addition, mobile phone scanning in the above process needs to be done through a client App, which is expensive to develop a mobile app [2].

This paper combines the advantages of the above method and its own application scenarios to design an intelligent inspection system [3] of coal preparation plant based on dynamic two-dimensional code.

3   System Design

3.1   System Framework Design

The system consists of three parts: server-side PC, board terminal and mobile phone. The board end consists of an Arduino main control board integrating ATmega32u4 and W5500 Ethernet chips and an LCD 12864 display expansion board. The server side and the board side are connected via Ethernet. As shown in Fig. 1:

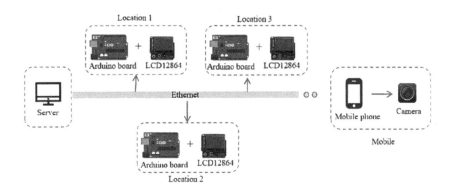

Fig. 1. Block diagram of the system.

3.2   System Function Design

The functional requirements of the system are that the server side assigns inspection tasks to the mobile terminal and display a dynamic QR code on the board card at each inspection address. Users scan the corresponding QR code through the WeChat mini program on the mobile phone to obtain the corresponding QR code information. After that, users fill in the relevant inspection information on the WeChat mini program and take photos on the spot. After completion, users? upload the QR code information together with the inspection information. After successful verification with the local QR code information, the user's inspection location and inspection time can be determined, and the inspection record can be generated. First, the server generates the QR code, communicates through the Web socket, sends the timestamp and QR code to the board display on a regular basis, and saves the time information, sending location and QR code information to the database. The QR code and timestamp are displayed through the LCD 12864 screen. Through the WeChat mini program, the mobile terminal identifies the QR code of the board display end, fills in the inspection information after success, and finally sends the inspection information and the QR code information to the server. Compare the two-dimensional code information in the information with the time and the information in the server-side database, and the inspection record can be generated after the successful completion of the comparison. The system function diagram is shown in Fig. 2. The system data flow is shown in Fig. 3.

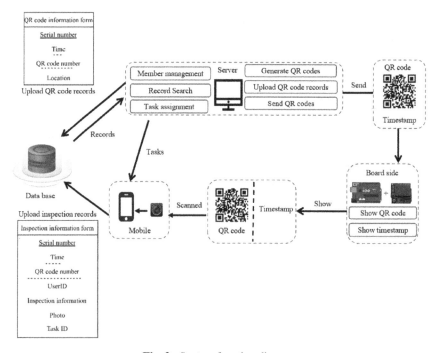

Fig. 2. System function diagram

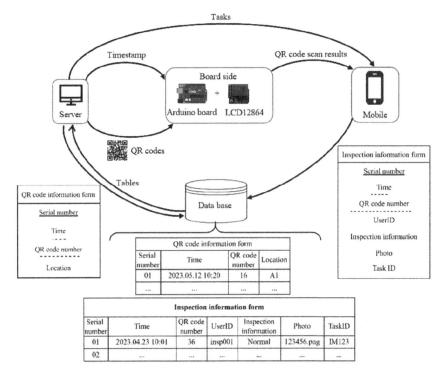

Fig. 3. System data flow

4 Server-Side Software Design

The functions to be realized by the system server include: (1) Inspection personnel management. (2) Assignment of inspection tasks. (3) Inspection record inquiry. (4) QR code generation. (5) When sending the QR code to the board end, save the location, time and QR code information sent to the database. (6) Send the QR code and timestamp to the board end through the web socket.

4.1 Inspection Personnel Management

This system designs the inspection personnel management module to realize the addition and deletion of inspection personnel and the modification of personnel information. The inspection personnel information includes user ID, name, and phone number, and the system can edit the personnel information through the relevant buttons. The inspection personnel management interface is shown in Fig. 4 below:

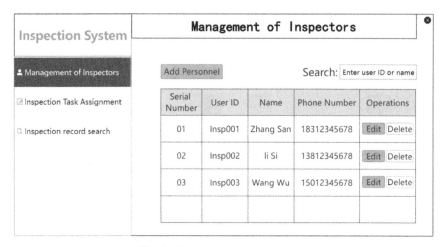

Fig. 4. Inspector management page

4.2 Inspection Task Assignment

This system realizes the formulation and allocation of server-side inspection tasks. Each inspection task is an inspection route, including the location, arrival time and task cycle of the required arrival, and each inspection task has a unique task ID. Figure 5 shows the inspection task assignment interface.

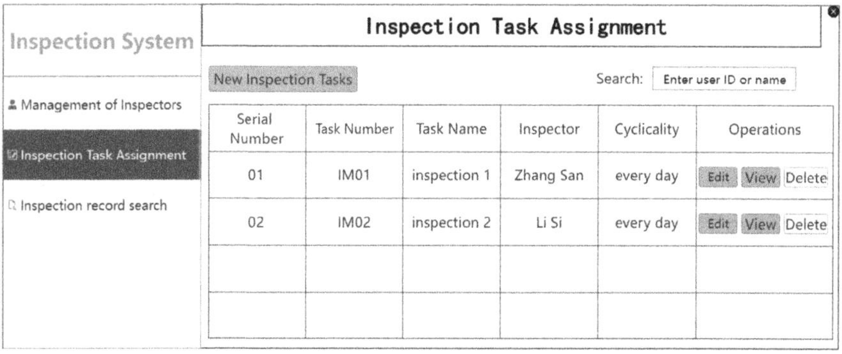

Fig. 5. Task assignment page

In addition, when the "New Task Route" button is selected, you can visually develop inspection routes. To create a new inspection route, you need to specify the period, inspection personnel, and inspection route. The table on the left can specify the inspection time and location, and the route will eventually be marked on the inspection map on the right; The inspection map on the right can also be determined by clicking the circular icon on the map successively. Figure 6 shows the interface of the new inspection task.

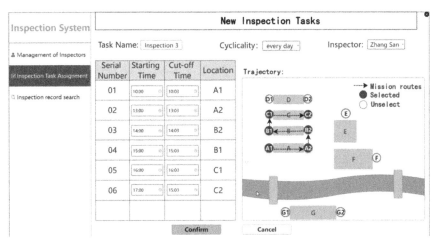

Fig. 6. New task route page

4.3 Query Inspection Records

After uploading the inspection information on the mobile terminal, the server generates a new inspection record locally [4]. This system provides two ways to view inspection records. The first way is presented in a tabular form, which includes the task name, inspection time, name of inspection personnel, and inspection information and photos. The page is shown in Fig. 7 below.

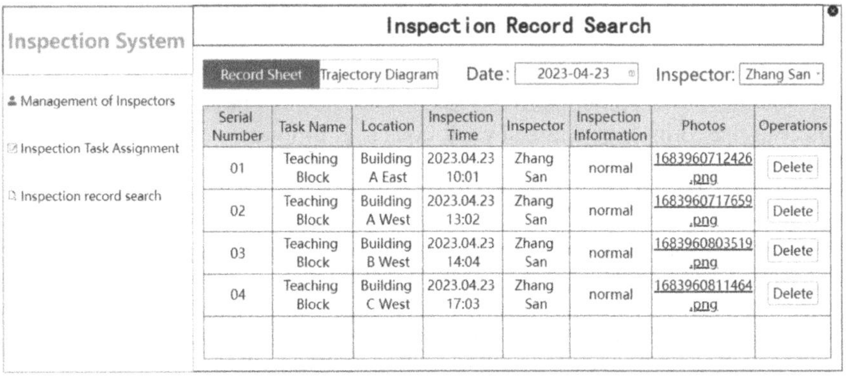

Fig. 7. The page of the inspection record sheet

The second method is presented in the form of a trajectory diagram, as shown in Fig. 8 below. The inspection time and location are displayed in the table on the left. The task route and the actual route are marked on the track chart on the right. Three types of inspection results are represented in different colors: punctuality, delay and absence.

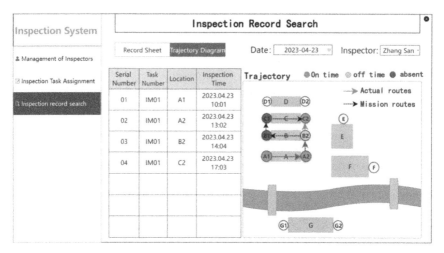

Fig. 8. The page of the track chart of the inspection record

4.4   Code Generator Module

The QR code of this system is generated using the QR code library in Python. Due to its own characteristics, QR code has a limited number under the same version and error correction level. To ensure the absolute uniqueness of the QR code received by the board display terminal at the same time, this article uses a fixed QR code generated locally and numbered it (the number of QR codes is much greater than the number of Arduino terminals), and the information contained in the QR code is its number. According to this, when the server sends the QR code to the board display end every minute, the QR code is randomly selected to be sent, and the dynamic QR code display can be completed.

Since the width and height of the LCD 12864 display screen on the board end is 128bit by 64bit, the maximum size of the QR code that can be displayed is 64 bits by 64 bits. The size formula generated by QR code is as follows:

$$qr\_size = (21 + (version - 1) \times 4 + border \times 2) \times box\_size \tag{1}$$

Among them, the qr_size is the size of the QR code, the $version$ is the error-corrected version, the number of pixels contained in each module by border, and the number of modules contained in the border box_size. In the maximum range that the display can display, version takes three, border takes one, and box_size takes two, then the size of the generated QR code is 62 bits by 62 bits.

QR code generated by the server? and its number (from 0 to 49) are shown in Fig. 9 below.

Fig. 9. QR codes generated by the Server

4.5   Module for Sending QR Codes with Timestamps

The network model of this system is C/S structure, that is, client/server structure. The Arduino terminal is used as the client side, and the host computer is used as the server side [5]. There are three types of sockets: streaming sockets, datagram sockets, and raw sockets. This system uses streaming sockets to transmit image data.

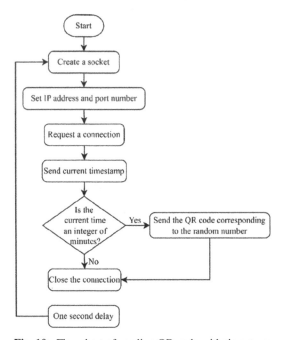

Fig. 10. Flowchart of sending QR code with timestamp

Since the width of the picture displayed by the LCD 12864 display screen on the board end needs to be the number of whole bytes, it is necessary to insert two blank columns on the right side of the QR code before sending the QR code, so that the size of the picture becomes 64 bits by 62 bits. The host computer sends a QR code to the board end every minute and the current timestamp to the board end every second. The flowchart is shown in Fig. 10.

4.6 The Storage Module for QR Code Sending Information

Server-side databases are implemented using SQLite. When the host computer sends the QR code to the display end of the board, upload the QR code number, the location of the transmission, and the time of sending the QR code to the database table. Table 1 describes the results of exporting database tables.

Table 1. Table of information to send QR codes

Serial number	Time	QR code number	Location
1	2023-04-23 10:00	11	A1
2	2023-04-23 10:01	36	A1
3	2023-04-23 10:03	48	A1
4	2023-04-23 13:00	15	A2
5	2023-04-23 13:01	22	A2
6	2023-04-23 13:02	14	A2
7	2023-04-23 14:00	39	B2
8	2023-04-23 14:01	45	B2
10	2023-04-23 14:02	2	B2
11	2023-04-23 14:03	18	B2
12	2023-04-23 14:04	6	B2
13	2023-04-23 17:00	31	C2
14	2023-04-23 17:01	27	C2
15	2023-04-23 17:02	49	C2
16	2023-04-23 17:03	23	C2

5 Software Design on the Board Side

The board end consists of an Arduino main control board integrating ATmega32u4 and W5500 Ethernet chips and an LCD 12864 display expansion board [6]. Arduino terminals will be deployed in various locations that need to be clocked in. The host computer is connected to the Arduino main control board through an Ethernet cable, and the board display can complete the receipt of the QR code picture and timestamp through the Web socket and display the picture and timestamp through the LCD 12864 display. As shown in Fig. 11.

Fig. 11. Physical diagram of the board side

The flow chart of the board display program is shown in Fig. 12.

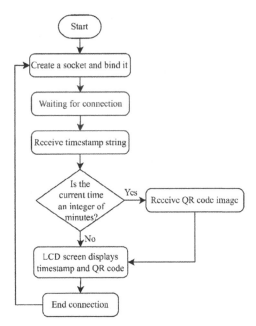

Fig. 12. Software flowchart on the board side

6 Software Design for Mobile Phones

The User ID of the same user's WeChat account corresponding to a WeChat Mini Program is unique, so the User ID can be used as the identity of the WeChat user. The system uses WeChat developer tools to develop a QR code scanning mini program. First, receive inspection tasks through the Mini Program. After that, scan the corresponding QR code on the QR code scanning page of the Mini Program and identify the QR code information. Finally, the obtained QR code information is uploaded to the cloud database with the scanning time, User ID and related inspection information.

The Mini Program has four pages, the initial login page, the inspection task page, the QR code scanning page, and the inspection information upload page. Figure 13 shows each interface of the Device Inspection Mini Program. On this page, users who log in for the first time need to manually complete the authorization login.

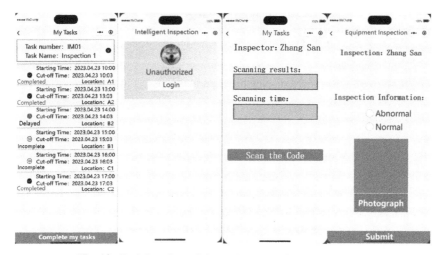

Fig. 13. Each interface of the device inspection mini program

After the authorization is successful, the system will enter the inspection task interface, which displays the inspection tasks sent by the server and visually displays the completed and uncompleted inspection tasks. Select "Go to finish", and the Mini Program will automatically jump to the QR code scanning interface. In this interface, call the wx.scanCode interface to call up the client code scanning interface to scan the code. After the code is successfully scanned, the scan result, scanning time and User ID are displayed on the interface, and then automatically jump to the inspection information filling interface. Under this interface, the user clicks "Submit" after filling in the information, and the WeChat Mini Program will upload the above-mentioned QR code scanning information and the inspection information and photos filled in on the page to the database.

The board terminals are deployed at each point, and the mobile phone logs in to the WeChat Mini Program to scan the QR code of four points, and the resulting database records are exported as shown in Table 2.

Table 2. Table of inspection information

Serial number	Time	QR code number	UserID	Inspection information	Photo	TaskID
01	2023-04-23 10:01	36	Insp001	Normal	1683960712426.png	IM01
02	2023-04-23 13:02	14	Insp001	Normal	1683960717659.png	IM01
03	2023-04-23 14:04	6	Insp001	Normal	1683960803519.png	IM01
04	2023-04-23 17:03	23	Insp001	Normal	1683960811464.png	IM01

7   Location Authentication Process

In order to determine the user's successful inspection at the location of the display end of a board, the system needs to compare the record of the QR code sent by the host computer to the display end of the board and the record of the QR code information uploaded after the mobile terminal successfully scans the code according to the time and QR code number, so as to determine whether the user is patrolling at the equipment set up at the display end of the board at that moment [7]. After the comparison is successful, a new record containing the time, location, task ID, User ID, and inspection information is generated in the server database, as shown in Fig. 14.

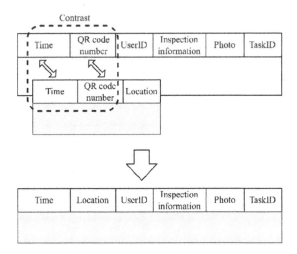

Fig. 14. Location Authentication Process

In this system, the server-side QR code sending database (see Table 1 above) is compared with the inspection information database (see Table 2 above), and the results generate a new record. In this system, the new database tables are shown in Table 3.

Table 3. Table of new records

Serial number	Time	Location	UserID	Inspection information	Photo	TaskID
01	2023-04-23 10:01	A1	Insp001	Normal	1683960712426.png	IM01
02	2023-04-23 13:02	A2	Insp001	Normal	1683960717659.png	IM01
03	2023-04-23 15:04	B2	Insp001	Normal	1683960803519.png	IM01
04	2023-04-23 17:03	C2	Insp001	Normal	1683960811464.png	IM01

8   Related Tests

Due to the limited screen brightness, size and clarity of the QR code displayed by the board, the success rate and scanning time of users scanning the QR code with their mobile phones will vary with the distance. In this article, the QR code is scanned five times with a mobile phone at each set distance, and the success of the scan and the scan time are recorded, as shown in Table 4.

Table 4. Relationship between the scanning result and the scanning distances

Distance (cm)	Number of failures (time)	Number of successes (time)	Interval of time t for successful code scanning (second)
10	0	5	$t < 2$
20	0	5	$t < 2$
30	0	5	$t < 2$
40	0	5	$t < 2$
50	0	5	$t < 2$
60	0	5	$3 \leq t \leq 5$
70	0	5	$6 \leq t < 10$
80	1	4	$14 \leq t \leq 20$
90	4	1	$t \geq 20$
100	5	0	∞

Through the simulation debugging and on-site testing of the system, the entire intelligent inspection system runs well, and the response time of system login and logout is maintained within 0.5 s; the time elapsed from the user submits the inspection information to the server generates the inspection record and returns "submission successful" is less than 3.0 s. The response time for inspector information modification is less than 1.0 s. The inspection task takes less than 2.0 s from publishing to sending it to mobile.

9 Conclusion

This paper designs and implements an intelligent inspection system scheme for coal preparation plants based on dynamic QR code, including server-side database management, inspection information upload based on WeChat mini program, and dynamic QR code display based on Arduino [8]. This inspection scheme has high security. However, this system is only suitable for the area covered by the network and is not suitable for working environments with weak network signals such as underground operations. In the future, consideration will be given to improving the applicability of the system based on extensive testing and improvement of system efficiency, stability and reliability [4].

References

1. Yang, Z., Yin, X., Chen, X., Cao, H., Su, H.: Design and application of intelligent management system for equipment inspection in coal preparation plant. Coal Preparation Technol. **2019**(03), 99–104 (2019)
2. Hu, Y., Zhang, J., Wang, D.: Research on intelligent inspection technology of distribution network based on RFID. IOP Conf. Ser. Earth Environ. Sci. **651**(2), 05–22 (2021)
3. Zhang, S., Zhang, Y., Cao, S., Li, B., Qi, X., Li, S.: Design and application of intelligent patrol system in substation. J. Phys. Conf. Ser. **2237**(1), 99–110 (2022)
4. Jin, C., Tong, D.: Wearable device-based intelligent patrol inspection system design and implementation. Int. J. Distrib. Syst. Technol. (IJDST) **14**(2), 1–10 (2023)
5. Long, D.: Application of mobile intelligent inspection system in substation equipment management. Energy Power Eng. **9**(04), 408–413 (2017)
6. Kirthima, M.A., Raghunath, A.: Air quality monitoring system using raspberry pi and web socket. Int. J. Comput. Appl. **169**(11), 28–30 (2017)
7. Zhang, Y., Zhang, Y., Fan, S., Wang, K., Jiang, P.: Indoor positioning and navigation based on two-dimensional code and database. J. Eng. **2019**(23), 8820–8823 (2019)
8. Zhang, J., Wang, H., Hu, Y., Liu, C., Zhang, D.: Design of intelligent inspection system for distribution network based on circular multi-agent structure. IOP Conf. Ser. Earth Environ. Sci. **218**(1), 12–18 (2019)

On the Construction of Information Management System for Railroad Grouping Station Operation

Yufei Nie, Xuefei Xu, and Cheng Cai[(✉)]

Shanghai DianJi University, Shuihua. 300, Shanghai 201306, China
216003010224@st.sdju.edu.cn, {xuxf,caic}@sdju.edu.cn

Abstract. In order to optimize the information management of railway freight car inspection technology operations and improve the efficiency of inspection shift personnel, this article analyzes the existing inspection technology workflow and conducts an in-depth study of the information management aspects of inspection operations. The information management system designed in this article utilizes the real-time collection and transmission of actual production data based on the RS485 bus Modbus communication protocol. The development language chosen is C#, and the MySQL database is selected for storing backend data. This system integrates functional modules such as operational fault uploading, detection data querying, and comprehensive alarm display. It achieves the goal of accessing and analyzing real-time operational data from the field through the network, enabling the real-time understanding of technical operations during train inspections and optimizing the information management of train inspection technology operations.

Keywords: Rail Wagon · Technical Operation · Information Management

1 Introduction

The railroad grouping station is called "the manufacturing factory of cargo trains" [1], where the cargo trains from all directions are assembled and arranged into new trains. The train inspection duty room is mainly responsible for the production organization and information coordination of the train inspection operation site. By understanding the stage plan of the station, the train inspection duty room gives the operation notice to the operation group, then remotely controls the rutting machine equipment to carry out the operation of derailer up or down, and organizes the operation group to complete the technical operation inspection and fault repair of the train. After the inspection, the operation group will inform the vehicle maintenance information and vehicle repair information to the inspection duty room. The inspection duty room will upload the vehicle information to the railroad truck technical management information system, and then contact the station duty officer [2].

The extensive deployment and utilization of various technological equipment such as the 5T system (Train Operation Dynamic Safety Inspection System), train rear-end wind

J. Li et al. (Eds.): 6GN 2023, LNICST 553, pp. 44–52, 2024.
https://doi.org/10.1007/978-3-031-53401-0_6

pressure detection devices, intelligent electric derails, and handheld devices have greatly enriched the means of inspection technology operations. However, various equipment are functionally independent and cannot be interrelated. The organization of train inspection technology operations has become increasingly complex, accompanied by a growing number of management forms and reports. Repetitive information entry has resulted in low efficiency for staff on duty and increased safety risks. This paper focuses on the construction of a railroad grouping station's inspection operation information management system, which is conducive to improve the operation quality and transportation level of the grouping station [3].

2 Demand Analysis

2.1 Function Analysis

Technical operation information management system is an important tool for dispatching and commanding the operation of the whole station [4]. The system uses Modbus protocol to communicate and interact with STM32 microcontroller system to realize the operation control of electric derailleur and display the operation status, and integrates video monitoring, image recognition function, fault alarm, fault upload, car inspection data query, protection light status monitoring and other functions [5, 6]. Its main functional structure is shown in Fig. 1.

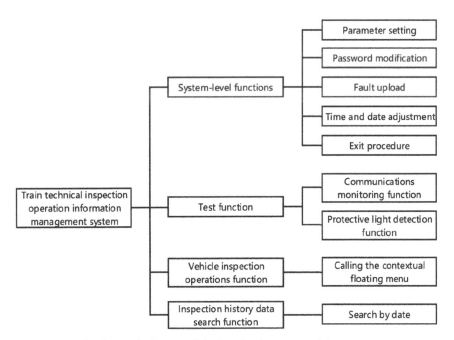

Fig. 1. Block diagram of the functional structure of the system.

2.2 Analysis of Inspection Process

The flow chart of inspection operation is shown in Fig. 2. The key link in the process is that when the derailer performs the upward and downward derailment, the 160 V power switch of the track should be turned off at first, then the option "perform upward derailment" from the floating menu is selected, and then the 160 V power switch will be turned off after the derailer is in place. If the up and down actions are in place, the inspector should be notified that the inspection process can be start, and the system starts the inspection operation timing. If the upper off actions are not in place, the duty officer should report to the duty leader, immediately organize the maintenance, after confirming the derailer on in place or install the mobile derailer, with the privilege operation to confirm in place, system will start the inspection timing. If the duty officer cannot report in time, the electric derailer of the strand will be deactivated.

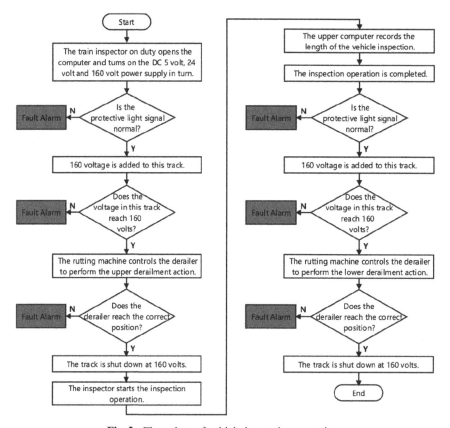

Fig. 2. Flow chart of vehicle inspection operation.

2.3 System Equipment Composition

The hardware components of the system are shown in Fig. 3. As shown in the figure, the host computer is connected to the control box through the bus, and there is a 160 V voltage detection board in the control box, which is used to detect whether the voltage at both ends of the rutting machine reaches 160 V. The upper computer is connected to the outdoor serial server and hard disk recorder through the switch.

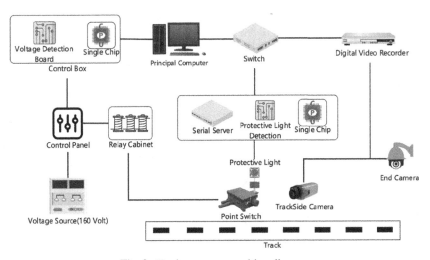

Fig. 3. Equipment composition diagram

3 System Design and Implementation

To increase the logic of the code and reduce the coupling of the code, the programming idea of this system adopts a three-layer architecture, which are data access layer, business logic layer, and interface display layer [7].

3.1 Data Access Layer

The data access layer consists of only one class named Database, which performs all operations for all database access. Its class diagram is shown in Fig. 4 (Table 1).

3.2 Data Access Interface Layer

The data access interface layer includes a class named SqlStringConstructor, which completes the function of constructing SQL statements, and its class diagram is shown in Fig. 5 (Table 2).

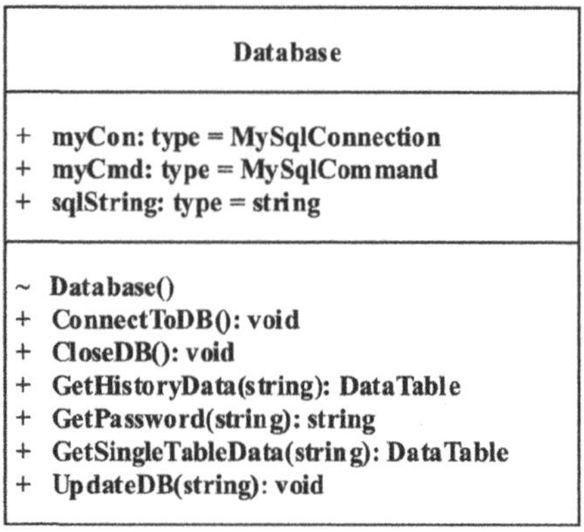

Database
+ **myCon: type = MySqlConnection** + **myCmd: type = MySqlCommand** + **sqlString: type = string**
~ **Database()** + **ConnectToDB(): void** + **CloseDB(): void** + **GetHistoryData(string): DataTable** + **GetPassword(string): string** + **GetSingleTableData(string): DataTable** + **UpdateDB(string): void**

Fig. 4. Class diagram of Database class

Table 1. Table Description of the class members of Database

Property/Method	Function Specification
myCon	Public properties, Database connection object
myCmd	Public properties, Database command object
sqlString	Public properties, Command string
Database	Constructor method without reference
ConnectToDB	Public method, Connect and open database
CloseDB	Public method, Close the database
GetHistoryData	Public method, Obtain historical vehicle inspection data
GetPassword	Public method, Obtain system password
GetSingleTableData	Public method, Get all the data of a particular data table
UpdateDB	Public method, Update the data of a particular data table

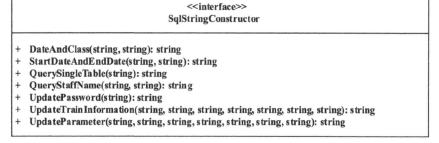

<<interface>> SqlStringConstructor
+ **DateAndClass(string, string): string** + **StartDateAndEndDate(string, string): string** + **QuerySingleTable(string): string** + **QueryStaffName(string, string): string** + **UpdatePassword(string): string** + **UpdateTrainInformation(string, string, string, string, string, string, string): string** + **UpdateParameter(string, string, string, string, string, string, string): string**

Fig. 5. Class diagram of SqlStringConstructor class

Table 2. Table Description of the class members of SqlStringConstructor

Method	Function Specification
DateAndClass	Public method, Statement that query data by date and shift
StartDateAndEndDate	Public method, Statement that query data by start and end date
QuerySingleTable	Public method, Construct a statement for single table queries
QueryStaffName	Public method, Statements for querying employee names
UpdatePassword	Public method, Construct the change password statement
UpdateTrainInformation	Public method, Statement to update inspection information
UpdateParameter	Public method, Constructing statements to change system parameters

3.3 Business Logic Layer

The business logic layer completes the data exchange among the monitoring terminal of each field of the formation station, the operation terminal and the column inspection operation information system to ensure the real-time accurate data. The business logic layer realizes the early warning function by comparing the data from the operation terminals of the operation yards of the formation stations with the safety data thresholds to ensure the safety of program operation [8]. As an example, the class diagram of the Track class is shown in Fig. 5 (Fig. 6 and Table 3).

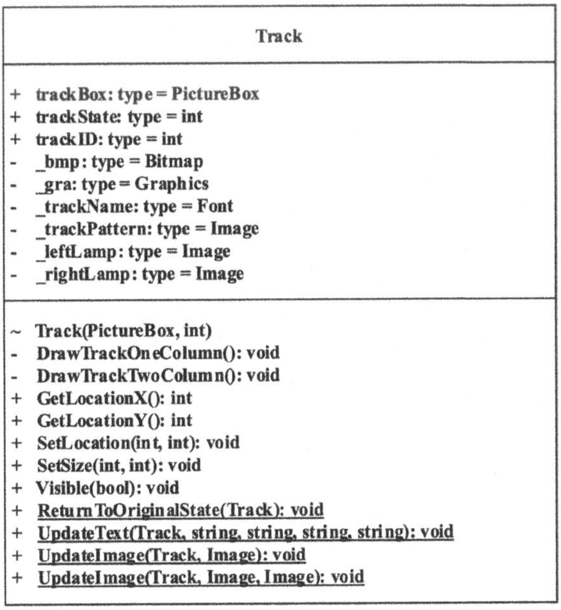

Fig. 6. Class diagram of Track class

Table 3. Table Description of the class members of Track

Property/Method	Function Specification
trackBox	Public properties, Track Object
trackState	Public properties, The working state of the track
trackID	Public properties, Numbering of the track
_bmp	Private properties, Drawing board for drawing track's information
_gra	Private properties, Paintbrush for drawing track's information
_trackName	Private properties, Text font of the track's name
_trackPattern	Private properties, Pictures of the track style
_leftLamp	Private properties, Left derailer status indicator
_rightLamp	Private properties, Right derailer status indicator
Track	Public method, Constructor method
DrawTrackOneColumn	Private method, Initialize each track by one column
DrawTrackTwoColumn	Private method, Initialize each track by two column
GetLocationX	Public method, Get the horizontal coordinates of this track's pictureBox
GetLocationY	Public method, Get the longitudinal coordinates of this track's pictureBox
SetLocation	Public method, Set the location of this track's pictureBox
SetSize	Public method, Set the size of this track's pictureBox
Visible	Public method, Set the visible attribute of this track's pictureBox
ReturnToOriginalState	Static Public method, Restore the track picture to its original form
UpdateText	Static Public method, Plot the train information for the track
UpdateImage(Track, Image)	Static Public method, Draw a picture of the left or right derailer status indicator for this track
UpdateImage(Track, Image, Image)	Static Public method, Draw a picture of the left and right derailer status indicator for this strand

3.4 Interface Display Layer

Main Interface

The main interface of the system is shown in Fig. 7.

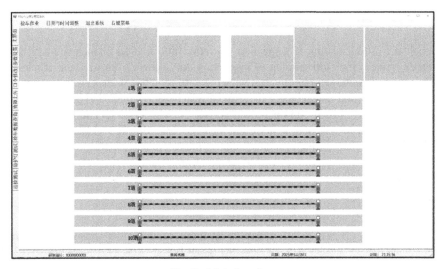

Fig. 7. Main interface

Train Information Interface

The duty officer clicks on the floating menu item "Train Arrived" in the operation unit, and then the train information window pops up. The interface design of this window is shown in Fig. 8.

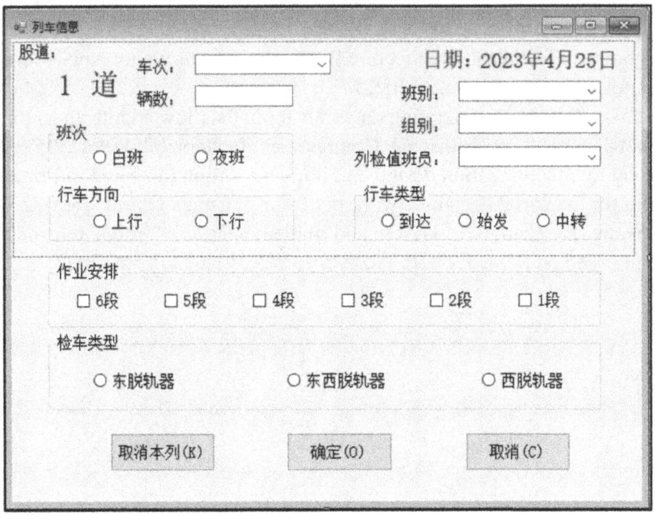

Fig. 8. Interface for entering vehicle inspection information

4   Conclusion

Through the establishment of inspection operation information management system, it greatly reduces the work burden of inspection duty officer, improves the efficiency of inspection technology operation, satisfies the requirements of inspection technology operation information management in terms of real-time, dynamic and actual situation responsiveness, etc. [9]. It can query and access the inspection technology operation information of any historical time in a combined way, and display it visually in the form of corresponding charts, which is convenient for the relevant personnel to make rational analysis of inspection technology operation situation [10].

References

1. Luan, D., Yang, H., Feng, J.: Research on intelligent technology of automatic brake duct disconnection for vehicles in formation stations. Rail Transp. Econ. **43**(08), 51–57+70 (2021)
2. Bharadwaj, D.: Integrated freight terminal and automated freight management system: a theoretical approach. Transp. Res. Procedia **48**, 260–279 (2020)
3. Marzena, K., Edyta, P., Maciej, W.: Reliability of the intermodal transport network under disrupted conditions in the rail freight transport. Res. Transp. Bus. Manag. **44**, 100686 (2022)
4. Smith, P., Kyriakidis, M., Majumdar, A.: Impact of European railway traffic management system on human performance in railway operations: European findings. Transp. Res. Rec. **2374**(1), 83–92 (2013)
5. Ahmad, A., Makarand, W.S., Vairavasundaram, I.: Design and implementation of SAE J1939 and Modbus communication protocols for electric vehicle. Machines **11**(2), 201 (2023)
6. Mu, X., Du, Y., Wang, H.: Research on cloud-edge AC power monitoring systems based on MODBUS. J. Phys. Conf. Ser. **2479**(1), 012007 (2023)
7. Tiko, I.: Creating a Technical Architecture Framework for m-voting Application. Afr. J. Sci. Technol. Innov. Dev. **14**(1), 86–93 (2022)
8. Feng, Y., Li, D., Tan, X.: Accelerating data transfer in dataflow architectures through a look-ahead acknowledgment mechanism. J. Comput. Sci. Technol. **37**(4), 942–959 (2022)
9. Qi, N., Zhang, X.: Optimization design and implementation of shared information management system for industrial design network platform. J. Comb. Optim. **45**(1), 29 (2022)
10. Chen, D., Wang, S., Zhang, W.: Design and implementation of bridge information management system based on BIM. J. Intell. Fuzzy Syst. **43**(3), 2973–2984 (2022)

Research on Short Term Power Load Forecasting Based on Wavelet and BiLSTM

Rongyang Liao, Juhui Ren[✉], and Chunlei Ji

Shanghai Dianji University, Shanghai 201306, China
renjh@sdju.edu.cn

Abstract. With the developing of big data and artificial intelligence, the application of smart grid has received widespread attention. Specifically, accurate power load forecasting plays an important role in the safety and stability of power system production dispatching. However, traditional load forecasting methods still has some limitations in processing large-scale nonlinear time series data. to accurately predict short-term power loads, a new forecasting method combining BiLSTM and Wavelet decomposition is proposed, named Wavelet-BiLSTM. First, the input time series data is decomposed into different sequences using wavelet decomposition. By comparing the prediction results of different decomposition levels, it is determined that a 2-level decomposition is the most appropriate, resulting in sequences A2, D1, and D2. Next, for each wavelet decomposition coefficient sequence, a separate BiLSTM model is constructed for training and prediction. Next, the prediction results of each Wavelet coefficient series are reconstructed by inverse Wavelet transformation to obtain the final load data prediction value. Experimental results show that the proposed Wavelet-BiLSTM method can improve the prediction precision and accuracy, Therefore, it's a promising approach for electricity load forecasting tasks.

Keywords: Load Forecasting · Neural Networks · BiLSTM · Wavelet Analysis

1 Introduction

Power Load (or active power) forecasting refers to the process of estimating or forecasting electricity demand over a period in the future. By analyzing historical data, weather conditions, social activities and other factors, power load forecasting can help power system operators and planners make sound decisions, including dispatching generator sets, optimizing energy supply and demand balance, and ensuring stable operation of the grid.

Given the dependence of power load on factors such as weather conditions and social activities, short-term load forecasting is more subject to randomness than medium and long-term load forecasting, Therefore, researchers generally agree that short-term forecasting is a challenging and interesting problem.

Short term prediction methods can generally be divided into the following three categories [1]: statistical methods, machine learning methods, and the hybrid method

© ICST Institute for Computer Sciences, Social Informatics and Telecommunications Engineering 2024
Published by Springer Nature Switzerland AG 2024. All Rights Reserved
J. Li et al. (Eds.): 6GN 2023, LNICST 553, pp. 53–65, 2024.
https://doi.org/10.1007/978-3-031-53401-0_7

which combines statistical and machine learning methods. Due to the nonlinear nature of power load time series, neural networks with nonlinear behavior learning capabilities are an ideal choice for power load forecasting. However, traditional neural networks also have some inherent drawbacks, such as slow learning speed, tendency to fall into local minimum solutions, and difficulty in determining the structure. To overcome these problems, researchers proposed the concept of wavelet neural networks [2], which combines wavelet transform with neural networks. In traditional neural networks, nonlinear sigmoid functions are replaced by nonlinear wavelet bases. Therefore, the wavelet neural network is formed by introducing the wavelet function with scaling and translation factors as the nonlinear function of the network.

Because Wavelet functions have scaling and translation factors, Wavelet neural networks provide higher degrees of freedom than traditional neural networks, allowing them to fit functions better. In order to analyze the time and frequency domain characteristics of power load data, Discrete Wavelet Transform (DWAVELET) is an applicable method that represents the load time series as multiple approximate and detail components with different frequency resolutions [3]. The DWAVELET method has been widely applied in load forecasting [4].

Recurrent neural networks (RNNs) are usually used to process time series data. However, traditional RNNs have limitations when dealing with long-term dependencies [5]. Long short-term memory (LSTM) is a special type of RNN, which does well to handle long-term dependence. Therefore, LSTM is widely used in load forecasting, in related studies such as [6]. BiLSTM is an extension of traditional LSTM, which can consider both past and future context information when processing sequence data. This structure allows the network to learn and model sequences from both directions, further improving the ability to represent sequence data [7].

To further improve load forecasting skill, a short-term load forecasting framework based Wavelet and BiLSTM networks is proposed in this paper. The proposed framework has following novelties. (1) Combining BiLSTM network with Wavelet decomposition to decompose time series data into subseries with different frequencies. This combination better captures the different frequency components in the time series data, providing more predictive information. (2) An independent BiLSTM model is established for training and prediction for each Wavelet coefficient sequence. Which can better adapt to the characteristics of different frequency components. (3) The prediction results of each Wavelet coefficient sequence are reconstructed by inverse Wavelet transform. In this way, the forecasts of the individual subseries can be combined to obtain the final load data forecast.

2   Related Methods and Models

2.1   Wavelet Transform

Wavelet transform is a mathematical tool for working with nonstationary and nonlinear signals. It converts time-domain signals into time-frequency domain representations, allowing us to analyze signals at different frequencies and time scales. In other words, Wavelet transforms decompose nonlinear and nonstationary data into a different set of frequency curves. These frequency components extract the appropriate statistical characteristics of the input data sequence. High-frequency components provide short-term variation information on data and help improve the accuracy of short-term predictions. Wavelet transforms can be divided into two types depending on the input signal: continuous Wavelet transform (CWT) and Discrete Wavelet transform (DWT) [8]. The continuous Wavelet transform can be mathematically expressed as:

$$W_{CWT}(a, b) = a^{-1/2} \int_{-\infty}^{\infty} PL(t) \psi\left(\frac{t-b}{a}\right) dt \tag{1}$$

Here, $PL(t)$ is the power load time series and $\psi(t)$ is the parent Wavelet, scaled and time-shifted by "a" and "b", respectively. The Wavelet basis function $\psi(t)$ needs to meet certain orthogonality and scale conditions.

The discrete Wavelet transform (DWT) of a time series can be expressed as a linear combination of Mother Wavelet, $\psi_{m,k}$ and Father Wavelet, $\psi_{M,k}$. This representation is based on the principle of multiscale analysis. Time series in DWT can be decomposed into Wavelet coefficients at different scales and translational downshifts. Specifically, for a time series of length N, Wavelet coefficients of multiple scales can be obtained by performing multi-level Wavelet decomposition. Mathematically, the discrete Wavelet transform is expressed as:

$$x(n) = \sum_k c_{m,k} \psi_{m,k}(n) + \sum_k d_{M,k} \psi_{M,k}(n) \tag{2}$$

where $c_{m,k}$. represents the mother Wavelet coefficient with scale m and translation k, and $d_{M,k}$ represents the parent Wavelet coefficient with scale M and translation k. The mother Wavelet and parent Wavelet are pairs of orthogonal function sets that represent the characteristics of a time series at different scales and translation.

By performing discrete Wavelet transforms, the decomposition representations of time series are obtained at different frequencies and time scales, to extract the time-frequency characteristics of the signal. This representation allows us to better analyze and process nonstationary signals and apply them to tasks such as signal compression, feature extraction, noise removal, and more in time series analysis.

Or in simple terms, the series of finite lengths for finite decomposition is:

$$PL(t)_W = D1 + D2 + D3 + \cdots + Dn + An \tag{3}$$

where "D1, D2,..., Dn" is the detailed component of the input time series, and "A1, A2, A3,..., An" is the approximate component of the input time series.

2.2 BiLSTM

BiLSTM (Bidirectional Long Short-Term Memory Network) is an extension of LSTM that introduces an additional inverse LSTM layer into the model. This allows the model to make predictions using both past and future contextual information, resulting in a more complete understanding of correlations in sequence data. The BiLSTM network consists of forward and backward LSTMs, where data can be processed forward and backward. Backward direction processing captures latent features and patterns within the data that are often overlooked by traditional LSTM models.

The architecture of the BiLSTM is shown in Fig. 1. Forward Hidden Layer for Updating the Network "L_f", Backward Hidden Layer "L_b" and output sequence "PL_o". The network is iteratively updated backwards, i.e., from "T" to "1", and forward, i.e., from "1" to "T". The update parameters can be mathematically expressed as:

$$L_f = \sigma \left(W_1 PL_i(t) + W_2 L_{f-1} \right) + b_{L_f} \tag{9}$$

$$L_b = \sigma \left(W_3 PL_i(t) + W_5 L_{b-1} \right) + b_{L_b} \tag{10}$$

$$PL_o = W_4 L_f + W_6 L + b_{PL_t} \tag{11}$$

where "L_f", "L_b" and "PL_o" represents the forward pass, backward pass, and final output layers respectively. "W" is the weight coefficient, "b_{L_f}" & "b_{L_b}" is the bias (Cheng et al., 2019), σ represents the Sigmoid function.

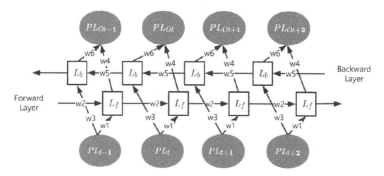

Fig. 1. The architecture of the BiLSTM

2.3 Other Models

A time series model is a statistical model used to analyze and forecast time series data. They consider temporal relationships between data points and autocorrelations within the series to extract patterns such as trend, seasonality, and periodicity, and make predictions. Following three common time series models will be selected for comparison.

ARIMA (autoregressive moving average model) [10]: The ARIMA model is a classic time series model for working with nonstationary data. It consists of three components: autoregressive (AR), differential (I), and sliding average (MA).

SARIMA (seasonal autoregressive moving average model) [11]: The SARIMA model is an extension of the ARIMA for working with time series data with seasonality. It adds seasonal differential and seasonal ARMA models to the ARIMA model.

Prophet [12]: The Prophet model is a time series forecasting model developed by Facebook for data with trends and seasonality. It employs additive decomposition models and utilizes nonlinear trends and seasonal components. The mathematical representation of the Prophet model is:

$$y(t) = g(t) + s(t) + h(t) + \varepsilon_t \tag{12}$$

where $g(t)$ represent the trend term, $s(t)$ represent the seasonal term, $h(t)$ represent the holiday effect, and ε_t represent the error term.

3 Proposed Wavelet-BiLSTM Model

The Wavelet-BiLSTM is proposed by combining BiLSTM and wavelet decomposition to integrate time and frequency domain information of time series, then to give more accurately predict trends and fluctuations in power loads. To construct the Wavelet-BiLSTM model, the following steps will be undertaken.

Step1. Wavelet decomposition: Use "db2" and other wavelet functions to perform two-layer wavelet decomposition on the time series data of the input electrical load. This results in three sets of subsequences, namely the approximation coefficient A2 and the detail coefficients D1 and D2.

Step2. Independent BiLSTM training and prediction: For each wavelet-decomposed sequence (A2, D1, D2), an independent BiLSTM is used for training and prediction. Each BiLSTM network will accept a wavelet factorization sequence as input and output the corresponding prediction results.

Step3. Subseries prediction reconstruction: In order to obtain the final load data prediction result, the predicted value of each subseries is reconstructed in the BiLSTM network. Specifically, the inverse wavelet transform is used to recombine the prediction results of each wavelet factorization sequence into a single load forecast sequence. Here is n-layer wavelet decomposition:

$$PL(t)_W = D1 + D2 + D3 + \ldots D_n + An \tag{13}$$

Independent BiLSTM training and prediction:

$$A_n = BiLSTM_{A_n}(PL(t)_W) \tag{14}$$

$$D_n = BiLSTM_{D_n}(PL(t)_W) \tag{15}$$

$$D_{n-1} = BiLSTM_{D_{n-1}}(PL(t)_W) \tag{16}$$

$$\cdots\cdots$$

$$D_1 = BiLSTM_{D_1}(PL(t)_W) \tag{17}$$

Subsequence prediction reconstruction:

$$\hat{X}(t) = waverec(A_n, D_n, D_{n-1}, \ldots\ldots D_1) \tag{18}$$

4 Experimental Schema

4.1 Data

The proposed prediction method is verified based on the actual power data. These data come from electricity load data in a region of China, The data spans from August 14, 2018, to August 31, 2021, The day in the data is divided into 96 time points, There are over 100,000 data points in total. To make the experiment more rigorous, the periodicity of the data was first verified, the results are shown in Figs. 2, 3 and 4.

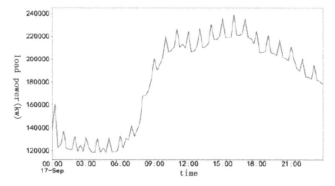

Fig. 2. Daily periodicity

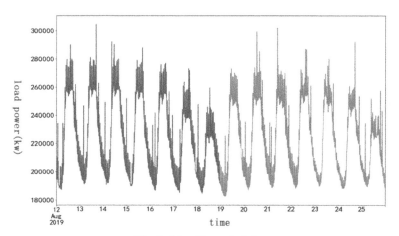

Fig. 3. Monthly periodicity

As can be seen from the Figures, the load data is cyclical on a daily, weekly, and yearly basis. In addition to checking the periodicity of data, it also should check the missing values, outlier, duplicate values, and the stability of data. These data processing parts will be described in detail in the chapter of model establishment.

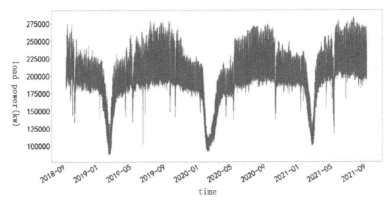

Fig. 4. Annual periodicity

4.2 Evaluation Indicators

To evaluate different models, four metrics are adopted in this paper. The first one is the MAPE (Average Absolute Percentage Error), as follows:

$$e_{MAPE} = \frac{\sum_{i=1}^{n} \frac{PL_i - \tilde{PL}_i}{PL_i}}{n} \tag{19}$$

where is PL_i represents the actual value, the predicted value is wide, and the total number of predictions is n. The smaller the MAPE, the higher the accuracy.

The second error metric used is the RMSE (Root Mean Square Error), as follows:

$$e_{RMSE} = \sqrt{\frac{1}{n} \sum_{i=1}^{n} \left(PL_i - \tilde{PL}_i\right)^2} \tag{20}$$

The third measurement factor used is R2 (Coefficient Of Determination), which can be used to represent the goodness of the model and is the square of the correlation coefficient range from 0 to 1, as follows:

$$R^2 = 1 - \frac{\sum_{i=1}^{n} \left(PL_i - \tilde{PL}_i\right)^2}{\sum_{i=1}^{n} (PL_i - mean(PL_i))^2} \tag{21}$$

The last one used is MAE (Absolute Mean Square Error), as follows:

$$e_{MAE} = \frac{1}{n} \sum_{i=1}^{n} |PL_i - \tilde{PL}_i| \tag{22}$$

where n represents the total number of predictions, PL_i represents the true value, $\tilde{PL}_i$ represents the predicted value, and the smaller the MAE, the higher the accuracy.

4.3 Model Construction

The Wavelet-based BiLSTM is constructed by combing Wavelet and BiLSTM. The overall flowchart is shown in Fig. 5. The specific steps are as follows.

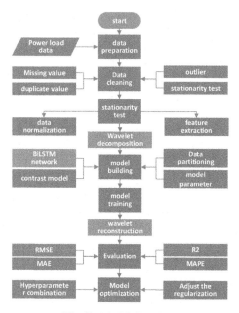

Fig. 5. Model flowchart

Step1: Data preparation. The collection of 15-min load data load data in a domestic region is the first step of data mining, including time series and corresponding load values. Python data mining is used to mine the 15-min load data of the region.

Step2: Data cleaning. Due to the large and messy data, it is impossible to directly analyze the data manually, but by using the relevant models and mathematical analysis, the data can become more regular, and more targeted data processing can be carried out to obtain the relevant information needed. In this experiment, the power load data collected from the site is in its original form, which will have a great impact on the model during model training, so the purpose of data cleaning is to obtain clean and reliable data to improve the accuracy and robustness of the load prediction model. Data cleansing is an important part of data preprocessing, which can help remove noise, outliers, and irregularities from the data, making the data more suitable for modeling and analysis. Specifically, we identified duplicates, outliers, missing values in the data, and imputed them with KNN neighbors. Data stationarity test: Most load data collected from a site is usually random, nonlinear, periodic, seasonal, and unexpected events. Therefore, in order to improve the efficiency of the model, the data should be preprocessed to meet the stationarity requirements before feeding it to the model. This preprocessing may involve removing trends, seasonality, or other influencing factors, thereby ensuring the data remains stable over time. Only when the data reaches a stable state can it be used for training and making predictions with the predictive model. This paper considers the root of unity test for verification, using the most used ADF test (Augmented Dickey-Fuller Test).

By performing a unit root test on the load data, its stationarity can be determined, and further smoothing processes such as differential, seasonally adjusted, etc., can be performed as needed to improve the accuracy and efficiency of the load forecasting model. In order to better understand and predict the characteristics of the load data, the load data is decomposed in time series, and the decomposition chart is shown in Fig. 6, by performing time series decomposition, we can gain a more comprehensive understanding of the composition of the load data.

This decomposition allows to separate the data into different components, such as the trend, seasonality, and residuals. By utilizing these components as features and inputting them into the model, we can enhance the accuracy and interpretability of load forecasting. This approach can capture the underlying patterns and variations in the data, resulting in more accurate predictions and better insights into the factors affecting the load behavior. In addition, the decomposed time series components can be used for deeper data analysis and mining, revealing hidden patterns and trends in load data.

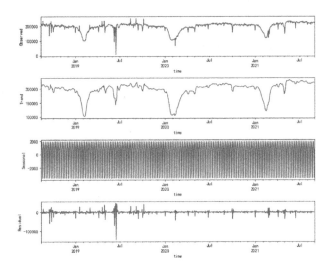

Fig. 6. Time series decomposition plot

Steps3: Feature Engineering. This step can be divided into two parts: (1) Wavelet decomposition: Firstly, wavelet decomposition is performed on the power load time series data after data governance, and it is decomposed into sub-sequences of different frequencies, including approximation coefficients and detail coefficients. This allows the extraction of feature information at different frequencies and captures short- and long-term changes in the data. Then, Wavelets are applied to input data series to decompose into approximate and detailed combinations. This work observes that the Daubechies (db7) wavelets provide the smoothness required for the data under consideration. However, wavelet decomposition up to levels 1–7 was observed experimentally, with stage

2 and DB7 found to be the most suitable dataset for use in the study. (2) Feature transformation. Transform features to meet the requirements of the model or improve the distribution of features. For example, normalization operations can be performed to make feature data more distributed or comparable.

Steps4: (1) Data Setup: Prepare data through feature engineering, then split into training, validation, and test sets. (2) Model Architecture: Design BILSTM-Wavelet model combining bidirectional LSTM for temporal dependencies and wavelet characteristics for frequency components. (3) Build comparative models: Build ARIMA, SARIMA, Prophet, BiLSTM as benchmark models.

Steps5: Model training. Use the training set to train the BILSTM-Wavelet model. The weights and parameters of the model are adjusted by backpropagation algorithms and optimizers (such as Adam, SGD, etc.) so that it can adapt to the characteristics of the power load data.

Step6: Prediction reconstruction. Predict each wavelet coefficient sequence to obtain corresponding prediction results. The IDWAVELET (Inverse Discrete Wavelet Transform) is used to combine the prediction results of each wavelet coefficient sequence with other wavelet coefficient sequences to reconstruct the prediction results of the load data, and the final prediction result is the predicted value of the reconstructed load data.

Step7: Model optimization. Utilize the validation set to verify the trained model and fine-tune it based on the validation results. Evaluate the model's performance using metrics like mean squared error, mean absolute error, and others, and make adjustments to hyperparameters or enhance the model structure as required. This iterative process allows for continuous improvement and ensures the model's effectiveness in accurately predicting the desired outcomes.

Step8: Model optimization. Use the validation set to verify the trained model, and tune the model based on the verification results. You can evaluate the performance of the model by monitoring metrics such as mean squared error, mean absolute error, etc., and adjust hyperparameters or improve the model structure as needed.

5   Results and Analysis

All experiments in this paper are conducted in a server, whose CPU is Intel i7-12700H, the memory is 32GB, the GPU is NVidia RTX2020.

Here, results from ARIMA, SARIMA, Prophet, LSTM, BiLSTM, and Wavelet-BiLSTM predictive models are compared. All network structures used are two-layered, the number of neurons per layer is set to 64, the activation function is selected tanh, the fully connected layer activation function is Relu, and the step of LSTM is 5. The dataset inputs are all historical loads and times. The day is divided into 96 moments, and all models output a time prediction at a time. In this paper, 99% of the total data in each dataset is divided into training sets, and the power load data for the next 10 days is predicted. Table 1 lists the values predicted by each model for ten days.

Table 1. Comparison of error predictions for different models over 10 days

Model	ARIMA	SARIMA	Prophet	LSTM	BiLSTM	Wavelet-Bilstm
e_{MAPE}	11.29%	9.73%	6.58%	2.46%	1.49%	0.91%
e_{RMSE}	35761.778	23759.121	17871.44	6301.531	4498.488	2950.830
R^2	−1.056	0.0931	0.3469	0.9360	0.9674	0.9860
e_{MAE}	28055.978	21211.736	13152.28	5473.8505	3491.8701	2119.2550

Prediction Accuracy: In terms of e_{MAPE} (Mean Absolute Percentage Error) and e_{RMSE} (Root Mean Square Error), the Wavelet-BiLSTM model performs the best with the lowest values of 0.91% and 2950.830, respectively. The BiLSTM model follows with an e_{MAPE} of 1.49% and an e_{RMSE} of 4498.488. The Prophet model also shows good performance with an e_{MAPE} of 6.58% and an e_{RMSE} of 17871.44. ARIMA and SARIMA models exhibit lower prediction accuracy.

Goodness of Fit: R^2 (Coefficient Of Determination) measures the goodness of fit of the models to the observed values, with a range of $[-\infty, 1]$. According to the R^2 values, the Wavelet-BiLSTM model demonstrates the highest goodness of fit with a value of 0.9860, followed by the BiLSTM model (0.9674) and the Prophet model (0.3469). The ARIMA and SARIMA models exhibit lower R^2 values, and some even show negative values, indicating a poor fit with the observed values.

Error Analysis: Based on the e_{MAE} (Mean Absolute Error) results, the Wavelet-BiLSTM model achieves the lowest error (2119.2550), followed by the BiLSTM model (3491.8701) and the Prophet model (13152.28). The ARIMA and SARIMA models show larger errors.

The results of the Wavelet-BiLSTM model demonstrate significantly better performance compared to other models. Deep learning networks outperform traditional time series analysis methods, and bidirectional LSTM models outperform unidirectional LSTM models. Models that incorporate Wavelet analysis outperform those without it, as bidirectional LSTM effectively captures features in the sequence, and the Wavelet analysis mechanism further enhances the prediction accuracy.

Figure 7 depicts a comparison between the actual and predicted data for the next 10 days using various models, including ARIMA, SARIMA, prophet, LSTM, BiLSTM, and Wavelet-BiLSTM. The true values are represented by the blue line, while the orange line represents the predicted values. The visual comparison illustrates the superior predictive ability of the Wavelet-BiLSTM model over the other methods, confirming its effectiveness in forecasting future data accurately.

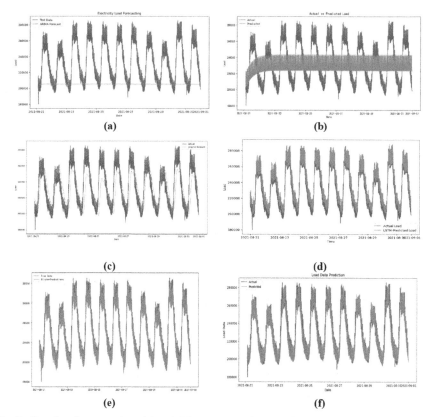

Fig. 7. Results of proposed model and Other comparative models. (a) ARIMA (b) SARIMA (c) Prophet (d) LSTM (e) BILSTM (f) Wavelet-Bilstm

6   Conclusion

A load aggregation model for predicting the next 10 days is proposed using Wavelet and BiLSTM networks in this paper. Wavelet is used to extract statistical features of the input time series using low- and high-frequency time series, while the BiLSTM network is used for forecasting, through power load forecasting, using different evaluation criteria to rigorously evaluate the performance of the proposed model. At the same time, the predictive performance of the proposed model is compared with traditional time series analysis, independent LSTM, BiLSTM and two different Wavelet-based BiLSTM models. The results show that this model outperforms other models in RMSE, MAE, R2 and MAPE. Finally, compared with BiLSTM and LSTM, the maximum percentile improvement of the proposed model on RMSE is 34.4% and 53.17%, respectively. Compared with the time series models SARIMA and Prophet, the maximum percentile improvement of RMSE was 83.49% and 87.58%, respectively. Therefore, considering all power load time series factors, the proposed model proves to be the best forecasting model for 10-day load forecasting. However, the study found some challenges when designing the model, such as selecting the exact appropriate Wavelet basis function and decomposition level, and feature engineering of parameters and data for BiLSTM networks. Therefore, more accurate and efficient results will be obtained in a fast simulation time in the future.

References

1. Nti, I.K., Teimeh, M., Nyarko-Boateng, O., et al.: Electricity load forecasting: a systematic review. J. Electr. Syst. Inf. Technol. **7**(1), 1–19 (2020)
2. Ghoushchi, S.J., Manjili, S., Mardani, A., et al.: An extended new approach for forecasting short-term wind power using modified fuzzy wavelet neural network: a case study in wind power plant. Energy **223**, 120052 (2021)
3. Gong, M., Wang, J., Bai, Y., et al.: Heat load prediction of residential buildings based on discrete wavelet transform and tree-based ensemble learning. J. Build. Eng. **32**, 101455 (2020)
4. Zhang, L., Alahmad, M., Wen, J.: Comparison of time-frequency-analysis techniques applied in building energy data noise cancellation for building load forecasting: a real-building case study. Energy Build. **231**, 110592 (2021)
5. Bouktif, S., Fiaz, A., Ouni, A., et al.: Multi-sequence LSTM-RNN deep learning and metaheuristics for electric load forecasting. Energies **13**(2), 391 (2020)
6. Mokarram, M.J., Rashiditabar, R., Gitizadeh, M., et al.: Net-load forecasting of renewable energy systems using multi-input LSTM fuzzy and discrete Wavelet transform. Energy **275**, 127425 (2023)
7. Siami-Namini, S., Tavakoli, N., Namin, A.S.: The performance of LSTM and BiLSTM in forecasting time series. In: 2019 IEEE International Conference on Big Data (Big Data), pp. 3285–3292. IEEE (2019)
8. El-Hendawi, M., Wang, Z.: An ensemble method of full wavelet packet transform and neural network for short term electrical load forecasting. Electric Power Syst. Res. **182**, 106265 (2020)
9. Yu, Y., Si, X., Hu, C., et al.: A review of recurrent neural networks: LSTM cells and network architectures. Neural Comput. **31**(7), 1235–1270 (2019)
10. Khan, S., Alghulaiakh, H.: ARIMA model for accurate time series stocks forecasting. Int. J. Adv. Comput. Sci. Appl. **11**(7) (2020). SARIMA
11. Dubey, A.K., Kumar, A., García-Díaz, V., et al.: Study and analysis of SARIMA and LSTM in forecasting time series data. Sustain. Energy Technol. Assess. **47**, 101474 (2021)
12. Bashir, T., Haoyong, C., Tahir, M.F., et al.: Short term electricity load forecasting using hybrid prophet-LSTM model optimized by BPNN. Energy Rep. **8**, 1678–1686 (2022)

Big Data Mining, D2D Communication, Security and Privacy for 6G Networks

Spectrum Allocation Algorithm Based on Improved Chimp Optimization Algorithm

Xingdong Huo, Kuixian Li, and Hang Jiang[✉]

Harbin Engineering University, Harbin 150001, China
jha@hrbeu.edu.cn

Abstract. In response to the rapid growth of spectrum resource demand and low utilization of spectrum, this paper proposes a spectrum allocation solution based on improved chimp optimization algorithm (ICHOA). First, opposition-based learning is applied to improve the quality of the initial solution by generating opposing solutions; Next, a proportional weight-based location update method is used to dynamically adjust the location update vector to play the leading role of the attacking chimp. Then, a nonlinear convergence factor is used for spatial search to enhance the local exploration ability and global exploitation ability; Finally, the improved algorithm is combined with the spectrum allocation model to maximize the efficiency of the system. Compared with genetic algorithm (GA), improved particle swarm optimization algorithm (IPSO), improved gray wolf algorithm (IGWO) and traditional chimp optimization algorithm (CHOA), the simulation results show that ICHOA algorithm has faster convergence speed and higher system benefits in spectrum allocation. It effectively improves the utilization rate of spectrum resources and the proportion fairness of cognitive users.

Keywords: Spectrum Allocation · Chimp Optimization Algorithm · Proportional Weight · Opposition-based Learning · Fairness

1 Introduction

Along with the rise of 6th generation wireless systems [1], radio communication has been widely applied in new wireless devices, playing an increasingly important role in human production and life. However, it also brings a series of problems. The development of emerging services further intensifies the gap of spectrum resources and seriously restricts the development and application of new technologies. As a non-renewable strategic resource, allocating spectrum resources reasonably to different users is an important goal to meet people's communication needs. Therefore, many experts have proposed cognitive radio [2, 3] technology to maximize spectrum utilization through dynamic allocation [4] of spectrum resources.

This work is supported by the National Natural Science Foundation of China (No: 62201172), the National Key Research and Development Program of China (2022YFE0136800). This work is also supported by Key Laboratory of Advanced Marine Communication and Information Technology, Ministry of Industry and Information Technology, Harbin Engineering University, Harbin, China.

J. Li et al. (Eds.): 6GN 2023, LNICST 553, pp. 69–78, 2024.
https://doi.org/10.1007/978-3-031-53401-0_8

At present, graph theory [5], game theory [6] and auction bidding models [7] are usually used to solve spectrum allocation problems. Spectrum allocation based on graph theory is a typical NP-hard optimization problem, and intelligent optimization algorithm is a common method to solve such problems. In recent years, scholars have paid more attention to solving problems in the field of spectrum allocation by intelligent optimization algorithms, and have achieved many results. In literature [8], a particle swarm optimization is proposed to adjust parameters by using sawtooth mode. To a certain extent, this algorithm improves the resource allocation efficiency of cellular mobile networks. Literature [9] proposes a slime mold algorithm combining the addition of unselected factors, and the transfer function with the best performance is selected to solve the spectrum problem. The spectrum allocation scheme obtained can realize the effective use of network resources. In order to realize efficient resource scheduling of satellite wireless signals, an artificial bee swarm algorithm combined with differential evolution strategy in literature [10], is improved to improve the fairness and throughput of cognitive users. A multi-agent Cuckoo algorithm is proposed in literature [11], which reduces the computational complexity of 5G network resource allocation and effectively improves the speed of spectrum access.

However, there is no good balance between the optimization ability and convergence speed of the above algorithms. Therefore, a new spectrum allocation algorithm based on the improved chimp optimization algorithm (ICHOA) is proposed. Compared with genetic algorithm (GA), improved particle swarm optimization algorithm (IPSO), improved gray Wolf algorithm (IGWO) and chimp optimization algorithm (CHOA), ICHOA has higher system benefits and faster convergence rate. The experiment results demonstrate the feasibility and superiority of ICHOA algorithm.

2   Graph Coloring Model

This paper will consider two spectrum sharing modes: overlay and underlay. In overlay mode, the system can only allocate the unused authorized channel of the primary user to the sub-user. In underlay mode, the system can assign any authorized channel to the sub-user, but the transmitting power of sub-user should be controlled below the interference threshold of the given primary user. No matter in which mode, it is to consider whether the channel is available to sub-users. Therefore, the spectrum allocation problem is modeled as a graph theory coloring problem, which is described by the following parameters [12, 13].

Channel availability matrix L:

$$L = \left\{ l_{n,m} \middle| l_{n,m} \in \{0, 1\} \right\}_{N \times M} \tag{1}$$

where N is the number of sub-users and M is the number of non-overlapping channels, if $l_{n,m} = 1$, sub-user n can use channel m; Otherwise, m cannot be assigned to n.

Channel reward matrix B:

$$B = \{b_{n,m}\}_{N \times M} \tag{2}$$

where B denotes the benefit obtained by n using m. If $l_{n,m} = 0$, $b_{n,m} = 0$.

Interference constraint matrix C:

$$C = \left\{ c_{n,k,m} \middle| c_{n,k,m} \in \{0, 1\} \right\}_{N \times K \times M} \tag{3}$$

where C denotes whether there is interference between sub-users n and k when they use m together. If $c_{n,k,m} = 0$, n and k will not interfere with each other and can jointly use m. In practice, spectrum resources should be allocated to as many devices as possible. Therefore, in the presence of frequency conflict, priority should be given to the equipment with less spectrum resources.

Conflict free channel assignment matrix A:

$$A = \left\{ a_{n,m} \middle| a_{n,m} \in \{0, 1\} \right\}_{N \times M} \tag{4}$$

where A denotes the channel allocation result. If $a_{n,m} = 1$, n can use m. At the same time, the matrix satisfies the constraint conditions of L and C.

User system benefits R:

$$R = \{\alpha_n\}_{N \times 1} \tag{5}$$

where R denotes the benefit obtained by sub-user n on all channels.

$$\alpha_n = \sum_{m=1}^{M} a_{n,m} \times b_{n,m} \tag{6}$$

Maximum total system benefit U_R:

$$U_R = \sum_{n=1}^{N} \sum_{m=1}^{M} a_{n,m} \times b_{n,m} \tag{7}$$

where U_R denotes the total system benefit for all sub-users on the available spectrums within a certain time slot.

$$U_F = \left(\prod_{n=1}^{N} \left(\sum_{m=1}^{M} a_{n,m} b_{n,m} + 10^{-6} \right) \right)^{1/N} \tag{8}$$

where U_F denotes the fairness among cognitive users, which is used to maximize the fairness of spectrum resource allocation.

According to the above definition, spectrum allocation problem is equivalent to finding matrix A, when the constraints of L and C are satisfied and U_R is maximized.

3 Spectrum Allocation Based on Improved Chimp Optimization Algorithm

3.1 Traditional Chimp Optimization Algorithm

M. Khishe et al. [14] proposed CHOA based on individual behavior and sexual motivation in 2020. In a chimp group, four types of chimps lead the other chimps to hunt for the prey (optimal solution), in which Attacker is the leader who plays a decision-making role and is responsible for predicting the location of the prey, Barrier play a decision-making role, Chaser is responsible for chasing the prey, and Driver and other chimps assist in driving the prey. During the hunt, the chimps position changes with the position of the prey. Therefore, the authors model the algorithm by mimicking chimp behavior:

$$D = |G \cdot X_p(t) - H \cdot X_c(t)| \tag{9}$$

$$X_c(t+1) = X_p(t) - K \cdot D \tag{10}$$

$$K = 2 \cdot f \cdot r_1 - f \tag{11}$$

$$G = 2 \cdot r_2 \tag{12}$$

$$H = \text{Chaotic}_{\text{value}} \tag{13}$$

where X_c and X_p are the position of the current chimp and the current prey respectively, K, G and H are the coefficient vector, f is the convergence factor, r_1 and r_2 take random values from 0 to 1, H is the chaotic vector based on Gaussian distribution.

$$\begin{cases} D_{\text{attacker}} = |G_1 \cdot X_{\text{attacker}} - H_1 \cdot X^*| \\ D_{\text{aarrier}} = |G_2 \cdot X_{\text{barrier}} - H_2 \cdot X^*| \\ D_{\text{chaser}} = |G_3 \cdot X_{\text{chaser}} - H_3 \cdot X^*| \\ D_{\text{driver}} = |G_4 \cdot X_{\text{driver}} - H_4 \cdot X^*| \end{cases} \tag{14}$$

$$\begin{cases} X_1^* = X_{\text{attacker}} - K_1 \cdot D_{\text{attacker}} \\ X_2^* = X_{\text{barrier}} - K_2 \cdot D_{\text{barrier}} \\ X_3^* = X_{\text{chaser}} - K_3 \cdot D_{\text{chaser}} \\ X_4^* = X_{\text{driver}} - K_4 \cdot D_{\text{driver}} \end{cases} \tag{15}$$

$$X^*(t+1) = (X_1^* + X_2^* + X_3^* + X_4^*)/4 \tag{16}$$

where X_{attacker} means the optimal solution, X_{barrier} denotes the suboptimal solution, X_{chaser} denotes the third solution, X_{driver} denotes the fourth solution, and $X^*(t+1)$ denotes the updated position of the chimp.

3.2 Improved Chimp Optimization Algorithm

Population Initialization based on Oppositional Learning
CHOA algorithm makes the chimp locations evenly distributed in the search space through random initialization. Since the initialization of the population affects the search efficiency of the algorithm, if the initial solution position generated by random initialization is poor, the convergence rate of the algorithm will be directly affected. Therefore, the initialization method based on opposition learning [15] is adopted to increase the diversity of the initial population and enhance the search efficiency of the algorithm by generating opposition solutions. The detailed steps are as follows:

a) In the D-dimensional search space, randomly initialize chimps as the initial population $Q = [q_1, q_2, \cdots q_d]$, where d is the number of chimps in the population.
b) Use the method of opposite learning to generate the opposite population Q' of the initial population Q. Suppose the individual position of the chimp is $Q = (0, 1, 1, ..., 0)$, then the position of the opposite individual of this chimp is $Q' = (1, 0, 0, ..., 1)$.
c) Compare the fitness values of all the chimps in the original population Q and the opposing population Q', and the chimps with the top N fitness values are selected as the new initial population.

Position Update based on Proportional Weights
The traditional chimp optimization algorithm only calculates the average of the four chimp positions through formula (16) for position update, without considering the hierarchical relationship between chimp groups or the contribution of the attacking chimp as the decision maker. Therefore, a position updating formula based on proportional weights is proposed in this paper.

$$\gamma_1 = F_{it1}/(F_{it1} + F_{it2} + F_{it3} + F_{it4}) \tag{17}$$

$$\gamma_2 = F_{it1} + F_{it2}/(F_{it1} + F_{it2} + F_{it3} + F_{it4}) \tag{18}$$

$$\gamma_3 = F_{it1} + F_{it2} + F_{it3}/(F_{it1} + F_{it2} + F_{it3} + F_{it4}) \tag{19}$$

$$X(t+1) = \begin{cases} X_1, \ rand(1) < \gamma_1 \\ X_2, \ \gamma_1 < rand(1) < \gamma_2 \\ X_3, \ \gamma_2 < rand(1) < \gamma_3 \\ X_4, \ \gamma_3 < rand(1) \end{cases} \tag{20}$$

where $F_{it1}, F_{it2}, F_{it3}$ and F_{it4} are the fitness values of the four types of chimps respectively, γ_1, γ_2 and γ_3 are the weight factors. Obviously, in each iteration, the value of the weight factor changes with the fitness value. Therefore, the probability of the four chimps being selected in each iteration also changes dynamically, and the higher the social rank, the higher the chance of being selected when the position is updated, which improves the speed of algorithm convergence.

Spatial Search based on Nonlinear Convergence Factor
In CHOA, if $|K| > 1$, the population will expand the range of hunting, search for prey, and global search; If $|K| < 1$, the population will narrow down the area of contraction, attack the prey and conduct a local search. Formula (11) shows $|K|$ dynamically changes with the change of f. The convergence factor f decreases linearly in CHOA algorithm, resulting in poor global search ability in the later stage and easily falls into the local optimal. Therefore, a nonlinear convergence factor f is shown below:

$$f = S_M \cdot e^{-(S_N \cdot \frac{iter}{iter_{max}})^4} \tag{21}$$

where S_M and S_N are constant, $iter$ is the number of iterations, $iter_{max}$ is the maximum number of iterations.

According to formula (21), convergence factor f changes slowly in the early stage, which makes $|K| > 1$ take up a larger proportion in the iterative process and enhances the global exploration ability; f changes rapidly in the late stage, which causes the proportion of $|K| < 1$ in the iterative process to decrease and the algorithm performs local search rapidly in the late stage. This allows the algorithm to find the optimal solution more quickly and accurately.

3.3 Spectrum Allocation Based on Improved Chimp Optimization Algorithm

For the binary discrete optimization problem, the variables of spectrum allocation are all 0 or 1 in the discrete space. Therefore, the transfer function proposed in the literature [16] is introduced in this paper to discretize the algorithm, and the channel allocation problem is then transformed into the problem of finding the optimal solution for chimp population. Different chimp locations [16] represent different spectrum allocation schemes. The purpose is to maximize the overall benefit of the system without interference and find out the best spectrum allocation scheme. The detailed steps are as follows:

Step 1: Set the population size $Nagent$ and $iter_{max}$, initialize L, B, C and the positions of chimps X, adopt the method of opposing learning to generate opposing solutions and form initial population with higher fitness value.

Step 2: Check whether the individual position meets the constraint conditions. Determine sub-user n and sub-user k that satisfy the condition $c_{n,k,m} = 1$, and then verify that the values $a_{n,m}$ and $a_{k,m}$ of A are the same. When both values are 1, set the sub-user with more spectrum resources to 0. Finally, the solution treated with interference constraints is encoded in binary.

Step 3: Update algorithm parameters according to Eqs. (11) (12), (13) and (21), and update algorithm position according to Eqs. (14), (15) and (20).

Step 4: The corresponding fitness value is obtained according to B, and the best fitness value and corresponding position are saved as the optimal solution and the optimal position.

Step 5: If $iter < iter_{max}$, skip to step 3 to execute the loop, otherwise print out the maximum system benefit and the optimal solution.

4 Experimental Evaluation and Results

4.1 Parameter Setting

To verify the performance of the proposed ICHOA algorithm in cognitive radio spectrum allocation, a comparison test was conducted with GA, IPSO, GWO [18] and CHOA. Population size $Nagent = 30$. L, B and C of the simulation experiment in different scenarios are different and the same in the same experiment, which can be initialized in the simulation program. The test results are obtained by the algorithm repeated 30 times to obtain the average value.

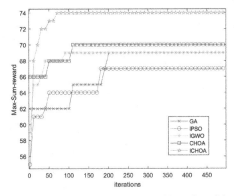

Fig. 1. When M = 10, N = 20, relationship between total benefits of the system and the number of iterations.

4.2 Simulation Results and Performance Analysis

Figure 1 shows the convergence curve of each algorithm in a random experiment when N = 20 and M = 10. As seen, the system benefit increases with the number of iterations. The final system benefit of ICHOA algorithm is much higher than that of other algorithms. ICHOA algorithm also has quicker convergence speed and can quickly find the optimal solution in the spectrum allocation optimization process. This is because ICHOA enhances the exploration ability in the early stage and the local search ability in the late stage, which can effectively promote the population to move to the optimal position and show a good effect in spectrum allocation.

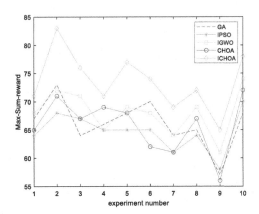

Fig. 2. When M = 10, N = 20, benefit of the system.

Figure 2 shows the relationship between the total system benefits and the number of experiments. In 20 experiments, N = 20 and M = 10 remained unchanged. It can be seen that the total system benefit changes dynamically in each experiment, but the system benefit generated by ICHOA algorithm is always higher than other algorithms.

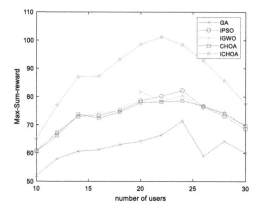

Fig. 3. When M = 10, N changes from 10 to 30, benefit of the system.

Figure 3 and Fig. 4 show the relationship curves between system benefit and sub-users, the relationship curves between average system benefit and sub-users respectively. M = 10 is fixed and the number of sub-users increases from 10 to 30. As can be seen, the benefits of the system first increase and then gradually decrease. This is because the system benefit is the sum of the benefits of all sub-users. As the number of sub-users increases, multiple sub-users will use the same spectrum, and the system benefit gradually increases. However, with the further increase of sub-users, due to the limited number of available spectrum and the increasing interference between sub-users and primary users, and between sub-users and sub-users, so when the number of sub-users increases to a certain extent, the benefit of the system declines. Nevertheless, the system efficiency achieved by the ICHOA algorithm is still the highest. The average system benefit refers to the average benefit obtained by all sub-users participating in the spectrum allocation. As the number of sub-users grows, the competition for sub-users to acquire a small amount of spectrum resources becomes more and more fierce, and the average system efficiency becomes lower and lower.

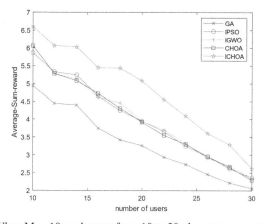

Fig. 4. When M = 10, n changes from 10 to 30, the average system benefit.

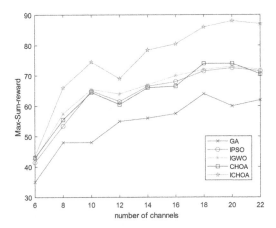

Fig. 5. When N = 20, M changes from 10 to 30, benefits of the system.

Figure 5 shows the relationship between system benefit and the amount of spectrum available. When N = 20 remains unchanged, M increases from 10 to 30, the system benefit increases with the amount of spectrum, because under the same conditions, the more the number of spectrums, the more channel resources available to sub-users, the more system benefits. The ICHOA algorithm has the highest total system benefit compared to other algorithms.

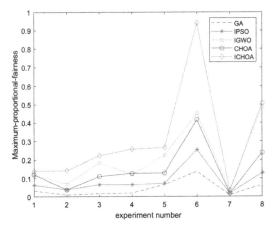

Fig. 6. When M = 10, N = 20, the maximum of the sub-user's proportional fairness.

Figure 6 shows the relationship between sub-user access fairness and the number of experiments. In 8 experiments, N = 20 and M = 10 remained unchanged. It is known that ICHOA has the greatest network fairness, which is better than other algorithms, and reflects the effectiveness of ICHOA algorithm.

5 Conclusion

In this article, in order to improve the efficiency and quality of spectrum allocation, the chimp algorithm is improved by adding population initialization based on opposition learning, positions update based on proportional weights, and space search based on nonlinear convergence factors. Compared with other representative algorithms, simulation results show that ICHOA can achieve faster convergence speed and higher system benefits when solving the spectrum allocation problem, and spectrum resources can be better utilized and allocated.

References

1. Wang, M., Lin, Y., Tian, Q., et al.: Transfer learning promotes 6G wireless communications: recent advances and future challenges. IEEE Trans. Reliab. **70**(2), 790–807 (2021)
2. Lin, Y., Wang, C., Wang, J., et al.: A novel dynamic spectrum access framework based on reinforcement learning for cognitive radio sensor networks. Sensors **16**, 1675 (2016)
3. Ya, T., et al.: Large-scale real-world radio signal recognition with deep learning. Chin. J. Aeronaut. **35**(9), 35–48 (2022)
4. Hei, Y., Qiu, Z., Liu, J., et al.: Efficient Taguchi algorithm for cognitive radio spectrum allocation. Trans. Emerg. Telecommun. Technol. **27**(5), 640–647 (2016)
5. Wang, Y., Ye, Z., Wan, P., et al.: A survey of dynamic spectrum allocation based on reinforcement learning algorithms in cognitive radio networks. Artif. Intell. Rev. **51**(3), 493–506 (2019)
6. Teng, Z., Xie, L., Chen, H., et al.: Application research of game theory in cognitive radio spectrum allocation. Wireless Netw. **25**(7), 4275–4286 (2019)
7. Sofia, D.S., Edward, A.S.: Auction based game theory in cognitive radio networks for dynamic spectrum allocation. Comput. Electr. Eng. **86**, 106734 (2020)
8. Zhang, X., Zhang, X., Wu, Z.: Utility- and fairness-based spectrum allocation of cellular networks by an adaptive particle swarm optimization algorithm. IEEE Trans. Emerg. Top. Comput. Intell. **4**(1), 42–50 (2020)
9. Li, L., Pan, T.S., Sun, X.X., et al.: A novel binary slime mould algorithm with AU strategy for cognitive radio spectrum allocation. Int. J. Comput. Intell. Syst. **14**, 1–18 (2021)
10. Xiao, Y., Chen, D., Zhang, L.Y.: Research on spectrum scheduling based on discrete artificial bee colony algorithm. J. Phys. Conf. Ser. **1856**(1), 012059 (2021)
11. Ai, N., Wu, B., Li, B., et al.: 5G heterogeneous network selection and resource allocation optimization based on cuckoo search algorithm. Comput. Commun. **168**, 170–177 (2021)
12. Cao, Y., Li, Y., Ye, F.: Improved fair spectrum allocation algorithms based on graph coloring theory in cognitive radio networks. J. Comput. Inf. Syst. **7**(13), 4694–4701 (2011)
13. Zhang, H., Peng, S., Zhang, J., Lin, Y.: Big data analysis and prediction of electromagnetic spectrum resources: a graph approach. Sustainability **15**, 508 (2023)
14. Khishe, M., Mosavi, M.R.: Chimp optimization algorithm. Expert Syst. Appl. **149**, 113338 (2020)
15. Cao, C., Li, K.: Spectrum allocation algorithm based on improved wolf swarm algorithm. In: 2022 9th International Conference on Dependable Systems and Their Applications (DSA), Wulumuqi, China, pp. 1000–1001 (2022)
16. Emary, E., Zawba, H.M., Hassanien, A.E.: Binary grey wolf optimization approaches for feature selection. Neurocomputing **172**, 371–381 (2016)
17. Du, N., Zhou, Y., Deng, W., et al.: Improved chimp optimization algorithm for three-dimensional path planning problem. Multimedia Tools Appl. **81**(19), 27397–27422 (2022)
18. Al-Tashi, Q., Abdulkadir, S.J.A., Rais, H.M., et al.: Binary optimization using hybrid grey wolf optimization for feature selection. IEEE Access **7**, 39496–39508 (2019)

Research on Model Evaluation Technology Based on Modulated Signal Identification

Songlin Yang⊙, Menghchao Wang⊙, and Yun Lin$^{(\boxtimes)}$⊙

Harbin Engineering University, Harbin, Heilongjiang, China
`linyun@hrbeu.edu.cn`

Abstract. Nowadays, the electromagnetic space is complex and variable, and the accurate identification of modulated signals is becoming increasingly difficult. And model quality can directly affect signal recognition. Therefore, this paper constructs a model performance evaluation system based on the recognition effect of modulated signals, and builds a hierarchical model of evaluation indexes from the classification performance, complexity performance, noise robustness and adversarial robustness of the model, to make a comprehensive and credible evaluation of the model quality from multiple dimensions. Through the experiment, we found that the complexity and classification performance of the model can affect the robustness of the model to a certain extent. The results of the evaluation show that the evaluation system can make a comprehensive and reasonable assessment of the quality of the model under the modulation-based signal recognition task.

Keywords: Evaluation System · Modulation Identification · Evaluation Metrics · Model Performance

1 Introduction

With the development of the electromagnetic field, the number of electromagnetic signals existing in the electromagnetic space and the modulation of these electromagnetic signals are becoming more and more diverse, which makes the electromagnetic environment increasingly complex. Initially, electromagnetic signal identification was performed by manually analyzing the characteristics of electromagnetic signals, and this method required huge workload and high labor cost. Currently, deep learning models are able to perform better recognition of signal samples.

In the electromagnetic signal recognition task, there is no uniform standard for the evaluation of model performance, so a comprehensive evaluation system is of great significance both for the evaluation of signal recognition performance in the complex electromagnetic environment and for the objective evaluation and optimization of model performance.

Deep learning has yielded better results in electromagnetic signal recognition tasks. In [1], the signal is converted into a graph and a graph convolutional

© ICST Institute for Computer Sciences, Social Informatics and Telecommunications Engineering 2024
Published by Springer Nature Switzerland AG 2024. All Rights Reserved
J. Li et al. (Eds.): 6GN 2023, LNICST 553, pp. 79–90, 2024.
https://doi.org/10.1007/978-3-031-53401-0_9

network is used for classification. In [2], the complex-valued network is used for modulation recognition. In [3], a binarized complex neural network is combined with a deep complex neural network for modulated signal recognition. In [4], the signals of eight species are identified by fusing image features and handcrafted features. In [5], an ADS-B signal dataset was created to measure the quality of the deep learning model.

In terms of model evaluation, in [6], the classification effect of the model for signals under the influence of some channel losses is studied. In [7], the performance of deep learning models in signal recognition tasks was evaluated by a model quality evaluation system. In [8], a comprehensive evaluation metric is proposed to assess the impact of non-target and target attacks on convolutional neural network-based device recognition. In [9], the model recognition performance is evaluated for signal constellation maps with different signal-to-noise ratios. In [10], an evaluation metric based on the difference in sample fit is proposed, which can effectively quantify the strength of electromagnetic signals to perturbations. In [11], the modulation classification performance in complex environments is evaluated.

2   Introduction

Modulated signal recognition is to recognize electromagnetic modulated signals in a certain electromagnetic environment and get recognition results. This paper selects the typical neural network extraction to learn the features of modulated signals, then identify different categories of modulated signals. The recognition results are input to the evaluation system, the evaluation index and evaluation dimension are selected, and the evaluation results are finally obtained. The relationship between modulated signal identification and evaluation system is shown in Fig. 1.

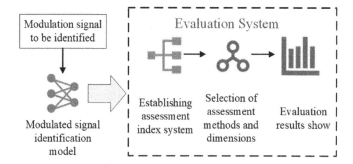

Fig. 1. Diagram of the relationship between modulated signal identification and evaluation system.

Under the task of electromagnetic signal identification, a reasonable evaluation of model performance requires the selection of reasonable evaluation indexes.

When constructing the evaluation system, it is necessary to consider the performance evaluation under different dimensions. The evaluation indexes under each dimension need to satisfy scientificity, typicality, representativeness and reliability. Each Indicator needs to be able to intuitively reflect the relevant performance of the model and the system is able to reflect the existence of certain correlation effects between indicators. Through the determination of the index system, a reliable evaluation system that can reflect the characteristics of the electromagnetic domain and the performance of the model is constructed. The evaluation system is constructed as shown in Fig. 2.

As shown in the figure, the main objective of the electromagnetic modulated signal recognition effect evaluation system is to provide a comprehensive evaluation of the modulated signal recognition results, and the evaluation dimensions are categorized into classification performance, complexity, noise robustness, and adversarial robustness. The index layer is the end of the evaluation system, including the evaluation indexes of different dimensional recognition effects. The evaluation indexes chosen in this paper can provide scientific and reliable assessment of the characteristics under different dimensions and can reflect the characteristics of the model under different domains.

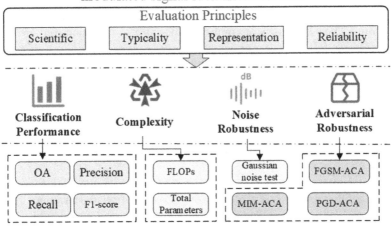

Fig. 2. Evaluation system chart.

3   Evaluation Index System

3.1   Classification Performance Evaluation Metrics

a) Original Sample Classification Accuracy: Normal sample classification accuracy refers to the number of normal samples correctly classified as a percentage of the overall sample size.

$$OA = \frac{1}{N} \sum_{i=1}^{N} count\, (F(x_i) = y_i) \tag{1}$$

where N is the number of natural samples; $count(\cdot)$ indicates 1 when the model prediction category is consistent with the true label, and 0 otherwise.

b) Precision: In model classification performance evaluation, Precision is the ability of the model to classify the truly positive signal samples as positive.

$$Precision = TP/(TP + FP) \tag{2}$$

where, in the multi-classification problem, TP denotes a sample with a true label of L_A and the predicted outcome is also a sample of L_A. FP denotes a sample with a true label of L_B and a predicted label of not L_B.

c) Recall: The recall is the percentage of the number of samples that are actually predicted to be positive out of the number of signal samples that are truly positive. It is shown by Eq. (3).

$$Recall = TP/(TP + FN) \tag{3}$$

FN indicates that the actual label of the sample is L_A, but the predicted label results in L_B.

d) F1-score: The F1 score is a combined measure of precision and recall in a classification model. It is a more comprehensive representation of how good the system model is. This is shown by the following equation.

$$F1 = 2(R_{pre} \times R_{rec})/(R_{pre} + R_{rec}) \tag{4}$$

where R_{pre} denotes precision; R_{rec} denotes recall.

3.2 Complexity Performance Assessment Metrics

Model complexity is one of the dimensions to evaluate the performance of a model. A good model needs to have good classification performance and low computational complexity. This section presents the model complexity evaluation metrics chosen in this paper.

a) Floating Point Operations: FlOPs reflects the model computational complexity, and the metric reflects the number of floating-point operations the model performs during forward propagation. The following equation demonstrates the calculation of FLOPs in the convolutional layer of the model:

$$FLOPs = [M_{in} + A_{in} + 1] \times E_{out} \tag{5}$$

where, M_{in} represents the multiplication operation amount, A_{in} represents the addition operation amount, $+1$ represents the error between the predicted label and the real label, E_{out} represents the number of elements in the output feature map. Various calculation methods are shown below:

$$M_{in} = C_{in} \times K_w \times K_h \tag{6}$$

$$A_{in} = M_{in} - 1 \tag{7}$$

$$E_{out} = C_{out} \times H \times W \tag{8}$$

where C_{in} denotes the number of input channels, C_{out} denotes the number of output channels, $K_w \times K_h$ denotes the size of the convolution kernel.

In the fully connected layer, the FLOPs are calculated as shown below:

$$FLOPs = (2N_{in} - 1)N_{out} \tag{9}$$

where N_{in} and N_{out} represent the number of input and output nodes in the fully connected layer, multiplying the two represents the multiplication operation. $(N_{in} - 1)N_{out}$ represents the addition operation in the fully connected layer.

b) Total model parameters: For model complexity assessment, the total model parameters can directly reflect the complexity of the model, which includes all the parameters of convolutional layers, fully connected layers, etc.

3.3 Noise Robustness Evaluation

In signal recognition, modulated signals may be accompanied by effects such as noise. In this paper, we refer to this test of how a model performs in the face of signal data anomalies asnoise robustness. In this paper, Gaussian white noise (AWGN) is added to the test sample. This is used to simulate the interference damage to the electromagnetic signal caused by the complex and variable channel.

Signal-to-noise ratio, abbreviated as SNR, represents the power ratio of signal to noise in a system.

$$SNR = 10 \lg \left(\frac{P_s}{P_n} \right) \tag{10}$$

where P_s indicates the effective power of the electromagnetic signal and P_n indicates the effective power of the noise in dB.

3.4 Adversarial Robustness Evaluation

The data distribution does not change, but the data changes slightly, such as constructing an adversarial sample, and this type of robustness is called adversarial robustness.

Machine learning models are easily influenced by adversarial examples, which can easily make model decisions wrong by simply adding carefully constructed, imperceptible perturbations to the input samples to the human eye. This section describes the algorithms and indicator used to evaluate the adversarial robustness of the model.

a) *Adversarial examples classification accuracy:* Adversarial example classification accuracy (ACA) is a metric for assessing the classification capability of the evaluation model under adversarial robustness. The metric represents the ratio of correctly classified adversarial samples to all samples. Its formula is shown below:

$$ACA = \frac{1}{N} \sum_{i=1}^{N} count \left(F(x_i^{adv}) = y_i \right) \tag{11}$$

N is the total number of adversarial exsamples, x_i^{adv} represents the adversarial example i.

b) *Fast gradient sign method(FGSM)* FGSM is an method for targetless attacks that generates adversarial samples in the A-paradigm limit of the original samples [12]. The method performs a one-step update in the gradient direction only. The adversarial sample can be expressed as:

$$\tilde{x} = x + \varepsilon \cdot sign(\nabla_x J_\theta(x, l)) \tag{12}$$

where the initial sample is x, l is the identified label value for, ρ is the perturbation value, such that ρ is small enough to satisfy $||\rho||_\infty < \varepsilon$, where ε generates the maximum level of perturbation. J is the loss function.

c) *Basic Iterative Method:* The adversarial examples in the Basic Iterative Method (BIM) are generated by multiple iterations and the output of each iteration is intercepted to ensure that these output values lie within the appropriate range [13]. The formula is shown below.

$$x'_{n+1} = Clip_{x,\varepsilon}\{x'_n + \varepsilon \cdot sign(\nabla_x J(x'_n, l)) \tag{13}$$

where x'_{n+1} denotes the number of iterations of the adversarial sample at this time is n+1. $Clip_{x,\varepsilon}\{z\}$ denotes the cropping of z to range $[x - \varepsilon, x + \varepsilon]$, which is defined as shown below:

$$Clip_{x,\varepsilon}(a) = \min\{1, x + \varepsilon, \max(0, x - \varepsilon, a)\} \tag{14}$$

d) *Projected Gradient Descent:* Projected Gradient Descent (PGD) is an iterative attack that makes the adversarial examples achieve high loss and minimizes the adversarial loss function [14]. The method is formulated as follows:

$$\min_\theta \rho(\theta), \quad \rho(\theta) = \mathbb{E}_{(x,y)\sim D} \max_{\delta \in S} L(\theta, x + \delta, y) \tag{15}$$

where $\mathbb{E}_{(x,y)\sim D}$ is the defined overall risk, D is the sample distribution, and S is the allowed perturbation.

e) Momentum Iterative Method: Momentum Iterative Method(MIM) is similar to PGD, but in this algorithm, the perturbation under each round is also related to the direction of the previous gradient [15]. The gradient is calculated as follows:

$$g_{t+1} = \mu g_t + \frac{\nabla_x J_\theta(x'_t, l)}{\|\nabla_x J_\theta(x'_t, l)\|_1} \tag{16}$$

x'_t represents the input to the objective function; $\nabla_x J_\theta(x'_t, l)$ represents the gradient. g_{t+1} is updated by accumulating the velocity vector in the gradient direction. Thus, x'_{t+1} is updated as shown in the following equation.

$$x'_{t+1} = x'_t + \varepsilon \cdot sign(g_{t+1}) \tag{17}$$

4   Experimental Simulation

In this section, 11 classes of modulated signals in the RadioML 2016.10a dataset are identified and classified using the above models, and the model is tested under different evaluation dimensions to illustrate the effectiveness of this evaluation method.

4.1   Dataset

In this paper, we use the RadioML2016.10a dataset, which includes 11 commonly used modulation methods. The number of data is 220,000. These signals include SNR from -20 dB to +18 dB. The signal sampling rate is 200 kHz and the in-phase and quadrature components (I/Q) of the signals are represented as a (2 × 128) tensor. In this experiment, the RadioML2016.10a dataset is divided into training set and test set by 1:1.

4.2   Mechanical Learning Model

In this paper, ResNet-18 and VGG-16 are selected to classify modulated signals, and the network parameters of this network are modified to make it applicable to the signal tensor. Meanwhile, this paper selects VT-CNN2 as the recognition model of modulated signals [16]. The convolutional and fully connected layers of this model are both two layers, and the number of convolutional kernels in the two convolutional layers is set to 64 and 16, respectively. The neurons in the fully connected layer are set to 128 and 11. This paper also evaluates the multichannel convolutional long short-term deep neural network (MCLDNN) [17]. MCLDNN consists of three input channels, and incorporates CNN, LSTM and fully connected deep neural network.

4.3   Classification Performance Testing

Table 1 show the test results of different classification metrics of the models at 0dB and 18dB, respectively. It can be seen that at 0dB and 18dB conditions, MCLDNN has the best classification capability, with the recognition accuracy reaching 82.52% and 87.67%, respectively, while VT-CNN2 has the worst recognition capability, with the recognition accuracy reaching only 68.98% and 72.46%. For the other classification metrics, the test results were similar.

Table 1. Model Classification Performance Test.

Model	OA_0dB	Pre_0dB	Rec_0dB	F1_0dB	OA_18dB	Pre_18dB	Rec_18dB	F1_18dB
ResNet-18	0.7012	0.7149	0.7012	0.6861	0.7689	0.7848	0.7689	0.7502
VGG-16	0.7786	0.7883	0.7786	0.7680	0.8295	0.8494	0.8295	0.8166
MCLDNN	0.8252	0.8296	0.8252	0.8252	0.8765	0.8958	0.8765	0.8709
VT-CNN2	0.6898	0.6844	0.6898	0.6747	0.7246	0.7355	0.7246	0.7035

4.4   Complexity Performance Test

In this section testing model complexity, the FLOPs size of the model and the total model parameter size are tested separately.

Table 2. Complexity Testing.

Complexity Metrics	ResNet-18	VGG-16	MCLDNN	VTCNN2
FLOPs	1945279	16117514	545042	3136653
Total Parameters	973739	8064267	406199	1568539

In Table 2, it can be seen that VGG-16 has the largest FLOPs and total model parameters, but its recognition classification performance is not the best. MCLDNN has the least complexity and its OA is the largest, indicating that this model has better performance. VT-CCN2 has more complexity, but it has poor OA, indicating that this model has poor performance. In summary, it can be seen that there is no direct relationship between the complexity of the model and the classification performance, and the two need to be considered together to evaluate the model performance.

4.5   Noise Robustness Testing

In this section, we mainly test the recognition and classification performance of the models for the test set signals under different SNR conditions. From Fig. 3, the recognition performance of the four models starts to decrease rapidly at SNR lower than 4 dB. The accuracy of the MCLDNN is the best at signal-to-noise ratios greater than −12 dB. At −14 dB, the VGG-16 model has the highest accuracy. All four models are susceptible to the effects of noise. In the overall view, MCLDNN has the highest recognition performance under the condition of certain noise.

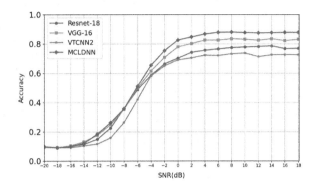

Fig. 3. Gaussian noise test.

4.6   Adversarial Robustness Testing

This section evaluates the adversarial robustness of the model. The results of the adversarial robustness test of the model under different adversarial attack algorithms at 0 dB are shown. The model recognition accuracy is plotted with different maximum perturbation values (eps). The test set in this section is 10%. The results of partial iterative and non-iterative attack tests are shown.

Table 3 shows the recognition accuracy of different models with different attack algorithms for modulated signals at 0 dB. Where the maximum perturbation is set to 0.0015. Where VGG-16 performs the best performance under each attack algorithm; MCLDNN has the smallest ACA. It indicates that the adversarial robustness of MCLDNN is poor under certain perturbations.

Table 3. Model Adversarial Robustness Testing Under 0 dB.

Model	FGSM-ACA	BIM-ACA	PGD-ACA	MIM-ACA
ResNet-18	0.4382	0.2800	0.2927	0.3309
VGG-16	0.4609	0.3282	0.3291	0.3636
MCLDNN	0.3073	0.1527	0.1554	0.1654
VT-CNN2	0.3373	0.1909	0.2018	0.1873

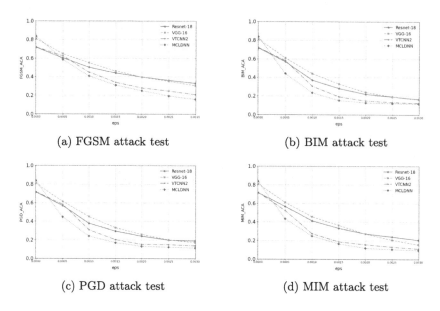

(a) FGSM attack test   (b) BIM attack test

(c) PGD attack test   (d) MIM attack test

Fig. 4. Adversarial Robustness Measurement

Figure 4 illustrates the ACA of the model under different perturbations. The ACA of MCLDNN decreases the fastest with increasing eps under the four adversarial attacks. At eps of 0.003, ResNet-18 has the highest ACA. For FGSM, PGD and BIM attacks, under low perturbation, VGG-16 has a greater advantage, followed by VT-CNN2 and ResNet-18. To summarize, MCLDNN has the best OA and its model complexity is the smallest, but its adversarial robustness is poor. Under low perturbation, the model complexity affects the adversarial robustness of the model to some extent.

4.7 Model Quality Performance Show

This section visualizes the quality of the model in terms of classification performance, complexity, noise Robustness, and adversarial robustness. Classification performance metrics are chosen as OA, precision and F1-score of the model at 0 dB. Complexity performance metrics are chosen as FLOPs metrics. Noise robustness performance and adversarial robustness were scored by observing the test curves and ranking the strengths and weaknesses of each model. Figure 5 is a visual presentation of the comprehensive evaluation of the models.

From the graph analysis, the MCLDNN has better classification performance, less model computational complexity, and less influence of Gaussian noise, but is vulnerable to attack algorithms. VGG-16 has better overall performance but more model complexity. ResNet-18 has slightly better performance than VT-CNN2 in terms of performance quality. In this paper, we argue that models with better classification performance are able to learn deeper model features.

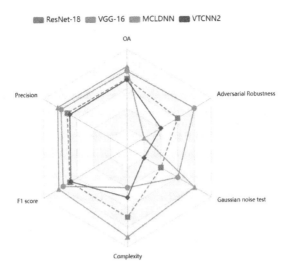

Fig. 5. Comprehensive evaluation chart of the model.

Therefore, when the model recognizes adversarial examples, the model is prone to misclassification because the deeper features have changed.

5   Conclusion

In this paper, a multidimensional fusion evaluation system is constructed for the comprehensive reliability assessment of models under the debug identification task, and a hierarchical model of evaluation metrics is established. The evaluation results show that the MCLDNN model has the best classification performance, but is vulnerable to the attack algorithm. The overall performance of VGG-16 is better, but the complexity of the model is larger. At the same time, there is a relationship between the metrics, and the complexity of the model and the classification performance can affect the robustness of the model to a certain extent. Models with good classification performance are easier to counteract the effects of adversarial examples. This paper proves that the evaluation system can provide a comprehensive and reasonable assessment of the quality of the model, and has a certain degree of scalability.

References

1. Liu, Y., Liu, Y., Yang, C.: Modulation recognition with graph convolutional network. IEEE Wirel. Commun. Lett. **9**(5), 624–627 (2020)
2. Tu, Y., Lin, Y., Hou, C., Mao, S.: Complex-valued networks for automatic modulation classification. IEEE Trans. Veh. Technol. **69**(9), 10085–10089 (2020)
3. Wang, J., Qi, L., Han, Y.: Modulation recognition based on complex binarized neural network. In: 2022 9th International Conference on Dependable Systems and Their Applications (DSA), pp. 126–132. IEEE (2022)

4. Zhang, Z., Wang, C., Gan, C., Sun, S., Wang, M.: Automatic modulation classification using convolutional neural network with features fusion of SPWVD and BJD. IEEE Trans. Signal Inf. Process. Netw. **5**(3), 469–478 (2019)
5. Ya, T., et al.: Large-scale real-world radio signal recognition with deep learning. Chin. J. Aeronaut. **35**(9), 35–48 (2022)
6. O'Shea, T.J., Roy, T., Clancy, T.C.: Over-the-air deep learning based radio signal classification. IEEE J. Sel. Top. Signal Process. **12**(1), 168–179 (2018)
7. Wang, J., Zha, H., Fu, J.: Evaluation of deep learning model in the field of electromagnetic signal recognition. In: IEEE INFOCOM 2022-IEEE Conference on Computer Communications Workshops (INFOCOM WKSHPS), pp. 1–6. IEEE (2022)
8. Bao, Z., Lin, Y., Zhang, S., Li, Z., Mao, S.: Threat of adversarial attacks on dl-based IoT device identification. IEEE Internet Things J. **9**(11), 9012–9024 (2021)
9. Doan, V.-S., Huynh-The, T., Hua, C.-H., Pham, Q.-V., Kim, D.-S.: Learning constellation map with deep CNN for accurate modulation recognition. In: GLOBECOM 2020-2020 IEEE Global Communications Conference, pp. 1–6. IEEE (2020)
10. Zhao, H., Lin, Y., Gao, S., Yu, S.: Evaluating and improving adversarial attacks on DNN-based modulation recognition. In: GLOBECOM 2020-2020 IEEE Global Communications Conference, pp. 1–5. IEEE (2020)
11. Cai, Z., Wang, J., Ma, M.: The performance evaluation of big data-driven modulation classification in complex environment. IEEE Access **9**, 26313–26322 (2021)
12. Shorten, C., Khoshgoftaar, T.M.: A survey on image data augmentation for deep learning. J. Big Data **6**(1), 1–48 (2019)
13. Kurakin, A., Goodfellow, I., Bengio, S.: Adversarial machine learning at scale, arXiv preprint arXiv:1611.01236 (2016)
14. Madry, A., Makelov, A., Schmidt, L., Tsipras, D., Vladu, A.: Towards deep learning models resistant to adversarial attacks, arXiv preprint arXiv:1706.06083 (2017)
15. Dong, Y., et al.: Boosting adversarial attacks with momentum. In: Proceedings of the IEEE Conference on Computer Vision and Pattern Recognition, pp. 9185–9193 (2018)
16. O'Shea, T.J., Corgan, J., Clancy, T.C.: Convolutional radio modulation recognition networks. In: Jayne, C., Iliadis, L. (eds.) EANN 2016. CCIS, vol. 629, pp. 213–226. Springer, Cham (2016). https://doi.org/10.1007/978-3-319-44188-7_16
17. Xu, J., Luo, C., Parr, G., Luo, Y.: A spatiotemporal multi-channel learning framework for automatic modulation recognition. IEEE Wirel. Commun. Lett. **9**(10), 1629–1632 (2020)

Parking Space Matching and Path Planning Based on Wolf Feeding Decision Algorithm in Large Underground Garage

Nan Dou[✉], Zhigang Lian, and Chunlei Guo

Shanghai Dianji University, Shanghai 201306, China
doudoudounan@163.com

Abstract. As cities grow, the number of complex underground parking garages with multiple entrances and exits is increasing. Randomly assigning parking spaces can lead to longer wait times for car owners during the parking and retrieval process and underutilization of resources. Therefore, allocating suitable parking spaces to car owners is crucial. This article proposes solutions for matching parking spaces and planning optimal routes for car owners in large underground parking garages based on their demands for shortest time, lowest price, and overall optimization. To reduce space and time complexity, a new heuristic decision optimization algorithm called Wolf Pack Search Algorithm (WSA) is proposed, which can solve multi-objective problems. WSA reduces redundant searches by using existing information to guide the search process. WSA simulates a wolf pack searching for food to find the optimal time cost for each intersection, reducing the space complexity of the data. Through information exchange and cooperative search, the wolf pack can find the optimal solution, abandoning redundant data. Compared with the ant colony algorithm, WSA has a significant advantage in space complexity and can meet car owners' requirements for parking space matching efficiency.

Keywords: Underground parking spot allocation · Scheduling optimization · Heuristic decision optimization algorithm · Wolf pack search algorithm · Multi-objective problem · Redundant search · A* algorithm · Intelligent computing

1 Foreword

1.1 Introduction

As an indispensable part of modern urban transportation system, underground parking garages provide convenient and fast parking places for urban residents. However, in large underground garages, the large number of parking spaces and complex distribution of entrances and exits, as well as the diverse combinations of entrances and exits chosen by car owners, pose challenges to parking space allocation and path planning. If parking

J. Li et al. (Eds.): 6GN 2023, LNICST 553, pp. 91–104, 2024.
https://doi.org/10.1007/978-3-031-53401-0_10

spaces are randomly assigned or chosen by car owners, it will result in longer parking and retrieval times for car owners and the inefficient use of garage resources. Therefore, allocating appropriate parking spaces to car owners and planning the optimal path have become important issues in managing large underground parking garages.

Traditional navigation algorithms can only satisfy the optimal search for a single point-to-point path and cannot flexibly meet car owners' optimal search requirements for time or price goals, nor can they effectively help car owners optimize the best parking space under a balance of time and price. Currently, there are two main types of algorithms to solve this problem: Label Setting Method [1] and Label Correcting Method [2]. Label Setting Method determines the optimal path from the starting node to a certain node in each search main loop, with typical algorithms such as A* algorithm and Dijkstra algorithm; Label Correcting Method can only ensure that the optimal path from the starting node to the last node has been found after traversing the entire road network, with representative algorithms including Graph Growth [3], Bellman [4], Ford, Moore [5], and Threshold algorithm [6]. The same problem can be solved by different algorithms, and the quality of an algorithm will affect the execution efficiency of the program.

With the rapid development of computer technology, the demand for solving various multi-objective optimization problems has become increasingly urgent. Traditional optimization algorithms often have high space complexity and large computational complexity, making them difficult to apply to large-scale problems. Therefore, designing an optimization algorithm with low space complexity and the ability to handle multi-objective problems has become particularly important. Among many heuristic optimization algorithms, ant colony algorithm and particle swarm algorithm have been widely used in multi-objective optimization problems. However, these algorithms often require maintaining a complete pheromone matrix or state matrix and other large amounts of data when dealing with large-scale problems, resulting in high space complexity. Car owners often have high requirements for matching timeliness.

The Wolf Search Algorithm (WSA) proposed in this article is a low space complexity and multi-objective optimization algorithm. Unlike the ant colony algorithm and particle swarm algorithm, the Wolf Search Algorithm does not require maintaining a complete pheromone matrix or state matrix and other large amounts of data, only requiring the maintenance of the optimal time cost for each exit. Therefore, it can adapt to large-scale problems, and has advantages such as fast screening of redundant data and fast iteration speed. In the parking decision problem, the Wolf Search Algorithm can find the optimal time cost for each intersection by simulating the behavior of wolves searching for food, thereby reducing the space complexity of the data. Through exchanging information and cooperative search, the wolf pack can abandon redundant data and find the optimal solution. Compared with the ant colony algorithm and particle swarm algorithm, the Wolf Search Algorithm has obvious advantages in space complexity. In summary, the Wolf Search Algorithm is a novel optimization algorithm with low space complexity and the ability to handle multi-objective problems. In solving large-scale problems such as parking decisions, the Wolf Search Algorithm can leverage its advantages to quickly find the optimal solution and allocate appropriate parking spaces to car owners and plan the optimal path by considering factors such as car owners' time and price. This article also introduces the mathematical model and implementation method of the algorithm and verifies the effectiveness of the algorithm through experiments.

2 Underground Garage Parking Space Matching and Optimization Problem

2.1 Problem Elaboration

Large underground parking lots have multiple entrances and parking spaces, and the combinations of entrances and parking spaces chosen by users are usually diverse. Randomly assigning parking spaces or allowing car owners to choose their own spaces would result in longer times for car owners to park and retrieve their cars, and the resources of the parking lot would not be used efficiently. Therefore, the system needs to allocate suitable parking spaces for car owners based on their location and their choice of entrances, while meeting their parking space requirements. Assuming that the GIS coordinates of the vehicles and the distribution of the location nodes of the parking lot are known in advance, car owners generally have the following requirements for parking spaces:

(1) Shortest time: The total driving time from the car owner's vehicle location to the destination parking space, as well as the walking time from the parking space to the target exit after parking the car, should be minimized.
(2) Lowest price: The parking lot management system selects the cheapest parking space within the car owner's price range.
(3) Best overall: The system takes into account both time and price, and assigns the car owner the best path and parking space, so that the combination of the driving time from the car owner's location to the destination parking space and the walking time from the parking space to the target exit, as well as the price of the parking space, is optimized.

2.2 Problem-Solving Approach

To match parking spaces for car owners and plan optimal paths in large underground parking lots with multiple entrances and parking spaces, the following strategies can be used:

(1) Shortest time

To achieve the shortest time possible, the GIS coordinates of the vehicle and parking lot can be used to calculate the shortest path from the car owner's current location to the relevant parking space, as well as from the parking space to the target exit, using the A* shortest path algorithm, in order to minimize the total driving and walking time.

(2) Lowest price

Firstly, the parking lot management system can select a group of parking spaces that meet the car owner's requirements and are distributed within the parking lot. Then, the cheapest parking space within this group can be selected. If multiple parking spaces have the same price, the A* algorithm can be used to select the one with the shortest time.

(3) Best overall

To achieve the best overall result, both time and price must be considered. Within the car owner's price range, suitable parking spaces can be identified, and the shortest path from the current location to the target parking space can be calculated using path planning algorithms, taking into account factors such as parking space price and location. The best path and parking space can be determined by synthesizing these factors.

To satisfy the car owner's best overall requirements, there needs to be a trade-off between parking space selection and navigation planning. A comprehensive evaluation function can be defined to evaluate parking spaces based on price and total driving and walking time. The parking space and navigation path with the best evaluation result can be selected.

The evaluation function can be defined as:

$$\text{Evaluation value} = \text{Price} + \lambda \times \text{Time} \tag{1}$$

When the car owner enters the parking lot, the current location can be determined using the vehicle's GIS coordinates. Then, an A* algorithm with time and price costs can be used to search for the best path for the car owner to reach the target parking space and provide navigation instructions.

From the above analysis, it is clear that we need to use the A* algorithm to calculate the time cost of the vehicle driving to the parking space and the time cost of walking from the parking space to the exit. Based on these time costs, we can make decisions and filter out parking spaces.

In large parking lots with multiple entrances and parking spaces, using multi-objective algorithms such as ant colony algorithms to analyze the time cost of each parking space is very complex, and the time cost is also related to the car owner's location and the expected exit.

The space complexity is equal to the number of parking spaces multiplied by the number of entrances multiplied by the number of exits. In order to reduce the space complexity of the data, this paper proposes a heuristic decision optimization algorithm called the "Wolf Pack Foraging Decision Algorithm" and presents a related mathematical model. The Wolf Pack Foraging Decision Algorithm is a heuristic optimization algorithm based on the foraging behavior of wolf packs and can be used to solve multi-objective optimization problems. The algorithm places more emphasis on using existing information to guide the search process and reduce redundant searching. The Wolf Pack Foraging Decision Algorithm does not require the maintenance of a complete pheromone matrix or other state matrix, resulting in low space complexity and the ability to handle large-scale problems.

When searching for the optimal time cost for each intersection, the Wolf Pack Foraging Decision Algorithm abandons paths with time costs greater than the known optimal time cost. The algorithm has the advantages of low space complexity, filtering redundant data, fast iteration speed, and the ability to handle multi-objective problems when solving problems.

2.3 Use of the A * Algorithm

The A* (A-Star) algorithm is an effective direct search method for finding the shortest path in a static road network and is also a commonly used heuristic algorithm for solving many other problems. The closer the distance estimation value in the algorithm is to the actual value, the faster the final search speed. Many preprocessing algorithms have emerged thereafter, such as ALT [8], CH [9], HL [10], and so on.

In actual road conditions, the distance between the starting and ending points of a road must be greater than or equal to the straight-line distance between the two points. Therefore, based on the selection principle of the distance estimation parameter $h(n)$ [11] in the A* algorithm, this paper uses the Euclidean distance [12] as the estimation function. For example, if there are a starting point A and an ending point B in a map with coordinates $(A.x, A.y)$ and $(B.x, B.y)$, respectively, the distance estimation between A and B can be calculated as follows:

$$
\begin{aligned}
h(n) &= \sqrt{(A.x - B.x)^2 + (A.y - B.y)^2} \\
g(n) &= \Sigma_{i=0}^{n} S_i \\
f(n) &= g(n) + h(n)
\end{aligned}
\tag{2}
$$

In the above formula, $h(n)$ is the minimum cost of the trajectory curve from the current state to the target, $g(n)$ is the minimum cost from the initial state to state n within the current state region, and $f(n)$ is the minimum cost from the initial state to the current state through state n. When using the Euclidean distance as the operation function, because $h(n)$ must be less than or equal to the actual distance of the target trajectory line, the A* algorithm can always plan the shortest path and operate efficiently.

2.4 Mathematical Model

To solve the optimization problem of matching parking spaces in underground garages, a mathematical model is constructed as follows: Assuming the location of the vehicle is point a, the parking space to be reached is b, and the exit of the garage is c. Both b and c are variables that are related to the real-time scheduling and planning of the system. The parking spaces are divided into different blocks based on their prices, numbered 1, $2, 3,..., i$.

p_{ab}: Cost of driving time from point a to parking space b;
q_{bc}: Cost of walking time from parking space b to exit c for the user;
l_{abc}: Sum of time cost from starting point a to destination parking space b and from leaving the garage to exit c;
M_i: Price cost of the i block;
t_i: Optimal time cost within the i block;
m: Block sequence number of the lowest-priced block in the garage, denoted by m;
t_m: Optimal time cost within the m block;
M_m: Price cost of the m block;
f_i: Unit time cost reduction for each unit price increase relative to block M in block i;
F: Overall optimal cost of parking.

Based on the above assumptions and parameter definitions, the model is established as follows:

Since t_m is the optimal time cost in the price-optimal section, when $t_m < t_i$, section i can be directly eliminated and is not part of the overall optimal solution. For the case where $t_m > t_i$, the optimal overall cost F can be selected using the ratio of price difference to time difference.

$$
\begin{aligned}
I &= minI_{abc} = minp_{ab} + minq_{bc} \\
M &= minM_i(i = 1, 2, 3...n) \\
f_i &= \frac{t_m - t_i}{M_m - M_i}(i = 1, 2, 3...n) \\
F &= \max f_i(i = 1, 2, 3, ...n)
\end{aligned}
\tag{3}
$$

3 Wolf Pack Feeding Decision Algorithm

3.1 The Explanation of the Wolf Pack Feeding Decision Algorithm

When hunting for food, wolf packs always send out members to explore various intersections and search for places near each intersection where prey may pass by, from closest to farthest. When the cost of the path to the prey's route is equal to the optimal known cost to reach the prey's den, the pack judges that the remaining path time cost at this intersection is greater than the known optimal time cost and abandons the remaining paths at this intersection. This allows for the rapid selection of the optimal time cost at each intersection, enabling the wolf pack to hunt down prey more quickly.

In the parking space decision problem, the wolf pack search algorithm can be used to find the optimal time cost at each intersection by simulating the behavior of a wolf pack searching for food, thus reducing the space complexity of the data. Specifically, each wolf can be viewed as a car owner, each intersection as a search state, and the behavior of the wolf in the search state can be transformed into the behavior of the car owner searching for a parking space in the garage. In each search state, the wolf pack can abandon redundant data and find the optimal solution through information exchange and cooperative search. After multiple rounds of iteration, the wolf pack search algorithm can output the optimal solution, i.e., the parking decision plan with the shortest time, lowest price, or best overall performance.

Compared to traditional computing methods, the wolf pack search decision algorithm is a heuristic optimization algorithm that does not require enumerating and analyzing the time cost of each parking space. Instead, it finds the optimal solution through the search behavior of the wolf pack, abandoning redundant data based on known data. The wolf pack search algorithm can greatly reduce the space complexity of the data, thereby reducing computation time and storage space.

3.2 Steps in Optimizing Path Time Cost for Wolf Feeding Decision

Problem: The system needs to select a parking spot and an exit for the car owner, such that the total time from the car owner's starting position to the destination parking spot, plus the time for the car owner to leave the garage from the selected exit is minimized.

Assuming there are $(a_1, a_2, a_3 \ldots a_i)$ exits in the underground garage, and $(b_1, b_2, b_3 \ldots b_j)$ corresponding parking spots, a set of optimal time costs $T\ (t_{a_1}, t_{a_2}, t_{a_3} \ldots t_{a_i})$ is created for each exit. The time cost for walking from the parking spot to the exit is q, the time cost from the entrance to the parking spot is p, and the overall cost is I, where $I = q + p$.

The algorithm steps are as follows:

Step 1: Take the parking spot closest to each exit in $(b_1, b_2, b_3 \ldots b_j)$ in turn, remove them from the set $(a_1, a_2, a_3 \ldots a_i)$, and calculate the overall time cost I, walking time cost q from the parking spot to the exit, and time cost p from the entrance to the parking spot using the mathematical model described in 1.2 and the A* algorithm described in 1.3. Mark the overall time cost I with the corresponding exit number and put them into a set T, denoted as $T\ (t_{a_1}, t_{a_2}, t_{a_3} \ldots t_{a_i})$.

Step 2: Continue to take the parking spots closest to each exit from $(a_1, a_2, a_3 \ldots a_i)$, remove them from the set, and calculate their I, q, p. Compare the walking cost q associated with these parking spots with the time cost in T one by one. If the q corresponding to a certain exit is greater than its corresponding time cost t_{a_i} in T, the optimal parking spot for that exit is the parking spot a_i corresponding to the overall time cost t_{a_i}. Otherwise, compare I with t_{a_i}. If I is less than t_{a_i}, replace the value of t_{a_i} in T with I.

Step 3: Continue to loop through steps 1 and 2 until the optimal time cost for each exit is calculated. Sort the time costs in T, and the minimum time cost corresponds to the optimal time cost that meets the requirements. The corresponding parking spot and exit are the optimal path for the minimum time cost.

Specific case is set up as follows:

Table 1. Cost of walking time from the exit

Exports/Spaces	b_1	b_2	b_3	b_4	b_5
a_1	2	4	10	15	20
a_2	14	3	9	2	19
a_3	4	5	8	13	18

Table 2. The cost of driving time from the parking space

car owner	b_1	b_2	b_3	b_4	b_5
Tom	8	5	6	7	8

According to Table 1, in Table 2, the data simulation is as follows:

Figure 1 in this article shows the path cost required to solve the minimum time cost problem using enumeration method, while Figs. 2, 3, and 4 show the path costs required for path planning using the wolf pack hunting optimization algorithm.

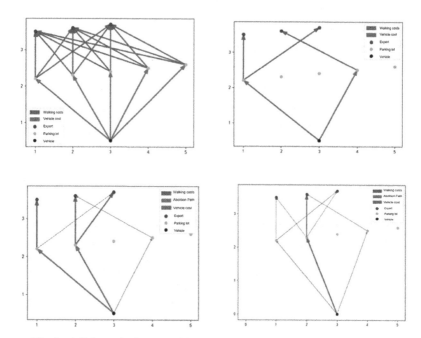

Fig. 1. A Schematic diagram of the small-scale dynamic path-finding process

In practical applications, different parking spot areas in the garage have different driving and walking costs, so the calculation of a single parking spot cannot meet the actual needs. Therefore, this article uses the wolf pack hunting decision algorithm to generate data for 200 parking spot areas with 10 exits, 1000 parking spot areas with 20 exits, and calculates the optimal path and parking spot using the path planning algorithm, and then uses small-scale pruning to select the specific optimal parking spot. Finally, the A* algorithm is used to calculate the specific path and the results are fed back to the car owner.

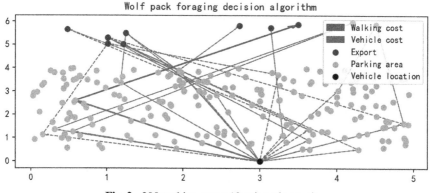

Fig. 2. 200 parking space 10 exit scale pruning

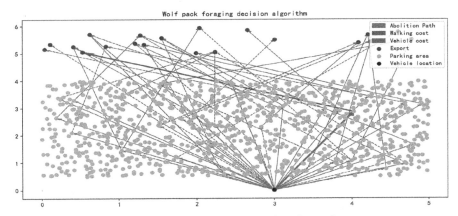

Fig. 3. 1000 parking space 20 exit scale pruning

As can be seen from the results in the medium and large-scale examples, the wolf pack hunting decision algorithm can effectively escape from local optimal values. Compared with classical multi-objective algorithms that are prone to local optimal solutions and computationally complex, the wolf pack hunting decision algorithm can maintain both global exploration and local search capabilities in a comparably fast manner.

3.3 Comparison of Data Scale Between Wolf Pack Feeding Decision and Ant Colony Algorithm

The main idea of the ant colony algorithm is to use pheromones to guide ants to search for the optimal solution. In the parking matching problem, pheromones can be represented as the attractiveness of each parking spot. When choosing a path, ants will prioritize parking spots with higher attractiveness. At the same time, ants will release pheromones during the search process, which increases the attractiveness of the path and attracts more ants to follow the same path, thereby forming a positive feedback effect. The advantage of this algorithm is that it can effectively avoid local optimal solutions and has good robustness. However, the disadvantage of the ant colony algorithm is that the search process takes some time, and the parameter settings of the algorithm are complex, requiring adjustments to the algorithm parameters, which also means that the ant colony algorithm needs to iterate and analyze the data repeatedly. Moreover, car owners often have high requirements for the timeliness of parking spot matching, and the wolf pack hunting decision algorithm can reduce the spatial complexity and screen out redundant data. The comparison of the selection efficiency between the wolf pack hunting decision algorithm and the ant colony algorithm is shown in Fig. 4. The time cost of parking spots and exits in the figure is determined by their respective driving and walking costs, and is unrelated to the path and straight-line distance between coordinates. It can be seen from the figure that the wolf pack hunting decision algorithm can exclude a large number of redundant paths and significantly improve the optimization efficiency.

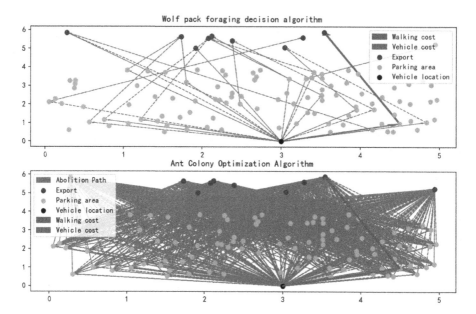

Fig. 4. Schematic diagram of data scale comparison between Wolf pack feeding decision and ant colony optimization algorithm

3.4 The Parking Fee is Optimized Based on the Wolf Pack Feeding Algorithm

The system needs to select a parking spot for the car owner that satisfies the cost optimality of the price and the cost. In underground parking lots, parking spots are generally divided into several blocks, where the time cost and price are similar. Table 3 shows that the parking garage is divided into 16 blocks based on different locations and prices, and the price of each block is listed in sequence according to the number, denoted as M_i *(10,20,15,30,25,40,45,33,44,59,43,29,39,17,20,18).*

The algorithm only needs to select the block with the lowest price and calculate the time cost optimization of the selected parking spots in the lowest-priced block using the wolf pack hunting decision algorithm in Sect. 2.2 to obtain the parking spot that is time-cost optimal under the price-cost optimal condition. From Table 3, block 0 has the lowest price, and the time-cost optimal parking spot in block 0 is the parking spot that the system needs under the condition of price-cost optimality. The specific optimal parking spot in the optimal parking spot area of block 0 is selected through the wolf pack hunting decision algorithm, and the time-cost data of the optimal parking spot in each block are shown in Table 3. The A* algorithm is then used to calculate the specific path of the parking spot and feed it back to the car owner.

3.5 Comprehensive Optimal Implementation Using the Valuation Function Based on the Results of the Wolf Feeding Decision

The algorithm needs to select the optimal parking spot for the car owner by considering both the price and time cost, so that the parking spot has the best cost-effectiveness. In

Table 3. Plate cost and the existing optimal parking space time cost in each plate

i	M_i	p	q	t_i
0	10	30	20	50
1	20	20	5	25
2	15	19	10	29
3	30	15	5	20
4	25	7	8	15
5	40	10	10	20
6	45	4	6	10
7	33	21	8	29
8	44	12	13	15
9	59	6	2	8
10	43	5	6	11
11	29	13	9	22
12	39	6	10	16
13	17	15	10	25
14	20	15	5	20
15	18	4	3	7

underground parking lots, parking spots are generally divided into several blocks, where the time cost and price are similar. The parking garage is divided into 16 blocks based on price, denoted as i (**0, 1, 2, 3, 4...15**), with the price of each block being Mi, and the time-cost data of the best parking spot in each block recorded as ti, as shown in Table 3.

The wolf pack hunting decision algorithm introduced in Sect. 2.2 is used to optimize the path cost of each block to select the best parking spot in each block. Then, the price difference between each of these parking spots and the reference parking spot (block 0) with the lowest price, $M_i - M_0$, and the ratio of the time difference between the reference parking spot and the best parking spot in each of the other blocks, $t_0 - t_i$, are compared to obtain the comprehensive optimality ratio F.

The formula for calculating F is:

$$F = maxf_i = \frac{t_0 - t_i}{M_i - M_0}, i = (0, 1, 2, 3, 4...) \tag{4}$$

The parking spot with the highest F value is the parking spot with the best cost-effectiveness that the algorithm selects for the car owner.

To obtain the optimal parking spot with the best cost-effectiveness, the following steps can be taken:

Step 1: Exclude negative ratios, as negative ratios indicate higher prices and greater time costs relative to the reference block.

Step 2: Next, select the ratio with the largest positive value in f_i, which is the optimal parking spot that the system is looking for. A larger F value indicates that more time cost is saved compared to the reference parking spot under the same price.

Step 3: If all ratios are less than zero, then the optimal parking spot is the reference parking spot with the lowest price.

Using the data from the small-scale underground parking lot in Table 3, the optimal ratio is calculated. To better show the comprehensive optimality ratio, the ratio is magnified by 10 times, as shown in the price and time comprehensive optimality ratio diagram in Fig. 5. It can be seen that the time-cost optimal parking spot in block 2 is the optimal parking spot that the algorithm searches for. The parking spot with the shortest dynamic path in block 2 is selected, and the A* algorithm is used to calculate the specific path and provide navigation feedback to the car owner.

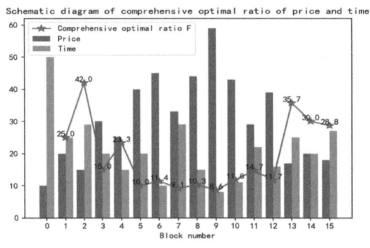

Fig. 5. Comprehensive optimal ratio of price and time

The simulation data shows that the wolf pack hunting decision algorithm can effectively identify the optimal parking spot with the best cost-effectiveness in multi-region parking spot decisions. Compared with ordinary traversal algorithms, the wolf pack hunting decision algorithm can effectively prune and dynamically select subsequent paths, improving overall efficiency.

The wolf pack hunting decision algorithm simulates the hunting behavior of wolves in the wild, which enables it to identify the optimal parking spot in a more efficient and intelligent way. By using the algorithm, the search space can be effectively reduced, and the search process can be optimized to find the optimal parking spot with the best cost-effectiveness.

In summary, the wolf pack hunting decision algorithm is an effective approach for multi-region parking spot decision-making, which can improve the efficiency of the decision-making process and help car owners find the best parking spot faster and with less effort.

4 Summary

The article explores various car navigation planning algorithms and proposes a parking spot search mode that considers both distance and time. To solve this problem, the wolf pack hunting decision algorithm is proposed based on the nature of the problem discussed in the article. Finally, the feasibility of the method is demonstrated through research on small, medium, and large-scale projects, solving the problem of comprehensive optimal selection of vehicle navigation and vehicle and pedestrian entrances and exits.

The model constructed in this article considers the needs of three user groups, namely time-optimal, parking cost-optimal, and comprehensive optimal, based on the attributes of the region and the distance between parking spots and exits, and considering the different time costs of walking and driving. The article proposes a parking spot search mode that considers price and time and a parking spot planning method based on the wolf pack hunting decision algorithm, solving the problem of comprehensive optimal selection of vehicle navigation and vehicle and pedestrian entrances and exits. Furthermore, this study considers the needs of different user groups and factors such as the attributes of the region and the distance between parking spots and exits, making the model more practical and operable. By conducting research on small, medium, and large-scale projects, the effectiveness and rationality of the method are demonstrated, providing new ideas and methods for research in the field of vehicle navigation.

The main contribution of this article is to propose a new method for solving the problem of parking spot allocation and path planning in large underground parking lots, providing car owners with a faster and more convenient parking experience, and also having important practical application value for parking lot management.

References

1. Lin, W.: Introducing an indoor path planning algorithm for the navigation grid. Surv. Mapp. Sci. **41**(2), 39–43 (2016)
2. Pan, S., Chen, Y., Gao, Y., Li, W.: Research on vehicle scheduling problem with multiple service requirements with soft time window. J. Wuhan Univ. Technol. (Transp. Sci. Eng. Ed.) **44**(06), 1123–1128 (2020)
3. Peng, L.: An analysis of the application of vehicle dispatching system in urban rail transit. Eng. Constr. Des. **20**, 90–91 (2020)
4. Liu, Y., Du, J., Zhang, Q.: Comparison of the performance of the A * and Dijkstra algorithms based on path optimization. Mod. Electron. Technol. **40**(13), 181–183+186 (2017)
5. Ran, D., Peng, F., Li, H.: Review of pathway planning studies based on the A * algorithm. Electron. Technol. Softw. Eng. **24**, 11–12 (2020)
6. Schiffer, M., Hiermann, G., Rüdel, F., Walther, G.: A polynomial-time algorithm for user-based relocation in free-floating car sharing systems. Transp. Res. Part B Methodol. **143**, 65–85 (2021)
7. Dellaert, N., Van Woensel, T., Crainic, T.G., Dashty, S.F.: A multi-commodity two-Echelon capacitated vehicle routing problem with time windows: Model formulations and solution approach. Comput. Oper. Res. **127**, 105154 (2021)
8. Groß, P.-O., Ehmke, J.F., Mattfeld, D.C.: Interval travel times for robust synchronization in city logistics vehicle routing. Transp. Res. Part E **143**, 102058 (2021)

9. Zhang, D., Li, D., Sun, H., Hou, L.: A vehicle routing problem with distribution uncertainty in deadlines. Eur. J. Oper. Res. **292**(1), 311–326 (2020)

10. Shi, Y., Zhou, Y., Ye, W., Zhao, Q.Q.: A relative robust optimization for a vehicle routing problem with time-window and synchronized visits considering greenhouse gas emissions. J. Clean. Prod. **275**, 124112 (2020)

11. Wang, H., Yin, P., Zheng, W., et al.: Mobile robot path planning based on the improved A~* algorithm and dynamic window method. Robotics **42**(3), 346–353 (2020)

12. Wang, H., Yin, P., Zheng, W., et al.: Path planning for mobile robots based on improved A~* algorithm and dynamic window method. Robotics **42**(3), 346–353 (2020)

Research on Physical Layer Authentication Method of Internet of Things Based on Contour Stellar Images and Convolutional Neural Network

Ying Lu[1], Jingchao Li[1(✉)], Yulong Ying[2], Bin Zhang[3], Tao Shi[4], and Hongwei Zhao[4]

[1] College of Electronic and Information Engineering, Shanghai Dianji University, Shanghai, China
lijc@sdju.edu.cn
[2] School of Energy and Mechanical Engineering, Shanghai University of Electric Power, Shanghai, China
[3] Department of Mechanic Engineering Kanagawa University, Yokohama, Japan
[4] Shandong Baimeng Information Technology Co., Ltd., Weihai, China

Abstract. Wireless communication networks are prone to information security issues due to the openness of radio transmission. User information security relies heavily on reliable identification and authentication methods. To address the channel differences between different users, this paper proposes a physical layer identity authentication method using contour stellar images. By converting one-dimensional signals into two-dimensional contour stellar images using measured channel information from two channel scenarios, and enhancing the data set with data enhancement technology, a convolutional neural network is employed to classify and recognize 800 contour stellar images in two scenarios. This achieves the goal of different user identity authentication. Additionally, the CNN model and SVM model are compared under the same conditions. The results indicate that the proposed CNN model achieves 98.3% recognition accuracy, demonstrating strong robustness and authentication efficacy in real-world scenarios.

Keywords: channel feature · contour stellar images · data enhancement · convolutional neural network

1 Introduction

Due to the rapid evolution of mobile communication technology, an increasing number of communication devices have been incorporated into wireless communication networks. As a result of the expansion of this vast industry, a large volume of confidential and valuable information is transmitted over wireless networks.

Sponsored by Shanghai Rising-Star Program (No. 23QA1403800), National Natural Science Foundation of China (No. 62076160) and Natural Science Foundation of Shanghai (No. 21ZR1424700).

J. Li et al. (Eds.): 6GN 2023, LNICST 553, pp. 105–116, 2024.
https://doi.org/10.1007/978-3-031-53401-0_11

Consequently, ensuring communication security has become a critical concern throughout the development of wireless communication technology. As a result, there is an urgent need to efficiently address the issue of wireless communication security, which holds significant research and practical value [1]. Physical layer authentication is a core technology that plays a crucial role in ensuring the security of wireless communication. Its fundamental principle involves utilizing the spatial and temporal characteristics of transmitting and receiving channels, as well as the signals transmitted, to verify the physical attributes of both communication parties. This process enables identity authentication at the physical layer. Compared to authentication technologies implemented in upper layers, physical layer authentication is highly effective in resisting copycat attacks. Additionally, it boasts several advantages, including fast authentication, low complexity, good compatibility, and a lack of requirement for considering various protocol executions [2]. In contemporary science and technology development, physical layer authentication technology is primarily categorized into three distinct types. (1) Watermark authentication technology based on embedded tags. As an illustration, document [3] incorporated authentication information in the form of a watermark into the transmission signal, and explored physical layer watermarking technology based on time-sharing long-term evolution. As a result, they achieved promising results in terms of anti-noise and anti-frequency offset performance. (2) Authentication technology based on RF fingerprint. In literature [4], the author investigated intelligent identification technology that utilizes signal multi-dimensional feature fusion. This technology was shown to enhance the accuracy of RFID in complex electronic station environments. Reference [5] presents a radio-frequency fingerprint identification method that effectively addresses the susceptibility of radio-frequency fingerprints to complex environments, particularly multipath fading. (3) Authentication technology based on channel characteristics. In literature [6], the author enhanced the traditional authentication method by incorporating deep learning and LSTM feature extraction technology. This was achieved by combining multiple channel estimation algorithms to analyze the wireless channel characteristics. In reference [7], a multi-feature fusion method based on short and long-term memory was used to study scene recognition in high-speed railway wireless channels. The results showed good recognition efficiency and provided an accurate method for recognizing transmission scenes. Among them, the authentication technology based on channel characteristics is currently a hot topic in the field of physical layer authentication. The most commonly used channel characteristics include channel state information, time domain, frequency domain, received signal strength, and channel impulse response. These characteristics change irregularly with time and the user's location, and are generally difficult to replicate, making them effective against external attacks.

In this paper, the channel state information characteristics of wireless channels are analyzed theoretically, and the channel characteristics are converted into two-dimensional images. At the same time, the convolutional neural network is used to train and classify the obtained image data set according to the deep learning technology's good classification and prediction performance of massive data. A physical layer authentication method based on contour stellar images and convolutional neural network is presented.

2 Wireless Channel Feature Authentication Model

In communication systems, signals can be compromised by channel instability during transmission, leading to signal interception and decryption by eavesdroppers, thereby undermining communication security. To address this issue, various encryption technologies have been developed; however, they cannot fully prevent eavesdropping.

In order to enhance communication security, Wyner [8] proposed a physical layer security authentication model that models the communication channel as consisting of two parts: the main channel and the eavesdropping channel. The main channel represents the transmission channel from sender to receiver, while the eavesdropping channel represents the channel used by eavesdroppers. Based on this, Wyner developed an encryption scheme where noise is added to the main channel to interfere with the eavesdropping channel, making it much harder for the eavesdropper to decrypt the communication. This model has been widely adopted in wireless communication and network security research.

Based on the eavesdropping channel model, this paper proposes a wireless channel feature authentication model that leverages the differences in channel features between legitimate users to achieve authentication, as illustrated in Fig. 1. In this model, Alice and Bob exchange signals as legitimate users, while Eve poses as a legitimate user and attempts to intercept and eavesdrop on Alice's signals. However, because Alice and Eve use different channels to transmit signals, the signals they receive have different channel characteristics. Bob needs to classify these signals to identify the legal identity of the sender. By analyzing the characteristics of different channels, identity authentication can be carried out more accurately, thereby enhancing communication security.

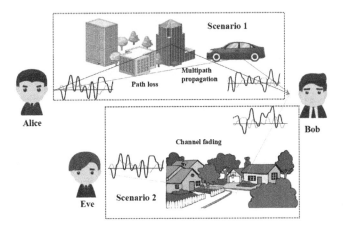

Fig. 1. Wireless channel feature authentication model.

3   Authentication Method

3.1   Channel State Information

The channel state information (CSI) of a communication system is a description of the characteristics of the communication link's channel. CSI encompasses the propagation of the signal within the channel, incorporating the effects of time delay, amplitude attenuation, and phase offset. In the frequency domain, it details the amplitude and phase of each subcarrier. Equation (1) is the received CSI:

$$CSI(f, t) = A_{\text{noise}} (f, t) e^{-j\theta_{\text{offset}} (f,t)} (H_S(f) + H_d(f, t)) \tag{1}$$

where A_{noise} is amplitude noise, θ_{offset} is random phase offset, $H_S(f)$ is static component, $H_d(f, t)$ is dynamic component.

CSI, which typically includes physical layer information in wireless communication protocols, is often difficult to obtain directly. In recent years, researchers have modified the firmware of some commonly used IEEE 802.11n standard wireless cards to provide detailed amplitude and phase information of different subcarriers in the form of CSI. In the IEEE 802.11n standard, Eq. (2) can be used to represent every CSI information in the received data packet:

$$H(k) = \|H(k)\| e^{j\angle H(k)} \tag{2}$$

where $\|H(k)\|$ for the first k is the carrier of amplitude; $H(k)$ is the channel state information of the kth subcarrier; $\angle H(k)$ is the phase of the kth subcarrier.

In this paper, Inter5300 wireless network card is used to collect data to obtain CSI information. The CSI matrix H can be waited until it is shown in Eq. (3):

$$H = [H(1), H(2), ..., H(k)]_{N_T \times N_R} \tag{3}$$

where, k is the number of subcarriers, N_R is the number of receiving antennas, and N_T is the number of transmitting antennas.

3.2   Contour Stellar Images

Contour stellar images are widely used in the field of communication. It serves as a tool to depict the transmission and reception of modulated signals in wireless channels, as well as to predict channel parameters. The different channel parameters exhibit distinct distribution shapes on the constellation maps. By examining the distribution characteristics of the constellation maps of the received signal, the phase shift, time delay, and other channel parameters can be inferred via the use of contour stellar images.

In general, the modulated signal in the signal can be expressed as:

$$S_N = A_m g(t) \cos(2\pi f_n t + a_k) \tag{4}$$

By expanding Eq. (4), we can obtain:

$$\begin{aligned} S_N &= A_m g(t) \cos\left(2\pi f_n t + a_k\right) \cos a_k - A_m g(t) \sin\left(2\pi f_n t + a_k\right) \sin a_k \\ &= I \cos\left(2\pi f_n t + a_k\right) - Q \sin\left(2\pi f_n t + a_k\right) \end{aligned} \tag{5}$$

among them,

$$I = A_m g(t) \cos a_k \tag{6}$$

$$Q = A_m g(t) \sin a_k \tag{7}$$

According to space theory, the following set of vectors can be selected:

$$\left[\sqrt{\frac{2}{\varepsilon_g}} g(t) \cos (2\pi f_n t), \sqrt{\frac{2}{\varepsilon_g}} g(t) \sin (2\pi f_n t) \right] \tag{8}$$

where, ε_g is the energy of the low-pass pulse signal, $\varepsilon_g = \int_0^T g^2(t) dt = T$.

The signal space vector represented by the modulated signal is shown in Eq. (9):

$$\left[\sqrt{\frac{\varepsilon_g}{2}} A_m \cos a_k, \sqrt{\frac{\varepsilon_g}{2}} A_m \sin a_k \right] \tag{9}$$

The constellation diagram is a two-dimensional image that represents the endpoints of vectors on a Cartesian coordinate system. It reflects the spatial distribution of amplitude and phase of the modulating signal and is used to visualize the modulation scheme.

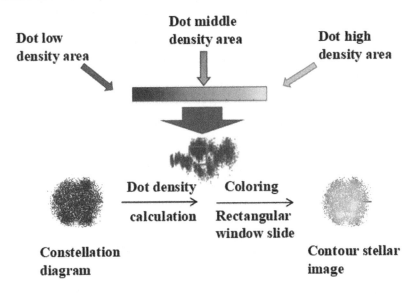

Fig. 2. Constellation map transform contour stellar image.

The constellation diagrams shows different densities of sampling points in different regions, which are represented by different colors, as depicted in Fig. 2. As the density window function moves across the two-dimensional coordinates, the number of points within the window is counted using the function, and the density is calculated accordingly. Different densities are then marked using different colors: yellow for the highest density range, green for the medium density range, and blue for the lowest density range. The rectangular window function calculates the point density within the rectangular box and converts it into contour stellar image using the point density formula.

3.3 Data Enhancement

When training deep learning models, having a sufficient amount of data is typically necessary to obtain a high-performing model. However, acquiring and labeling large amounts of data in the real world can be costly and time-consuming. To address this issue, researchers have developed a technique known as data augmentation. Data augmentation involves generating new training samples by applying a series of data transformations while preserving the distribution of the original data set. By increasing the sample size of the data set, the model can be better generalized to previously unseen data.

Data enhancement can be accomplished through various techniques. For instance, with image data, one can rotate, flip, crop, scale, add noise, adjust brightness and contrast, among other things. With audio data, the pitch can be altered, noise can be added, and the duration and speed of the audio can be modified. For natural language processing tasks, one can substitute synonyms, delete words, and randomly insert words, among other methods.

When implementing data enhancement, several factors must be considered, including the type, degree, and order of the transformations. It is important to note that data enhancement can introduce unnecessary noise and redundancy, which may negatively impact the model's performance. Thus, a careful design and evaluation process is necessary when applying data enhancement to ensure that the enhanced data set preserves the distribution of the original data set while minimizing the introduction of noise and redundancy.

3.4 Convolutional Neural Network Structure

Upon reviewing existing machine learning technologies, it has been found that object recognition is a highly accurate application of machine learning. Therefore, it is worth considering the use of machine learning to improve the efficiency of physical layer authentication. Convolutional neural network has been proven to be advantageous in processing two-dimensional images, allowing images to be directly used as input for the network, bypassing the complex feature extraction and data reconstruction processes in traditional recognition algorithms. The CNN can autonomously extract features such as color, texture, shape, and topological structure of the image, making it robust and efficient in recognition applications. The general structure of the network is illustrated in Fig. 3.

Compared to other neural networks, convolutional neural networks have a distinct feature that includes several convolutional units between the input and output layers. Each convolutional unit typically comprises a convolutional layer and a pooling layer, although the number of convolutional units may vary depending on the specific use case. The convolutional layer processes input data using convolutional operations, allowing it to more effectively extract important information from the sample data (such as the contour stellar images information used in this study). Meanwhile, the pooling layer reduces the dimensionality of the data via pooling operations and compresses the output from the convolutional layer. Finally, a fully connected layer maps the learned features and transmits them to the output layer.

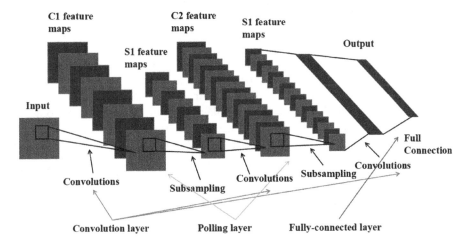

Fig. 3. Constellation map transform contour stellar images.

In this paper, convolutional neural network is used to construct a physical layer authentication scheme based on convolutional neural network. The specific architecture of the network is illustrated in Fig. 4 and consists of three convolutional layers and two pooling layers. Once the input layer receives the data, the network performs convolution and average pooling operations to extract information features and reduce dimensions. The output is then activated using the softmax activation function.

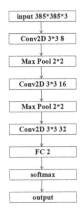

Fig. 4. General model of convolutional neural networks.

4   Experimental Results and Analysis

This experiment conducted channel data measurements in two different scenarios. Scenario 1 involved the same physical area where UEs within the area could belong to different cellular cells, while Scenario 2 involved the same cellular cell. Multiple samples were taken during the sampling process by varying the heights and directions of both the sending and receiving antennas. The resulting channel dimensions were 1000 samples, 4 receiving antennas, 8 transmitting antennas, and 52 resource blocks, with a carrier frequency of 2565 Hz and subcarrier interval of 30 kHz. The channel was divided into 13 subbands within the transmission bandwidth, and the corresponding CSI data was generated. The data dimension was 1000 samples × 8 feature vector length × 13 subbands.

The CSI data obtained were simulated using the MATLAB2020b environment. Specifically, 8000 data points were taken on each subcarrier and contour stellar images were generated for both scenarios. Figure 5 shows the contour stellar images for both scenarios, consisting of 12 maps in total.

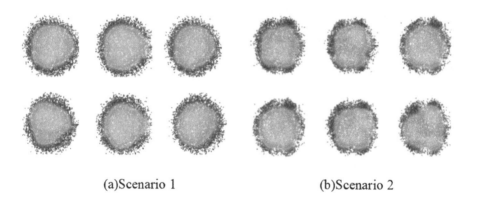

(a)Scenario 1   (b)Scenario 2

Fig. 5. Contour stellar images in each scenario.

To enhance the training effect, the data set was normalized to ensure the stability and reliability of the input data. Furthermore, data augmentation techniques were employed by rotating, flipping, cropping, and resizing the images to increase the diversity and quantity of the data set, thereby improving the model's generalization ability. Consequently, a total of 800 samples were obtained, with 400 samples from each scenario. Figure 6 shows 50 samples from each of the two scenarios.

In addition to processing the data set, this paper attempted various optimization algorithms such as Adam, SGD, and Adagrad, while adjusting the learning rate and regularization coefficient. Among these algorithms, Adam optimization algorithm was ultimately selected, with a learning rate of 0.001 and regularization coefficient of 0.001, in order to achieve the optimal training effect.

In the model verification test, the training set and test set were split in a 7:3 ratio. 560 samples were randomly selected for training the deep convolutional neural network, while the remaining 240 samples were used for recognition testing. Several experiments were conducted on the test set to verify the model's generalization ability and stability. The final accuracy of the model on the test set reached 98.3%, indicating that the system can effectively identify contour stellar images in different scenarios.

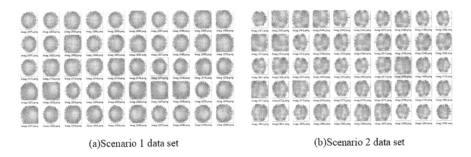

(a)Scenario 1 data set   (b)Scenario 2 data set

Fig. 6. Data set after data enhancement.

The training effect of the system model is illustrated in Fig. 7, where the smooth blue line represents the training curve. It continuously rises with the increase of the iteration number and eventually reaches a stable level. On the other hand, the black dashed line represents the verification curve, which also gradually rises with the iteration number. It can be observed that the system tends to stabilize after 50 iterations, and the final recognition rate reaches 98.3%, indicating a good recognition performance.

To demonstrate the effectiveness of the recognition methods proposed in this paper based on contour stellar images and convolutional neural network, a comparison is made between the constellation maps and contour stellar images

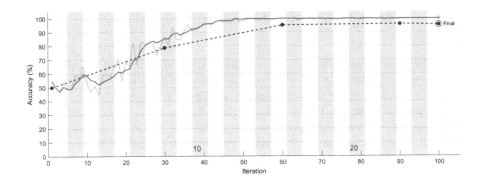

Fig. 7. Training effect of convolutional neural networks.

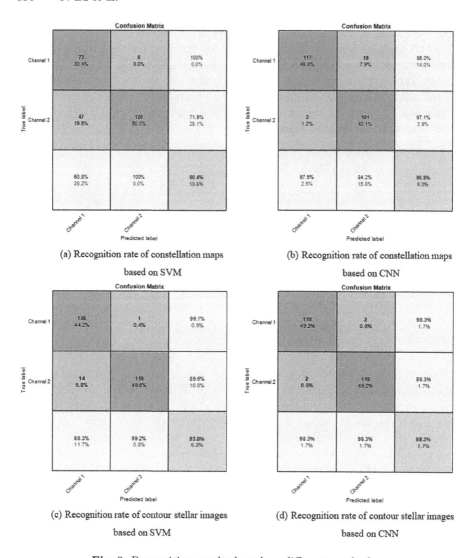

(a) Recognition rate of constellation maps
based on SVM

(b) Recognition rate of constellation maps
based on CNN

(c) Recognition rate of contour stellar images
based on SVM

(d) Recognition rate of contour stellar images
based on CNN

Fig. 8. Recognition results based on different methods.

under recognition methods based on support vector machine and convolutional neural network. The selected data sets are consistent, and confusion matrix analysis is carried out on the four models, as shown in Fig. 8.

The results shown in Fig. 8 indicate that the success rate of constellation maps recognition based on support vector machine is 80.4%, while that of constellation maps based on convolutional neural network is 90.8%. The success rate of support vector machine based on contour stellar images is 93.8%, and that of convolutional neural network based on contour stellar images is 98.3%. The confusion matrix demonstrates a very low classification error rate, indicating that the proposed method maintains high accuracy and robustness in the classification task of the two categories. This outcome provides a solid foundation for further research on physical layer authentication technology and can significantly improve recognition accuracy without compromising computational efficiency.

5   Conclusion

This paper proposes a physical layer authentication method based on contour stellar images and convolutional neural network. The CSI data is processed and analyzed to convert the one-dimensional signal into a two-dimensional contour stellar image, which is classified by a deep convolutional neural network to achieve physical layer authentication for users. Experimental results demonstrate high recognition accuracy and low error rate in different scenarios, and effectively identify the attack behavior of malicious devices. While the method in this paper achieves some achievements in physical layer authentication, further research is required to study users with similar channel characteristics in close proximity. In conclusion, the proposed physical layer authentication method provides a novel approach for device security authentication and holds significant value for developing secure and reliable Internet of Things applications. It is hoped that future scholars will continue to investigate Internet of Things security to foster the advancement and innovation of this technology.

References

1. Yong, H.: Research on Physical Layer Security Authentication Technology for Internet of Things. Huazhong University of Science and Technology (2022). https://doi.org/10.27157/d.cnki.ghzku.2022.000002
2. Jiang, H., Wang, S., Zhao, K.: Individual identification method of communication radiation source based on differential equipotential star map J. Jinan Univ. (Nat. Sci. Ed.) 35(05), 433–438+451 (2021). https://doi.org/10.13349/j.cnki.jdxbn.20210323.003
3. Wan, C.: Research on Watermarking technology of Physical Layer based on TD-LTE. Huazhong University of Science and Technology (2021). https://doi.org/10.27157/d.cnki.ghzku.2021.001167
4. Li, M., Xie, J., Yang, H.: RFID fingerprint recognition technology based on feature fusion of frequency hopping signal. Comput. Measur. Control 30(12), 319–325 (2022). https://doi.org/10.16526/j.cnki.11-4762/tp.2022.12.047
5. Jiajun, W.: Research and system implementation of wireless device Authentication Technology based on physical layer RF fingerprint. Nanjing University of Posts and Telecommunications (2022). https://doi.org/10.27251/d.cnki.gnjdc.2022.000982
6. Chaofan, X.: Research on Physical Layer Security Authentication based on Deep learning. Nanjing University of Posts and Telecommunications (2022). https://doi.org/10.27251/d.cnki.gnjdc.2022.000628
7. Wang, Y., Zhou, T., Tao, C.: Scene identification of high-speed rail wireless channel based on LSTM and multi-feature fusion. J. Radio Sci. 36(03), 453–459+476 (2021)
8. Cao, Y.: Research and design of space - time coding in MIMO eavesdropping Channel. Donghua University (2017)

9. Zhiyuan, Z.: Research on Physical Layer Authentication Technology based on Machine learning. Beijing University of Posts and Telecommunications (2021). https://doi.org/10.26969/d.cnki.gbydu.2021.002658

10. Gang, Z.: Remote sensing image scene classification based on deep Learning lightweight Convolutional Neural network. Nanjing University of Posts and Telecommunications (2022). https://doi.org/10.27251/d.cnki.gnjdc.2022.000193

11. Qiuping, H., Xiaofeng, L., Qiubin, G.: Research on feedback of large-scale antenna channel state Information based on Artificial Intelligence. Telecommun. Sci. **38**(03), 74–83 (2022)

12. Wang, Q., Hu, C., Li, F.: Scene recognition method based on deep transfer learning and multi-scale feature fusion. Electron. Technol. 1–9 (2023). https://doi.org/10.16180/j.cnki.issn1007-7820.2023.11.004

13. Zhanjian, W.: Research on indoor positioning technology based on channel state information and depth image. Nanjing University of Posts and Telecommunications (2022). https://doi.org/10.27251/d.cnki.gnjdc.2022.001088

14. Shuangde, L.: Research on wireless channel Measurement, simulation and modeling for 5G hotspot scene. Nanjing University of Posts and Telecommunications (2022). https://doi.org/10.27251/d.cnki.gnjdc.2022.000006

Double IRS-Aided Dual-Function Radar and Communication System

Zijia Chen$^{(\boxtimes)}$, Liu Yang, Ying Zhao, and Tiancheng Zhang

Shanghai Dianji University, Shanghai, China
2681758187@qq.com

Abstract. As an edge-cutting technology, Intelligent reflecting surface (IRS) can be well used in various fields. In this paper, we investigate to deploy double IRS in the dual-function radar and communication (DFRC) system, which we deploy double IRS to assist a base station (BS) with both radar sensing and communication capabilities with multiple single-antenna users. And we optimize the active beamforming at the base station and the passive beamforming of the IRS to maximize the achievable sum-rate of communication users. Because the optimization problems are non-convex, the fractional programming (FP), majorization-minimization (MM), and manifold optimization methods are used in an alternating algorithm to figure out these problems. Simulation results demonstrate the efficiency of the proposed algorithm.

Keywords: Intelligent reflecting surface (IRS) · Dual-function radar and communication (DFRC) · alternating optimization · beamforming design

1 Introduction

Thanks to the rapid development of wireless communication technology and 6G networks, the transmission quality and capacity of wireless signals have become a focus of attention [1]. However, due to factors such as buildings, obstacles and multipath propagation, the transmission of wireless signals is often hindered and interfered with, resulting in signal strength fading and transmission rate degradation. This forces us to find new solutions to overcome these problems and provide a more stable and efficient wireless communication experience.

Recently, dual-function radar and communication (DFRC) systems which contain both radar detection and communication functions in one system have attracted attention in various fields such as vehicular networks and multi-function radio systems, etc. In addition, DFRC systems reduce power consumption, electronic interference, and improve the efficiency of hardware. Therefore DFRC is considered as a key technology that can be applied in future wireless communication networks [2, 3].

Supported by National Natural Science Foundation of China (No. 62076160), Natural Science Foundation of Shanghai (No. 21ZR1424700), Shanghai Science and Technology Innovation Funds (No. 23QA1403800).

J. Li et al. (Eds.): 6GN 2023, LNICST 553, pp. 117–128, 2024.
https://doi.org/10.1007/978-3-031-53401-0_12

Dual-function radar transmit waveforms is one of the most important design in DFRC system. The application of multiple-input multiple-output (MIMO) technology to DFRC systems can result in better beam fugacity gain. However, channel degradation is a problem that cannot be ignored and leads to poor system performance. Thankfully, an innovative technique, IRS, can solve this nagging problem.

Intelligent Reflective Surface (IRS), an innovative and inexpensive technology, consists of multiple passive reflective elements. IRS has been used in many industries such as aerospace, satellite and in the wireless communication industry [4, 5]. Comparing to other existing technologies, IRS eliminates the need for a transmitter module, is cost effective and can operate in full duplex mode as an auxiliary component, and is less costly and less power hungry than AF relay.

In view of the above advantages, the application of IRS has attracted extensive research work [6–11], and in [6] the authors deployed an IRS with multiple targets and one user in a DFRC system. In [7], the article considers an IRS aided by a dual-function radar to detect a single target and multi-users. In [8], the article maximizes the communication data sum-rate by optimizing the beamforming vectors of the base station and reflecting surface. In [9], the paper uses IRS to reduce the interference to improve the total system rate. In [10], the paper uses discrete IRS phase shift design to minimize the interference of inter-user under Cramer-Rao bound (CRB) constraint. In [11], the paper deploys double IRS to assist communication and resist the interference of malicious users.

Thanks to the above findings, we assume that a realistic model where a dual- function BS detects multi-targets and serves multi-users simultaneously with the help of dual-IRS are considered in this paper. We corporately optimize the transmit beamforming of the BS and the passive beamforming of the IRS to enhance the target detection performance and achieve as many sum-rate as possible, under the constraints of reflection coefficient, transmission power and similarity of beampattern [12]. To solve the complex nonconvex optimization problem, we used the (FP) method to decompose the problem objective function into a manageable form, and also proposed an efficient iterative algorithm based on MM and manifold method to resolve the remaining issues. Numerical conclusions show the preponderances of employing double IRS in DFRC and the high efficiency of the proposed algorithm.

2   System Model and Problem Formula

The double IRS-aided DFRC system that includes a dual-function radar BS equipped with M antennas, radar targets, two N-element IRS and K single-antenna users. The BS serves users and detects radar targets simultaneously. Noteworthily, there is no connection between the users and the BS because of the obstruction. In order to improve the communication quality and detect tasks, the IRS is deployed to provide additional links and maintain good channel conditions.

We express the BS transmission signal as

$$\mathbf{Ws} = \mathbf{W_c}\mathbf{s_c} + \mathbf{W_d}\mathbf{s_d}, \tag{1}$$

where $\mathbf{W_c} \in \mathbb{C}^{M \times K}$ and $\mathbf{W_d} \in \mathbb{C}^{M \times M}$ denotes the matrixes regarding to beamforming of communication and radar sensing respectively. And $\mathbf{s_c} \in \mathbb{C}^{K \times 1}$ denotes the

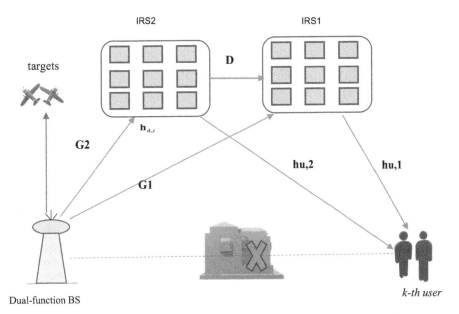

Fig. 1. Double IRS-aided DFRC system

communication vector with $E\{\mathbf{s_c s_c^H}\} = \mathbf{I_K}$, $\mathbf{s_d} \in \mathbb{C}^{M\times 1}$ denotes the detection vector with $E\{\mathbf{s_d s_d^H}\} = \mathbf{I_M}$. For simplicity, we write the whole beamforming matrix as $\mathbf{W} \triangleq [\mathbf{W_c}\ \mathbf{W_d}] \in \mathbb{C}^{M\times(K+M)}$ and the whole transmit vector as $\mathbf{s} \triangleq \begin{bmatrix} \mathbf{s_c^T}\ \mathbf{s_d^T} \end{bmatrix}^T \in \mathbb{C}^{K+M}$. Note the channel parameters of the IRS1 to the users and the IRS2 to the users as $\mathbf{h}_{u,1} \in \mathbb{C}^{N_1\times 1}$ and $\mathbf{h}_{u,2} \in \mathbb{C}^{N_2\times 1}$. And let the channel parameters of the BS to IRS1, the BS to the IRS2 and the IRS1 to IRS2 as $\mathbf{G}_1 \in \mathbb{C}^{N_1\times M}$, $\mathbf{G}_2 \in \mathbb{C}^{N_2\times M}$, $\mathbf{D} \in \mathbb{C}^{N_1\times N_2}$ respectively. The matrix of reflection coefficient of the IRS1 and the IRS2 is defined as $\mathbf{\Phi}_1 \triangleq diag(\phi_1) \in \mathbb{C}^{N_1\times N_1}$, $\mathbf{\Phi}_2 \triangleq diag(\phi_2) \in \mathbb{C}^{N_2\times N_2}$ separately, where $\phi_i \triangleq \begin{bmatrix} \phi_{i,1}, \cdots, \phi_{i,N} \end{bmatrix}^T \in \mathbb{C}^{N_i\times 1}$ with $|\phi_{i,n}| = 1$, $i \in (1, 2)$, $n \in (1, 2, \ldots, N)$. Then the cascade communication channel can be expressed as

$$\mathbf{T_d} = \mathbf{h}_{u,1}^H \mathbf{\Phi}_1 \mathbf{D} \mathbf{\Phi}_2 \mathbf{G}_2 + \mathbf{h}_{u,1}^H \mathbf{\Phi}_1 \mathbf{G}_1 + \mathbf{h}_{u,2}^H \mathbf{\Phi}_2 \mathbf{G}_2, \tag{2}$$

and the received signal at the k-th user can be expressed as

$$y_k = \mathbf{T_d W s} + n_k, \tag{3}$$

The scalar $n_k \sim \mathcal{CN}(0, \sigma_k^2)$ is additive AWGN. The SINR of the k-th user is given as

$$\gamma_k = \frac{|\mathbf{T_d w}_k|^2}{\sum_{j=1,j\neq k}^{K+M} |\mathbf{T_d w}_j|^2 + \sigma_k^2}, \tag{4}$$

where w_k denotes the k-th column of the $\mathbf{W}$, $k = 1, \ldots, K + M$. And the total sum-rate can be written as

$$R = \sum_{k=1}^{K} \log_2(1 + \gamma_k). \tag{5}$$

Considering radar detection, since the target detection probability is positively correlated with the sum rate of the radar, and in order to pursue better performance, a method is adopted to increase the signal power in the target direction and reduce the signal power in the other directions as well as to ignore the transmission delay. As shown in Fig. 1, the launch detection signal can reach the target via a straight blue line. In the later work, detection performance can be evaluated by the metric of similarity of beampattern which is used to match the ideal beampattern with the designed one. We define the vector of direction θ as $\mathbf{a}(\theta) \triangleq \left[1, e^{j2\pi \frac{\delta}{\lambda} \sin(\theta)}, \ldots, e^{j2\pi(M-1)\frac{\delta}{\lambda} \sin(\theta)} \right]^T$, where δ, λ are the space of antenna and the signal wavelength respectively. The transmission power can be formulated as

$$P_b(\theta, \mathbf{W}) = E\{|\mathbf{a}^H(\theta)\mathbf{W}s|^2\} = \mathbf{a}^H(\theta)WW^H\mathbf{a}(\theta). \tag{6}$$

The mean squared error (MSE) is given by

$$\mathcal{E}(\alpha, \mathbf{W}) \triangleq \frac{1}{L} \sum_{\iota=1}^{L} \left| \alpha P_d(\theta_\iota) - \mathbf{a}^H(\theta_\iota)\mathbf{WW}^H\mathbf{a}(\theta_\iota) \right|^2, \tag{7}$$

where $P_d(\theta_\iota)$ is the ideal transmission power, i.e., ideal beampattern, the angle of ι-th sample is represented by θ_ι and α is a ratio factor. With the factor α, the ideal beampattern can be approximated by designed beampattern.

In this paper, for maximizing the obtainable sum-rate, we corporately design the beamforming matrix $\mathbf{W}$ and the reflection coefficients ϕ_1, ϕ_2. Then the optimization problem is written as

$$\max_{\mathbf{W}, \phi_1, \phi_2, a} \sum_{k=1}^{K} \log_2(1 + \gamma_k) \tag{8.a}$$

$$\text{s.t.} \ \frac{1}{L} \sum_{\iota=1}^{L} \left| \alpha P_d(\theta_\iota) - \mathbf{a}^H(\theta_\iota)\mathbf{WW}^H\mathbf{a}(\theta_\iota) \right|^2 \leq \varepsilon, \tag{8.b}$$

$$||\mathbf{W}||_F^2 \leq P, \tag{8.c}$$

$$|\phi_{i,n}| = 1, \forall n, \tag{8.d}$$

where ε denotes the standard of the similarity of beampattern and P represents the transmission power. Obviously, it is hard to optimize the problem (8) because of the objection function (8.a) with log (.) terms, and the constraints (8.b) and (8.d). Hence, we cope with (8.a) by employing FP method in the next section. Then we first transform problem (8) into two easy-to-settle sub-problems then come up with efficient algorithms to settle them iteratively.

3 Joint Beamforming Design for Double IRS-Aided DFRC System

3.1 FP-Based Transformation

We convert the (8.a) problem into a tractable polynomial expression by using FP method. In particular, on the basis of Lagrangian dual reformulation [13], an instrumental variable c_k, $\forall k$ is introduced to transform (8.a) into

$$\sum_{k=1}^{K} \log_2(1 + c_k) - \sum_{k=1}^{K} c_k + \sum_{k=1}^{K} \frac{(1 + c_k)|\mathbf{T_d w}_k|^2}{\sum_{j=1}^{K+M} |\mathbf{T_d w}_j|^2 + \sigma_k^2}, \tag{9}$$

when the c_k has the appropriate value c_k^* as shown in (10), the problem (8.a) is equivalent to (9)

$$c_k^* = \frac{|\mathbf{T_d w}_k|^2}{\sum_{j=1, j \neq k}^{K+M} |\mathbf{T_d w}_j|^2 + \sigma_k^2}, \quad \forall k. \tag{10}$$

In (9), the direct solution cannot be obtained since the third term is difficult to figure out, thus we further convert it into

$$2\sqrt{1 + c_k}\Re\{g_k^* \mathbf{T_d w}_k\} - |g_k|^2 \sum_{j=1}^{K+M} |\mathbf{T_d w}_j|^2 - |g_k|^2 \sigma_k^2, \tag{11}$$

by quadratic transformation where the instrumental g_k has the optimal value

$$g_k^* = \frac{\sqrt{(1 + c_k)}|\mathbf{T_d w}_k|^2}{\sum_{j=1}^{K+M} |\mathbf{T_d w}_j|^2 + \sigma_k^2}, \quad \forall k. \tag{12}$$

Based on the transformations in (9) and (11), we reconstruct the (8.a) concerning $\mathbf{W}$ and ϕ_1, ϕ_2 as

$$\sum_{k=1}^{K} \left(2\sqrt{1 + c_k}\Re\{g_k^* \mathbf{T_d w}_k\} - |g_k|^2 \sum_{j=1}^{K+M} |\mathbf{T_d w}_j|^2\right). \tag{13}$$

Due to ϕ_1, ϕ_2 and $\mathbf{W}$ are coupled in the function (13), we iteratively solve $\mathbf{W}$ with block coordinate de-scent (BCD) method. The optimization of variables ϕ_1, ϕ_2 as shown in the following sub-sections.

3.2 Optimize W with Given ϕ_1, ϕ_2

Optimization problem $\mathbf{W}$ with given ϕ_1, ϕ_2 can be written as

$$\max_{\mathbf{W}, a} \sum_{k=1}^{K} \left(2\sqrt{1 + c_k}\Re\{g_k^* \mathbf{T_d w}_k\} - |g_k|^2 \sum_{j=1}^{K+M} |\mathbf{T_d w}_j|^2\right) \tag{14.a}$$

$$\text{s.t. } \frac{1}{L} \sum_{\iota=1}^{L} \left| \alpha P_{\rm d}(\theta_\iota) - \mathbf{a}^H(\theta_\iota)\mathbf{W}\mathbf{W}^H\mathbf{a}(\theta_\iota) \right|^2 \leq \varepsilon, \tag{14.b}$$

$$\|\mathbf{W}\|_F^2 \leq P. \tag{14.c}$$

We found that the (14.b) is a convex function with respect to the variable α. Hence, we decide to employ the first partial derivative to obtain its minimum, i.e. $\frac{\partial \mathcal{E}(\alpha, \mathbf{W})}{\partial \alpha} = 0$. And the optimal α can be written as

$$\alpha^* = \frac{\sum_{\iota=1}^{L} P_{\rm d}(\theta_\iota) {\rm vec}^H(\mathbf{A}_\iota) {\rm vec}(\mathbf{W}\mathbf{W}^H)}{\sum_{\iota=1}^{L} P_{\rm d}^2(\theta_\iota)}, \tag{15}$$

where we define $\mathbf{A}_\iota \triangleq \mathbf{a}(\theta_\iota)\mathbf{a}^H(\theta_\iota)$ for brevity. Then leverage some basic algebra transformations and substitute the optimal α^* (15) into $\mathcal{E}(\alpha, \mathbf{W})$, then the function $\mathcal{E}(\alpha, \mathbf{W})$ is degraded into $\mathcal{E}(\mathbf{W})$ which is an univariate function and (14.b) can be reformulated as

$$\mathcal{E}(\mathbf{W}) = {\rm vec}^H(\mathbf{W}\mathbf{W}^H)\mathbf{C}{\rm vec}(\mathbf{W}\mathbf{W}^H) \leq \varepsilon, \tag{16}$$

where we define

$$\mathbf{C} \triangleq \frac{1}{L} \sum_{\iota=1}^{L} b_\iota b_\iota^H, \tag{17.a}$$

$$b_\iota \triangleq \frac{P_{\rm d}(\theta_\iota) \sum_{\iota_1=1}^{L} P_{\rm d}(\theta_{\iota_1}) {\rm vec}(\mathbf{A}_{\iota_1})}{\sum_{\iota_1=1}^{L} P_{\rm d}^2(\theta_{\iota_1})} - {\rm vec}(\mathbf{A}_\iota). \tag{17.b}$$

Since beampattern MSE constraint (16) is a quartic and non-convex, we utilize the MM method to construct surrogate functions for $\mathcal{E}(\mathbf{W})$ which are easy-to-process. To be specific, after iterating t-th times, a more easy-processing solution $\mathbf{W}^t$, which approximates $\mathcal{E}(\mathbf{W})$ is written as

$$\mathcal{E}(\mathbf{W}) \leq \lambda_m {\rm vec}^H(\mathbf{W}\mathbf{W}^H)vec(\mathbf{W}\mathbf{W}^H) + \Re\{{\rm vec}^H(\mathbf{W}\mathbf{W}^H)b^u\} + c_1^t, \tag{18}$$

where in order to keep the formula short and easy to calculate, we define

$$b^u \triangleq 2(\mathbf{C} - \lambda_m\mathbf{I}){\rm vec}(\mathbf{W}^t(\mathbf{W}^t)^H), \tag{19.a}$$

$$c_1^t \triangleq {\rm vec}^H(\mathbf{W}^t(\mathbf{W}^t)^H)(\lambda_m\mathbf{I} - \mathbf{C}){\rm vec}(\mathbf{W}^t(\mathbf{W}^t)^H), \tag{19.b}$$

and λ_m is the largest characteristic value of $\mathbf{C}$. Owing to the function (14.c), the right-hand side of (18) is easily to acquire as

$$\lambda_m {\rm vec}^H(\mathbf{W}\mathbf{W}^H){\rm vec}(\mathbf{W}\mathbf{W}^H) = \lambda_m \left\| \sum_{j=1}^{K+M} \mathbf{w}_j\mathbf{w}_j^H \right\|_F^4 \leq \lambda_m \left(\left\| \sum_{J=1}^{K+M} \mathbf{w}_j\mathbf{w}_j^H \right\|_F^2 \right)^2 = \lambda_m P^2. \tag{20}$$

Put (20) into (18), $\mathcal{E}(\mathbf{W})$ can be expressed as

$$\mathcal{E}(\mathbf{W}) \leq \Re\{\text{vec}^H(\mathbf{W}\mathbf{W}^H)\mathbf{b}^u\} + c_1^t + \lambda_m P^2. \tag{21}$$

For simplicity, we re-write the term $\Re\{\text{vec}^H(\mathbf{W}\mathbf{W}^H)\mathbf{b}^u\}$ as

$$\Re\{\text{vec}^H(\mathbf{W}\mathbf{W}^H)\mathbf{b}^u\} = \sum_{j=1}^{K+M} \Re\{\text{vec}^H(\mathbf{w}_j\mathbf{w}_j^H)\mathbf{b}^u\} = \sum_{j=1}^{K+M} \Re\{\mathbf{w}_j^H\mathbf{B}^u\mathbf{w}_j\}, \tag{22}$$

where $\mathbf{B}^u \in \mathbb{C}^{N \times N}$ and in order to make the formula easier to calculate we take another way to represent the $\mathbf{b}^u$, i.e., $\mathbf{b}^u = \text{vec}(\mathbf{B}^u)$. On account of the formulas (17) and (19.a), we divide $\mathbf{B}^u$ into two distinct parts $\mathbf{B}^u \triangleq \mathbf{B}_1^u + \mathbf{B}_2^u$, and we define

$$\beta \triangleq \sum_{\iota=1}^{L} P_d^2(\theta_\iota), \tag{23.a}$$

$$\mathbf{B}_1^u \triangleq \frac{2}{L} \sum_{\iota_1=1}^{L} \frac{P_d^2(\theta_{\iota_1})}{\beta^2} \sum_{\iota_2=1}^{L} P_d(\theta_{\iota_2}) \text{vec}^H(\mathbf{A}_{\iota_2}) \text{vec}(\mathbf{W}\mathbf{W}^H) \times \sum_{\iota_3=1}^{L} P_d(\theta_{\iota_3})\mathbf{A}_{\iota_3}$$
$$+ \frac{2}{L} \sum_{\iota_1=1}^{L} \text{vec}^H(\mathbf{A}_{\iota_1}) \text{vec}(\mathbf{W}\mathbf{W}^H)\mathbf{A}_{\iota_1}, \tag{23.a}$$

$$\mathbf{B}_2^u \triangleq -\frac{4}{L}\Re\left\{\sum_{\iota_1=1}^{L} \frac{P_d(\theta_{\iota_1})}{\beta} \text{vec}^H(\mathbf{A}_{\iota_1}) \text{vec}(\mathbf{W}\mathbf{W}^H) \sum_{\iota_2=1}^{L} P_d(\theta_{\iota_2})\mathbf{A}_{\iota_2}\right\} - 2\lambda_m \mathbf{W}^t(\mathbf{W}^t)^H. \tag{23.c}$$

Obviously, the matrix $\mathbf{B}^u$ can be seen as $\mathbf{B}_1^u$ plus $\mathbf{B}_2^u$. The function $\mathbf{w}_j^H\mathbf{B}^u\mathbf{w}_j$ in (22) can be divided into $\mathbf{w}_j^H\mathbf{B}_1^u\mathbf{w}_j$ plus $\mathbf{w}_j^H\mathbf{B}_2^u\mathbf{w}_j$. Since the function $\mathbf{w}_j^H\mathbf{B}_2^u\mathbf{w}_j$ is non-convex, in order to turn it into a convex problem we employ the first-order Taylor expansion, and the function $\mathbf{w}_j^H\mathbf{B}_2^u\mathbf{w}_j$ can be constructed as

$$\mathbf{w}_j^H\mathbf{B}_2^u\mathbf{w}_j \leq \left(\mathbf{w}_j^t\right)^H \mathbf{B}_2^u\mathbf{w}_j^t + 2\Re\left\{\left(\mathbf{w}_j^t\right)^H \mathbf{B}_2^u(\mathbf{w}_j - \mathbf{w}_j^t)\right\}, \tag{24}$$

where $\mathbf{W}^t$ is the $\mathbf{W}$ in the t-th iteration and $\mathbf{w}_j^t$ is acquired from the j-th column of $\mathbf{W}^t$. . Hence, put the outcomes in (22)-(24) into (21), we can get $\mathcal{E}(\mathbf{W})$ and its expression is as follow

$$\mathcal{E}(\mathbf{W}) \leq \sum_{j=1}^{K+M} \Re\{\mathbf{w}_j^H\mathbf{B}_1^u\mathbf{w}_j + 2\mathbf{w}_j^H\mathbf{u}_j\} + c_2^t, \tag{25}$$

where we define

$$\mathbf{u}_j \triangleq \left(\mathbf{B}_2^u\right)^H \mathbf{w}_j^t, \tag{26.a}$$

$$c_2^t \triangleq -\sum_{j=1}^{K+M} \Re\{(w_j^t)^H (\mathbf{B}_2^u)^H w_j^t\} + c_1^t + \lambda_m \mathbf{P}^2 \tag{26.b}$$

According to the surrogate function (25), the optimization problem regarding to renovating $\mathbf{W}$ can be rewritten as

$$\max_{\mathbf{W}} \sum_{k=1}^{K} \left(2\sqrt{1+c_k} \Re\{g_k^* \mathbf{w}_k\} - |g_k|^2 \sum_{j=1}^{K+M} |\mathbf{T}_d \mathbf{w}_j|^2 \right) \tag{27.a}$$

$$\text{s.t.} \sum_{j=1}^{K+M} \Re\{\mathbf{w}_j^H \mathbf{B}_1^u \mathbf{w}_j + 2\mathbf{w}_j^H \mathbf{u}_j\} + c_2^t \leq \varepsilon, \tag{27.b}$$

$$\|\mathbf{W}\|_F^2 \leq \mathbf{P}. \tag{27.c}$$

Then the (27) is optimized to a convex problem and we use the CVX tool to settle it.

3.3 Optimize ϕ_1, ϕ_2 With Given $\mathbf{W}$

After optimizing $\mathbf{W}$, and for the sake of reducing the complexity of the optimization problem, we optimize ϕ_1 by giving ϕ_2 and optimize ϕ_2 by giving ϕ_1 after obtaining the optimized ϕ_1. we simplify the optimization problem as

$$\max_{\phi_1} \sum_{k=1}^{K} \left(2\sqrt{1+c_k} \Re\{g_k^* \mathbf{T}_{d,1} \mathbf{w}_k\} - |g_k|^2 \sum_{j=1}^{K+M} |\mathbf{T}_{d,1} \mathbf{w}_j|^2 \right) \tag{28.a}$$

$$\text{s.t} \ |\phi_{1,n}| = 1, \ \forall n, \tag{28.a}$$

where $\mathbf{T}_{d,1} = \mathbf{h}_{u,1}^H \boldsymbol{\Phi}_1 \mathbf{D} \mathbf{G}_2 + \mathbf{h}_{u,1}^H \boldsymbol{\Phi}_1 \mathbf{G}_1$.

Then, we use the equation $(\mathbf{h}_{u,1}^H \boldsymbol{\Phi}_1 \mathbf{D} \mathbf{G}_2 + \mathbf{h}_{u,1}^H \boldsymbol{\Phi}_1 \mathbf{G}_1)\mathbf{w}_j = \mathbf{h}_{u,1}^H \text{diag}[(\mathbf{D}\mathbf{G}_2 + \mathbf{G}_1)\mathbf{w}_j]\phi_1$, expression (28) can be re-formulated as

$$\min_{\phi_1} \ \phi_1^H \mathbf{Z} \phi_1 - 2\Re\{\phi_1^H \mathbf{z}\} \tag{29.a}$$

$$\text{s.t} \ |\phi_{1,n}| = 1, \ \forall n, \tag{29.b}$$

where we define

$$\mathbf{Z} \triangleq \sum_{K=1}^{K} |g_k|^2 \sum_{j=1}^{K+M} \text{diag}(\mathbf{w}_j^H (\mathbf{D}\mathbf{G}_2 + \mathbf{G}_1)^H)\mathbf{h}_{u,1}\mathbf{h}_{u,1}^H \text{diag}((\mathbf{D}\mathbf{G}_2 + \mathbf{G}_1)\mathbf{w}_j), \tag{30.a}$$

$$\mathbf{z} \triangleq \sum_{k=1}^{K} \sqrt{1+c_k} g_k \text{diag}(\mathbf{w}_k^H (\mathbf{D}\mathbf{G}_2 + \mathbf{G}_1)^H)\mathbf{h}_{u,1}. \tag{30.b}$$

We find that the non-convex unit-modulus constraint (29.b) is the main difficulty to handle the problem (29). Thanks to various existing algorithms which can solve the problem (29.b), we use the manifold-based algorithm to work out the problem (29). After optimizing ϕ_1, we can optimize ϕ_2 in the same way.

4 Simulation Results

In this section, we perform a numerical evaluation of the performance of double IRS-aided DFRC system. We adopt a scenario that double IRS assist a dual-functional BS equipped with $M = 10$ antennas servers $K = 5$ single antenna mobile communication users. IRS1 and IRS2 has the same number of reflection elements. The interference is set as $\sigma_k^2 = -80$ dBm.

The distances of BS to IRS1, BS to IRS2, IRS1 to user, IRS2 to user and IRS1 to IRS2 are set as 50 m, 54 m, 4 m, 8 m and 4 m, respectively. Set path-loss exponents of the BS-IRS2, BS-IRS1, IRS1-user, IRS2-user and IRS2-IRS1 as $\alpha_{B-I_2} = 2.5$, $\alpha_{B-I_1} = 4.5$, $\alpha_{I_1-user} = 3.5$, $\alpha_{I_2-user} = 5.5$ and $\alpha_{I_2-I_1} = 3.5$ separately. We consider the Rician fading channel in the scenario with the Rician factors being $\beta_{B-I_2} = 3$ dB, $\beta_{B-I_1} = 5$ dB, $\beta_{I_1-user} = 3$ dB, $\beta_{I_2-user} = 5$ dB, and $\beta_{I_2-I_1} = 3$ dB.

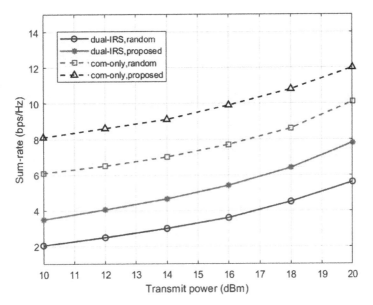

Fig. 2. Sum-rate versus transmit power

In Fig. 2, we illustrate the sum-rate versus the transmit power where the scenarios with double random phase-shift IRS ("dual-IRS, random") is included other than the proposed algorithm ("dual-IRS, proposed"). In addition, we include only-communication ("com-only") system for comparison. It is obvious to observe that the scenarios with double IRS in the proposed algorithm ("dual-IRS, proposed") achieve better performance than random phase-shift IRS, which demonstrates the effective-ness of proposed algorithm for double IRS-aided DFRC system. Moreover, it is obvious that the communication-only system and the DFRC system have a large difference because of the radar detection performance.

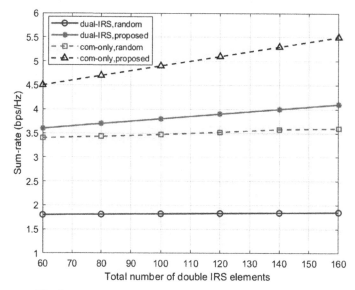

Fig. 3. Sum-rate versus total number of double IRS elements

Then we illustrate the sum-rate versus the total number of double IRS elements in Fig. 3.It shows the sum-rate curve in different IRS reflecting elements. Obviously, the more reflecting elements the IRS has, the larger sum-rate can achieve, which explain the importance of optimizing larger-scale IRS in pursing excellent performance. Besides, when the total quantity of reflective elements increases, the disparity between the "dual-IRS, random" scheme and "dual-IRS, proposed" scheme becomes larger, which further indicates the significance of the proposed algorithm.

Finally, we illustrate the relation curve regarding to the sum-rate versus reflecting elements of the IRS1 in Fig. 4. And the total quantity of reflective elements is 120, i.e. $N_1 + N_2 = 120$. We can observe that under the constraint of $N_1 + N_2 = 120$, the sum-rate can reach the maximum when the values $[N_1, N_2] = [60, 60]$, which shows that there is an optimal $[N_1, N_2]$ that maximum the sum-rate of double IRS-aided DFRC system.

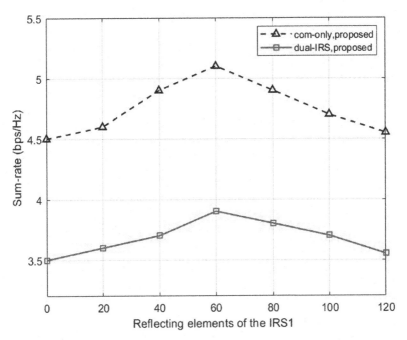

Fig. 4. Sum-rate versus reflecting elements of double IRS elements

5 Conclusion

We investigated the double IRS-aided DFRC system and the relationship between sum-rate and each index. The sum-rate was maximized and the efficiency of user communication was maintained under the restrictions of the IRS phase-shift, the beampattern similarity, and the transmit power and. An excellent alternating optimization algorithm is used to resolve the non-convex problem. And numerical results presented the superiority of employing double IRS in DFRC system. We will consider more practical scenarios, as well as crucial issues, in our future research. And We will continue to explore new technologies to better solve realistic problems, and explore new research areas.

References

1. Rasti, M., Taskou, S., Tabassum, H., Hossain, E.: Evolution toward 6G wireless networks: A resource management perspective (2021). arXiv:2108.06527
2. Liu, R., Li, M., Luo, H., Liu, Q., Swindlehurst, A.L.: Integrated sensing and communication with reconfigurable intelligent surfaces: opportunities, applications, and future directions, June 2022. https://arxiv.org/abs/2206.08518
3. Liu, F., Masouros, C., Petropulu, A.P., Griffiths, H., Hanzo, L.: Joint radar and communication design: applications, state-of-the-art, and the road ahead. IEEE Trans. Commun. **68**(6), 3834–3862 (2020)
4. Wu, Q., Zhang, S., Zheng, B., et al.: Intelligent reflecting surface aided wireless communications: a tutorial. IEEE Trans. Commun. **69**, 3313–3351 (2021)

5. Wu, Q., Zhang, R.: Towards smart and reconfigurable environment: intelligent reflecting surface aided wireless network. IEEE Commun. Mag. **58**(1), 106–112 (2020)
6. Song, X., Zhao, D., Hua, H., Han, T.X., Yang, X., Xu, J.: Joint transmit and reflective beamforming for IRS-assisted integrated sensing and communication. In: IEEE Wireless Communications and Networking Conference (WCNC), Austin, TX, April 2022, pp. 189–194 (2022)
7. Li, Y., Petropulu, A.: Dual-function radar-communication system aided by intelligent reflecting surfaces. In: Proceedings of IEEE Sensor Array Multichannel Signal Process. Workshop (SAM), Trondheim, Norway, June 2022, pp. 126–130 (2022)
8. Huang, C., et al.: Holographic MIMO surfaces for 6G wireless networks: opportunities, challenges, and trends. IEEE Wirel. Commun. **27**(5), 118–125 (2020)
9. Shtaiwi, E., et al.: Sum-rate maximization for RIS-assisted radar and communication coexistence system. In: Proc. IEEE Global Communication Conference (GLOBECOM), Madrid, Spain, December 2021, pp. 1–6 (2021)
10. Wang, X., et al.: Joint waveform and discrete phase shift design for RIS-assisted integrated sensing and communication system under Cramér– Rao bound constraint. IEEE Trans. Veh. Technol. **71**(1), 1004–1009 (2022)
11. Lu, Y., Hao, M., Mackenzie, R.: Reconfigurable intelligent surface based hybrid precoding for THz communications. Intell. Converg. Netw. **3**(1), 103–118 (2022)
12. Jian, L., Stoica, P.: MIMO radar with colocated antennas. IEEE Signal Process. Mag. **24**(5), 106–114 (2007)
13. Shen, K., Yu, W.: Fractional programming for communication systems—part I: power control and beamforming. IEEE Trans. Signal Process. **66**(10), 2616–2630 (2018). https://doi.org/10.1109/TSP.2018.2812733

Channel Estimation for RIS Communication System with Deep Scaled Least Squares

Tiancheng Zhang[✉], Liu Yang, Ying Zhao, and Zijia Chen

Shanghai Dianji University, Shanghai, China
135724953@qq.com

Abstract. Reconfigurable intelligent surfaces (RIS) are widely used in auxiliary millimeter wave communication systems due to their low energy consumption and low cost, accurate channel state information (CSI) is very important for channel estimation. However, the complexity of channel estimation is significantly increased by the fact that RIS are typically used as passive reflectors, that RISs are not capable of signal processing, and that the high complexity of RISs is caused by the abundance of reflective elements. This work suggests a deep systolic least squares channel estimation approach to solve this issue by lowering the guiding frequency overhead and increasing the channel estimation precision. We transform the problem of cascade channel estimation into the problem of noise elimination. By using scaled least square (SLS) algorithm, we can get the channel matrix containing noise, using a ResU-Net network to reduce the noise. The simulation results show that compared with the existing channel estimation method, the ResU-Net network algorithm proposed in this paper reduces the pilot frequency overhead and can significantly improve the accuracy of channel estimation.

Keywords: RIS · Channel estimation · SLS · ResU-Net

1 Introduction

With the development of technology, reconfigurable intelligent surfaces (RIS) play a key role in wireless communication systems [1, 2]. RIS are electromagnetically active surfaces with a multitude of passive elements [3]. Specifically, RIS facilitates wireless communication by integrating a large quantity of economical passive reflective components on a planar surface [4], which can independently modify the magnitude or phase of the communication channel [5]. This change in phase shift or amplitude is performed by the backhaul control link to the Base Station (BS), which allows the received signal from the BS to be adjusted accordingly, helping to improve the channel coverage and flexibly respond to different wireless communication channel transmission environments. Therefore, compared to traditional wireless communication transmission methods, RIS

Supported by National Natural Science Foundation of China (No. 62076160), Natural Science Foundation of Shanghai (No. 21ZR1424700), Shanghai Science and Technology Innovation Funds (No. 23QA1403800).

J. Li et al. (Eds.): 6GN 2023, LNICST 553, pp. 129–139, 2024.
https://doi.org/10.1007/978-3-031-53401-0_13

can break through the traditional wireless channel uncontrollable characteristics and reshape the wireless propagation environment. However, RIS is a passive device with no way to independently estimate the state information of the channel and unable to actively process the information.

The rise of RIS technology has brought about the emerging concept of "smart wireless environment". In the traditional wireless network, the environment is not controlled by the operator, in contrast, in the intelligent wireless environment to transform the environment into an intelligent reconfigurable space, can play a positive role in information transmission and processing. The future application scenario outlook of RIS is mainly divided into two categories, one is the application of traditional communication scenarios, and the other is the new application in vertical industries [6]. The potential application scenarios and requirements of RIS are as follows: Signal coverage supplement and expansion, BS edge coverage enhancement, indoor coverage enhancement, hot spot increase and line-of-sight multistream transmission, transmission stability enhancement, new large-scale antenna transceivers, high-precision positioning, vehicle networking communication, drone communication, communication security, energy collection and transmission, electromagnetic pollution reduction, mobile edge network delay reduction, etc. Despite the great potential of RIS, there are still many practical problems and challenges that need to be solved in RIS-assisted wireless communication. In particular, it is necessary to obtain accurate Channel State Information (CSI) in the RIS-assisted wireless communication system in order to realize the high control of RIS over the wireless communication environment [7]. In particular, in order to achieve effective RIS passive beamforming and reflection phase shift optimization for high rate and ultra-reliable communication, accurate CSI estimation of the communication environment reconfigured by RIS is critical.

First, because active units for baseband signal processing are expensive, low-cost passive RIS units can only reflect incident signals, but cannot transmit and receive pilot signals like active transceivers in traditional wireless communication systems. However, there are a large number of passive units on RIS, which generally have a lot of channel coefficients to estimate in RIS related channels, resulting in a large increase in the system overhead of RIS channel estimation. Therefore, different RIS architectures (semi-passive RIS, fully passive RIS, fully active RIS) put forward different requirements for channel estimation schemes. Second, with the accelerated deployment of 5G, High Speed Train (HST) application scenarios should also be considered in RIS auxiliary communication systems. Due to the characteristics of Doppler shift, fast time-varying channel and relatively short coherence time in high-speed railway system, the need for channel estimation and tracking in multiple coherence times is increasing urgently. Based on the above problems, many researchers have designed many effective channel estimation methods for different RIS architectures and different moving speed scenarios, aiming to achieve low training cost and high channel estimation accuracy.

A number of effective design ideas have been proposed to decrease the guide frequency overhead and enhance the accuracy of channel estimation. These include semi-passive channel estimation [8], sparse channel estimation based on compressed sensing (CS) [9] and channel decomposition based on channel variation characteristics [10]. The traditional CSI mainly uses the least squares (LS) algorithm and the least mean square error (MMSE), but it is difficult to realize in practical application because of its large computation, especially the matrix inversion process is quite complicated. However, due to the existence of a considerable number of antennas and RIS units in place, the dimension of cascade channel estimation is increased, and the CE calculation complexity of traditional methods is high. Convolutional neural networks estimate direct channels and cascades by training deep learning networks with different channel characteristics [11–13]. Based on CNN [13, 14], this paper introduces an SLS deep residual network, aiming to enhance the estimation performance of RIS auxiliary systems.

As a classical neural network framework, encoder-decoder architecture has been widely used in machine translation in NLP, image understanding in CV, and channel feedback in communication. Encoder-decoder architecture actually simulates the human cognitive process of information, that is, the encoder extracts input data into a low-rank vector or matrix through a large number of network layers for downsampling, which can be considered as the process of human understanding and memory of effective information, and the extracted information usually forms a low-rank vector. On the contrary, the decoder recovers the low-rank information of the encoder by upsampling and decodes it into the target data, which can be considered as the process of human recall and use of effective information. Therefore, the encoder-decoder architecture is applied in this paper. Firstly, the encoder is used to compress the input data into a smaller but more channel feature map, and then the desired cascade channel matrix is recovered from the compressed feature map by decoding it. For communication channels, the use of encoders to compress input data is not only beneficial to extract effective features, but also to suppress communication noise and other interference to a certain extent. We can obtain a channel matrix containing noise by using scaled least squares (SLS) algorithm, and propose a U-Net codec structure, which uses frequency hopping to process the input noise matrix, and constructs a new cascade channel matrix. In addition, in this study, we incorporate residual learning into the original U-Net [15], there by achieving enhanced accuracy in estimating the desired channel.

2 System Mode

We examine a millimeter-wave (mmWave) communication system assisted by an RIS, where the RIS board is utilized for channel estimation support, and multiple antennas are configured for the user and the BS. BS is equipped with a parameter controller for adjusting RIS components. The controller can control the amplitude and phase of RIS components, so that the incident signal can be cleverly reflected. In this article, the RIS assisted user communication system is shown in Fig. 1, where the RIS board has N reflective elements, the base station (BS) is equipped with M receive antennas, and there are K single-antenna users. $\mathbf{h}_D \in \mathbb{C}^{M \times 1}$ denotes the channel matrix from the user to the BS, $\mathbb{C}$ is the complex vector set, $\mathbf{h}_k \in \mathbb{C}^{N \times 1}$ denotes the channel matrix from the user to the RIS and $\mathbf{G} \in \mathbb{C}^{M \times N}$ denotes the channel matrix from the RIS to the base station. The RIS plays a passive role in this paper, so it is not possible to estimate the channel matrices $\mathbf{h}_k$ and $\mathbf{G}$ separately, and our aim is to estimate cascaded channels in RIS assisted multi-user systems $\mathbf{h}_G = \mathbf{G}diag(\mathbf{h}_k)$ and $\mathbf{h}_D$. Let $\mathbf{\Phi} = diag(\mathbf{r}) = diag\left(\left[e^{j\varphi_1}, \ldots, e^{j\varphi_N}\right]^T\right) \in$

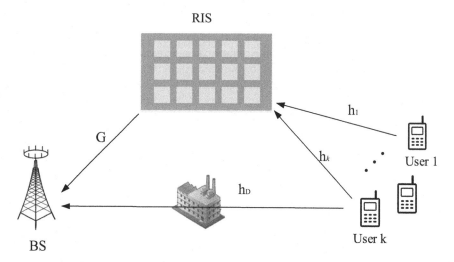

Fig. 1. RIS assisted multi-user system model.

$\mathbb{C}^{N \times N}$ be the reflection phase shift matrix at RIS and $\varphi_n \in [0, 2\pi]$ denote the phase of the nth reflection element of the RIS.

Firstly, it is assumed that the RIS may generate $D = N + 1$ different phase shift patterns, all K users have their guide frequency sequences orthogonal to each other. For the *Kth* user, its guide sequence is denoted as $\mathbf{x}_k$, the guide length is L, and the transmit power of each guide symbol is P. In the dth phase shift mode, the pilot signal received at the BS can be expressed as

$$\mathbf{y}_d = \sum_{k=1}^{K} \mathbf{H}_k \mathbf{p}_d \mathbf{x}_k^H + \mathbf{V}_d \in \mathbb{C}^{M \times L} \tag{1}$$

where $\mathbf{H}_k = [\mathbf{h}_D, \mathbf{h}_G]$ is the channel matrix to be estimated for the *kth* user, $\mathbf{p}_d = [1, \mathbf{r}_d]^T \in \mathbb{C}^{(N+1) \times 1}$ denotes the reflected phase shift matrix in d phase shift modes, and $\mathbf{V}_d$ is the Gaussian white noise. we rewrite the guide frequency signal received in phase shift mode in dth as

$$\mathbf{Y}_{d,k} = \frac{1}{PL} \mathbf{y}_d \mathbf{x}_k = \mathbf{H}_k \mathbf{p}_d + \mathbf{V}_{d,k} \tag{2}$$

At the base station in all phase shift modes can be expressed as follows

$$\mathbf{Y}_k = \mathbf{H}_k \mathbf{P} + \mathbf{V} \tag{3}$$

where $\mathbf{P}$ denotes the reflected phase shift matrix with D phase shift modes, as represented by Eq. (4), and we can determine this according to the discrete Fourier transform (DFT)

[16], $\mathbf{V} \in \mathbb{C}^{M \times D}$ is the noise matrix in which all elements obey a Gaussian distribution.

$$
\mathbf{P} = \begin{bmatrix} 1 & 1 & \cdots & 1 \\ 1 & w & \cdots & w^{D-1} \\ \vdots & \vdots & \ddots & \vdots \\ 1 & w^N & \cdots & w^{N(D-1)} \end{bmatrix} \in \mathbb{C}^{(N+1) \times D} \tag{4}
$$

3 SLS Initial Channel Estimation

3.1 Least Squares Channel Estimation

The least squares method is relatively easy to implement for channel estimation. We base our least squares criterion on the known reflection phase shift matrix $\mathbf{P}$ and the received signal $\mathbf{Y}_k$, the channel matrix obtained by LS algorithm can be expressed as

$$
\widehat{\mathbf{H}}_{k,\mathrm{LS}} = \mathbf{Y}_k \mathbf{P}^\dagger \tag{5}
$$

where $\mathbf{P}^\dagger = \mathbf{P}^H \left(\mathbf{P}\mathbf{P}^H \right)^{-1}$ stands for the pseudo-inverse of $\mathbf{P}$ and $(\cdot)^H$ means Hermitian transpose. The channel estimation error J_{LS} is represented as

$$
J_{\mathrm{LS}} = \mathrm{E}\left\{ \|\mathbf{H}_k - \widehat{\mathbf{H}}_{k,\mathrm{LS}}\|_F^2 \right\} = \sigma^2 M \, \mathrm{tr}\left\{ \left(\mathbf{P}\mathbf{P}^H \right)^{-1} \right\}. \tag{6}
$$

3.2 Scaled Least Square Channel Estimation

The LS channel estimation method is easy to implement, but it has many disadvantages, and the performance of our estimation degrades quickly in low SNR environment. Therefore, we improve the traditional LS estimation method, and enhancing the accuracy of channel estimation further holds significant importance. Based on the LS channel estimation algorithm, SLS channel estimation further reduces noise. The channel error for the LS channel estimate is shown below

$$
\begin{aligned}
& \mathrm{E}\left\{ \left\| \mathbf{H}_k - \gamma \widehat{\mathbf{H}}_{k,\mathrm{LS}} \right\|_F^2 \right\} \\
& = (J_{\mathrm{LS}} + \mathrm{tr}\{\mathbf{R_H}\}) \cdot \left(\gamma - \frac{\mathrm{tr}\{\mathbf{R_H}\}}{J_{\mathrm{LS}} + \mathrm{tr}\{\mathbf{R_H}\}} \right)^2 + \frac{J_{\mathrm{LS}} \mathrm{tr}\{\mathbf{R_H}\}}{J_{\mathrm{LS}} + \mathrm{tr}\{\mathbf{R_H}\}}
\end{aligned} \tag{7}
$$

The equation $\mathbf{R_H}$ is the autocorrelation matrix of $\mathbf{H}_k$ and γ is the scaling factor. For $\gamma = \gamma_0 = \frac{\mathrm{tr}\{\mathbf{R_H}\}}{J_{\mathrm{LS}} + \mathrm{tr}\{\mathbf{R_H}\}}$. The estimated channel according to the SLS algorithm can be rewritten as

$$
\widehat{\mathbf{H}}_{k,\mathrm{SLS}} = \gamma_0 \widehat{\mathbf{H}}_{k,\mathrm{LS}} = \frac{\mathrm{tr}\{\mathbf{R_H}\}}{\sigma^2 r \mathrm{tr}\left\{ \left(\mathbf{P}\mathbf{P}^H \right)^{-1} \right\} + \mathrm{tr}\{\mathbf{R_H}\}} \mathbf{Y}_k \mathbf{P}^\dagger \tag{8}
$$

The received signal under SLS channel estimation can be represented as

$$\mathbf{Y}_{k,\text{SLS}} = \widehat{\mathbf{H}}_{k,\text{SLS}}\mathbf{P} \tag{9}$$

where $\text{tr}\{\mathbf{R}_{\text{H}}\} = tr\left\{\widehat{\mathbf{H}}_{k,\text{LS}}^{H}\widehat{\mathbf{H}}_{k,\text{LS}}\right\}$ is a trace of $\mathbf{R}_{\text{H}}$, SLS channel estimation takes the effect of noise further into account compared to SL channel estimation, and SLS estimation achieves high accuracy even in low S/N regions. In addition to this, SLS estimation has many other benefits in that it does not require knowledge of the statistical properties of the a priori channel.

$$J_{\text{SLS}} = \frac{J_{\text{LS}}\text{tr}\{\mathbf{R}_{\text{H}}\}}{J_{\text{LS}} + \text{tr}\{\mathbf{R}_{\text{H}}\}} < J_{\text{LS}} \tag{10}$$

According to formula (10), the accuracy of SLS channel estimation is greater than that of traditional LS estimation. However, SLS estimates that the channel also contains a lot of noise, and the accuracy obtained is not high enough, and there is still room for improvement. Therefore, we designed a deep learning network to eliminate noise. Through this network, by utilizing the SLS algorithm, there by mitigating the impact of noise.

4 ResU-Net Based Channel Estimation

We convert the intricate cascaded channel estimation problem into a noise cancellation challenge. First, we build a general data set so that the neural network can be used at different lead lengths. On this basis, we propose a ResU-Net based network, which improves the extraction and elimination of noise features by ResU-Net to obtain higher channel accuracy.

4.1 Data Preprocessing

In channel estimation, considering that the channel matrix is usually a matrix containing real and imaginary parts, we usually input the real and imaginary parts to facilitate feature extraction. When we perform channel estimation in different communication environments, we need to change the number of nerve points in the input layer accordingly to accommodate different pilot lengths. To extract features from a complex noisy channel matrix. We take the real and imaginary parts of the pilot signals $\mathbf{Y}_{k,\text{SLS}}$ and $\mathbf{H}_k$ processed by SLS as inputs to the network, respectively.

4.2 Network Architecture

In response to the characteristics of RIS cascaded channels and the impact of data preprocessing, the dimensionality of the received pilot signal is increased through mathematical operations, which also makes our calculation more difficult. Therefore, we propose a ResU-Net architecture as shown in Fig. 2. From the overall network model, in the denoising block, we designed an encoder and decoder architecture. Using an encoder decoder

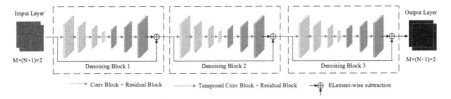

Fig. 2. The ResU-Net network architecture.

network skeleton to explore the potential sparsity of input data and suppress the noise contained in the input data, combining the advantages of U-Net and ResNet networks, a ResU-Net structure is formed to reduce information loss in the feature compression process of ResU-Net.

The encoder extracts input data into a low rank vector matrix through a large number of network layers for down-sampling, which can be considered as a process of human understanding and memory of effective information. Extracting information usually forms a low rank vector. On the contrary, the decoder performs up-sampling to recover the low-rank information of the encoder and decode it into target data, which can be considered as a process of human conference and utilization of effective information.

In the ResU-Net network, we use three denoising blocks to gradually remove noise in the channel, which share the same structure. Specifically, In the encoder stage, the input data from $M \times N \times 2$ is compressed to $M/2^E \times N/2^E \times 256$ using E encoding blocks, and in the channel count, the input data is expanded from a channel count of 2 to 256, resulting in an encoding matrix to $M/2^E \times N/2^E \times 256$. Each of which is down-sampled by a convolutional layer with a convolutional kernel sliding step of 2.The number of filters in the convolutional layer from the outside to the inside of the encoder is $f_i^e = 64 \times 2^{i-1}$, $(i = 1, 2, 3)$ where i indicates the sequence number of the encoder block from the outside to the inside. The complexity of the network and the channel estimation accuracy need to be considered when designing the depth and width of the ResU-Net, the size and number of convolutional kernels of the residual block in each encoding block are the same as the convolutional block. ResNet is widely used as a deep network, and residual learning can be represented as $F(x) = R(x) + x$, where $R(x)$ represents the result of the convolutional operation and x represents the initial input data.

The decoder of ResU-Net has a network body structure symmetrical to the encoder, and the decoder blocks recover the compressed features of dimension $M/2^E \times N/2^E \times 256$, each decoder block comprises a transposed convolution block and a residual block. Each transposed convolutional layer has a step size of 2 to up-sample the feature map, and the number of filters is $f_i^d = 512/2^{i-1}$, $(i = 1, 2, 3)$, where i is the decoder block number from inside to outside. The channel matrix needs to be output in the last layer of the encoder to meet the dual-channel output.

5 Simulation Results and Analysis

We do a number of experiments to prove the effectiveness of the proposed algorithm. In our experiments, BS has $M = 8$ receiver antennas, RIS is equipped with $N = 64$ reflector elements, and $K = 6$ single antenna users, $E = 4$. A path loss model is adopted in the simulations and the path losses of each channel can be modeled as $\alpha_k^{UB} = \alpha_0(\lambda_k^{UB}/\lambda_0)^{-\gamma_1}, \alpha_k^{UI} = \alpha_0(\lambda_k^{UI}/\lambda_0)^{-\gamma_2}$, and $\alpha_k^{IB} = \alpha_0(\lambda_k^{IB}/\lambda_0)^{-\gamma_3}$, where $\gamma_i, i \in \{1, 2, 3\}$, is the path loss exponent, $\lambda_0 = 15$ m is the reference distance, $\alpha_0 = -10$ dB is the path loss at the reference distance, and $\lambda_k^{UB}, \lambda_k^{UI}, \lambda_k^{IB}$ represent the distances of U_k-BS, U_k-IRS, and IRS-BS respectively. Specifically, we set $\lambda_k^{UB} = 90$m, $\lambda_k^{UI} = 15$ m, and $\lambda_k^{IB} = 80$ m. In addition, gradient fading models for all channels were considered to characterize small-scale fading. To assess the accuracy of our proposed scheme and validate its reliability, we employed the normalized mean square error (NMSE) as the evaluation metric for the experiments. The NMSE expression in this article is NMSE = $\mathbb{E}\left\{\|\widehat{\mathbf{H}}_k - \mathbf{H}_k\|_F^2 / \|\mathbf{H}_k\|_F^2\right\}$, where $\widehat{\mathbf{H}}_k$ is the estimated cascade channel. In the experiment, we generated 30,000 samples, randomly divided into training set accounting for 60%, verification set accounting for 20%, and test set accounting for 20%. The signal-to-noise ratio is 0 to 20 dB, the interval is 5 dB, and the training set lead length is set to L = 32. We test at different signal-to-noise ratios and L. Subsequently, we verify the channel estimation performance of our proposed ResU-Net method through a large number of experiments, and verify its relationship with SNR and pilot length respectively.

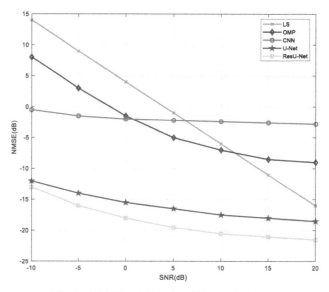

Fig. 3. NMSE and SNR for different algorithm.

We compare the NMSE performance of the SLS-based ResU-Net network presented in this paper with traditional LS estimators and OMP, and other DL models CNN and U-Net. As can be seen from Fig. 3, ResU-Net is superior to other algorithms in a certain range of SNR. In a low SNR environment, the performance of LS estimation diminishes, the LS estimator can obtain ideal performance at high signal-to-noise ratio. ResU-Net allows us to obtain more accurate channels and better remove noise through the superposition of residual blocks. At a signal-to-noise ratio of 5 dB, the ResU-Net network improves by about 3dB compared with the U-Net estimation method. Compared with other estimation schemes, the ResU-Net network has the best performance. ResU-Net can reduce more pilot overhead to achieve better NMSE performance.

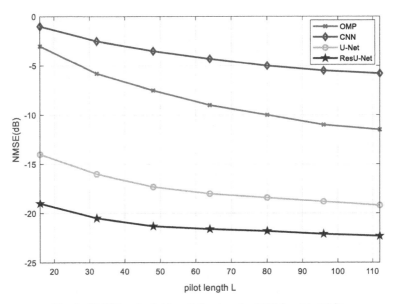

Fig. 4. NMSE and pilot length L when the SNR is set to 10dB.

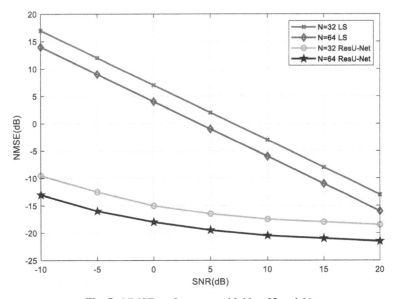

Fig. 5. NMSE performance with N = 32 and 64.

In Fig. 4, we compare NMSE performance at different pilot lengths. As L increases, the estimated results become more accurate, while the performance of the codec architecture does not change significantly with different lead lengths. With the increase of pilot length, the value of NMSE decreases. In Fig. 5, we perform tests with different RIS reflection element numbers (N = 32 and 64) and compare the proposed method with the traditional LS method. From the simulation results, it becomes evident that as the number of reflection elements increases, the channel matrix we need to estimate becomes more sparse, and the performance of NMSE obtained by experiment becomes more accurate. In addition, due to the addition of residual network in the ResU-Net scheme, when N = 32 and N = 64, the ResU-Net scheme is better than the traditional LS estimation, this observation strongly validates the ResU-Net scheme across different values of N.

6   Conclusion

This paper in order to solve the complexity of cascade channel estimation, improve the accuracy of the channel. The accuracy of traditional estimation methods is too low and the pilot frequency overhead is very high. In this paper, we propose SLS method for initial cascade channel estimation, and then send our initial estimated channel into ResU-Net network to further remove noise and improve the accuracy of the estimated channel. The algorithm has a low pilot frequency overhead and a good initial channel estimation accuracy. In the ResU-Net residual network, channel features are further extracted to eliminate noise. Finally, the experimental results demonstrate that the proposed algorithm achieves superior estimation accuracy compared to other channel estimation methods. In the future work, we hope to design a more simplified network structure to decrease the complexity of channel estimation and obtain higher estimation accuracy.

References

1. Zhang, J., Bjornson, E., Matthaiou, M., Ng, D.W.K., Yang, H., Love, D.J.: Prospective multiple antenna technologies for beyond 5G. IEEE J. Sel. Areas Commun. **38**(8), 1637–1660 (2020)
2. Wu, Q., Zhang, R.: Towards smart and reconfigurable environment: intelligent reflecting surface aided wireless network. IEEE Commun. Mag. **58**(1), 106–112 (2020)
3. He, Z., Héliot, F., Ma, Y.: Machine learning for IRS-assisted MU-MIMO communications with estimated channels. In: Proceedings of 2022 IEEE 95th Vehicular Technology Conference. Helsinki, Finland, pp. 1–6. IEEE (2022)
4. Huang, C., Zappone, A., Alexandropoulos, G.C., et al.: Reconfigurable intelligent surfaces for energy efficiency in wireless communication. IEEE Trans. Wirel. Commun. **18**(8), 4157–4170 (2019)
5. Dai, L., Wang, B., Wang, M., et al.: Reconfigurable intelligent surface-based wireless communications: antenna design, prototyping, and experimental results. IEEE Access **8**, 45913–45923 (2020)
6. Li, L., et al.: Enhanced reconfigurable intelligent surface assisted mmWave communication: a federated learning approach. China Commun. **17**(10), 115–128 (2020)
7. Xu, X., Zhang, S., Gao, F., Wang, J.: Sparse Bayesian learning based channel extrapolation for RIS assisted MIMO-OFDM. IEEE Trans. Commun. **70**(8), 5498–5513 (2022)

8. Chen, X., Shi, J., Yang, Z., Wu, L.: Low-complexity channel estimation for intelligent reflecting surface-enhanced massive MIMO. In: Proceedings of IEEE ICASSP, Brighton, U.K., May 2019, pp. 4659–4663 (2019)
9. Wang, P., Fang, J., Duan, H., Li, H.: Compressed channel estimation for intelligent reflecting surface-assisted millimeter wave systems. IEEE Signal Process. Lett. **27**, 905–909 (2020). https://doi.org/10.1109/LSP.2020.2998357
10. Liu, H., Yuan, X., Zhang, Y.-J.A.: Matrix-calibration-based cascaded channel estimation for reconfigurable intelligent surface assisted multiuser MIMO. IEEE J. Sel. Areas Commun. **38**(11), 2621–2636 (2020)
11. Liu, C., Liu, X., Ng, D.W.K., Yuan, J.: Deep residual learning for channel estimation in intelligent reflecting surface-assisted multi-user communications. IEEE Trans. Wirel. Commun. **21**, 898–912 (2021). https://doi.org/10.1109/TWC.2021.3100148
12. Kundu, N.K., McKay, M.R.: Channel estimation for reconfigurable intelligent surface aided MISO communications: from LMMSE to deep learning solutions. IEEE Open J. Commun. Soc. **2**, 471–487 (2021)
13. Taha, A., Zhang, Y., Mismar, F.B., et al.: Deep reinforcement learning for intelligent reflecting surfaces: Towards standalone operation. In: Proceedings of International Workshop Signal Processing Advances in Wireless Communications (SPAWC), Atlanta, GA, USA, August 2020, pp. 1–5. IEEE (2020)
14. Liu, M., Li, X., Ning, B., Huang, C., Sun, S., Yuen, C.: Deep learning-based channel estimation for double-RIS aided massive MIMO system. IEEE Wirel. Commun. Lett. **12**(1), 70–74 (2023)
15. He, K., Zhang, X., Ren, S., Sun, J.: Deep residual learning for image recognition. In: Proceedings of IEEE CVPR, pp. 770–778 (2016)
16. Jensen, T.L., Carvalho, E.D.: An optimal channel estimation scheme for intelligent reflecting surfaces based on a minimum variance unbiased estimator. In: International Conference on Acoustics, Speech, and Signal Processing, April 2020, pp. 5000–5004 (2020)

Algorithmic Protection Study Based on a Virtual Location

Zehui Wen and Yiqun Zhu[✉]

Shanghai DianJi University, Shuihua. 300, Shanghai 201306, China
zhuyiq@sdju.edu.cn

Abstract. With the development of big data technology and wireless communication technology, Internet-of-Vehicles (IOV) has been widely used. In order to reduce the probability of traffic accidents and provide different functional services for vehicle operation. Location-based services (LBS) have become an important part of our daily lives and are gradually spreading to vehicle terminals, bringing convenience to people's travel. However, while users enjoy the convenience provided by LBS, their location privacy information may be leaked, because the untrusted LBS server has all the information of the user in LBS, and may track the user in various ways or leak the user's personal data to a third party, which is easy to bring unexpected losses. In today's research, researchers have proposed schemes such as k-anonymity and Mix-zone in order to protect the user's location privacy and trajectory routes. However, these programs also have some deficiencies and shortcomings. To better address these issues, this paper proposes a new location privacy protection scheme. Vehicles on the road can dynamically generate a false position based on moving vehicles around them to blur driving routes, and realize the location privacy protection. Through the experimental analysis, the performance of the two schemes is compared, and the proposed algorithm can guarantee the user's location privacy while having a lower time cost.

Keywords: Location Privacy · Virtual Location · Route Disarray

1 Introduction

Vehicle networking [1] also known as vehicle AD hoc networking, refers to the use of various advanced technologies for road and traffic environment perception, to better achieve a wide range of data interaction. In the past five years, the Internet of Vehicles (IOV) has made rapid development, which can well serve the communication between vehicles (V2V) and vehicles and other infrastructure in the road network (V2I). The Internet of vehicles can not only make road traffic more harmonious, but also provide various entertainment services for in-vehicle users. The application field that is expected to make great use of this advanced technology is location-based services (LBS), which can provide corresponding application services based on the current location of in-vehicle users [2]. However, the location-based service system in the vehicle network often requires users to submit their current service request location information and query

J. Li et al. (Eds.): 6GN 2023, LNICST 553, pp. 140–151, 2024.
https://doi.org/10.1007/978-3-031-53401-0_14

content to the LBS service provider. Once this information is speculated and attacked by attackers, the privacy of the vehicle users will be leaked. Therefore, when vehicle users enjoy the services provided by the location-based service system of the Internet of vehicles, their real location and query content information need to be protected.

Communication in vehicle networking can be classified into two categories: vehicle-to-infrastructure (V2I) communication and vehicle-to-vehicle (V2V) communication [3]. As the main force of the future intelligent transportation system, the Internet of Vehicles has two main goals. First of all, Through Vehicle Ad-Hoc Networks (VANETs), vehicles can access up-to-the-minute data regarding the nearby traffic conditions and the status of other vehicles. This facilitates a decrease in traffic accidents while enhancing the overall efficiency of traffic flow. Secondly, as vehicles traverse the roads, they have the capability to connect to the Internet. Consequently, VANETs enable drivers to access a range of online services, such as news updates, social interactions, and entertainment options. This integration enhances the overall driving experience. By initiating location-based service (LBS) inquiries, users can avail themselves of the convenience offered by LBS technology [4]. Nonetheless, the situation arises where an untrusted LBS server possesses a comprehensive record of the user's location history, query submissions, activities, and more. This potential scenario raises concerns of unauthorized tracking or the inadvertent exposure of the user's private information to external entities. As a result, it becomes imperative to place increased emphasis on safeguarding the privacy of users. To achieve location privacy, numerous communication protocols and security strategies have been put forth by researchers. Among these, the K-anonymity concept stands out as a prominent example. K-anonymity entails a scenario where, within a given data query involving k users, no individual user can be discerned from the remaining K-1 users [5]. Drawing upon the principles of K-anonymity and employing techniques like virtual location and route obfuscation, a novel approach to safeguarding vehicle location privacy is introduced. This approach centers on the vantage point of vehicle routing. Illustrated in Fig. 1 are two distinct pathways taken by separate vehicles:

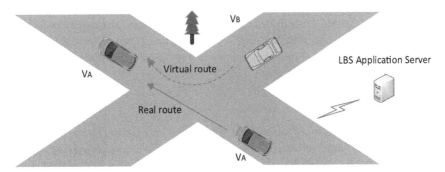

Fig. 1. Two routes for different vehicles

Xing et al. [6] proposed a double-k anonymity based on location privacy preservation approach by adding a cloud server as a trusted third party, which is able to isolate specific information from user requests, arrange and combine the user's location information and request information to reduce the correlation between identity and request. Added matrix algorithm and encrypted the matrix so that the new encrypted matrix is used to request location based on services instead of direct user requests. Again, adding cloud servers to the normal service flow will inevitably cause delays.

Liu et al. [7] designed a location privacy protection scheme (VVS) based on virtual vehicles, which firstly changed the previous way of certificate generation by a trusted authority, constructed a new network structure for distributing pseudonyms, and issued the function of generating pseudonym certificates to certificate centers, which realized distributed management at a time, and thus improved the efficiency of pseudonym generation. Second, virtual vehicles are added to the vehicle network to assist vehicles that want to perform pseudonym transformation. At the same time, the pseudonyms of the vehicles are to be replaced in a timely manner and the pseudonyms of the vehicles with offensive behavior are to be revoked. Compared with the existing schemes, the security of this scheme is significantly strengthened, and there is an advantage in the computational complexity. Finally, according to the analysis of the results, it is concluded that the VVS scheme can effectively protect the privacy of vehicle location and has a certain degree of feasibility.

Chen et al. [8] in trajectory analysis research has greatly facilitated the release of trajectory data, but any method has privacy issues, what he proposed is a flexible trajectory delineation scheme Mix-STC. This scheme is to find more and better candidate mixing zones by 3D spatiotemporal cube computation, he is to provide privacy preservation by restricting the lengths of the sub trajectories and the exact location of the interference endpoints. The study shows that we have to continue to optimize the distance between the length and Mix-STC, improve the method of data perturbation, increase the data availability and ensure more unlink ability.

2 Research Work

Preserving the confidentiality of location information holds significant significance within location-based service (LBS) frameworks. With the advancement of localization technologies within both in-vehicle and mobile devices, the matter of location privacy within Vehicle Ad-Hoc Networks (VANETs) has garnered considerable interest among researchers. As a result, numerous privacy-preserving strategies have been devised, grounded in LBS paradigms.

2.1 LBS Basic Architecture

Ensuring the confidentiality of location data is a crucial prerequisite for the proper functioning of location-based service (LBS) systems [9]. In the recent period, the advancement in localization capabilities of in-vehicle and mobile devices has sparked significant interest among researchers regarding the matter of location privacy within Vehicle Ad-Hoc Networks (VANETs). Consequently, a multitude of privacy-centric strategies, rooted in the framework of Location-Based Services (LBS), have been put forth by these researchers. A typical location-based service (LBS) system consists of four basic elements: user equipment, positioning system, network infrastructure, and external LBS provider. As shown in the Fig. 2:

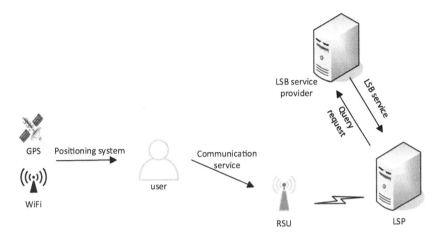

Fig. 2. Location-based service processes in the Internet of Vehicles

2.2 The LBS Query Process

Each user uses his or her real identity and device number when communicating with the network infrastructure, or choose a pseudonym to protect the confidentiality of their personal information. The user's location data will be periodically refreshed within the network, either through passive or active means. When utilizing LBS, the initial step for the user involves acquiring their location from the positioning system. Subsequently, this location data, along with the LBS inquiry, is transmitted to the LBS application server; The LBS application server has the capability to retrieve certain user details, encompassing pseudonyms, request messages, location coordinates, and timestamps. Furthermore, in typical scenarios, the locations associated with request messages from a single user tend to be contiguous. While the use of pseudonyms effectively prevents the exposure of a user's actual identity, the LBS application server still possesses the potential to deduce the user's whereabouts or travel patterns from this data. This phenomenon subsequently gives rise to concerns regarding location privacy, which in turn triggers the location privacy problem [10]; The wireless network creates a communication channel connecting the user with the LBS server, facilitating the transmission of location query requests initiated by the user and the subsequent delivery of outcomes provided by the LBS server; The LBS server responds to query requests in real time and returns results.

2.3 LBS Threat

Within the framework described in this paper, both the Roadside Unit (RSU) and the LBS application server are not considered entirely reliable, leaving privacy exposed to potential compromise by malicious actors aiming to illicitly obtain the vehicle's location data. However, the assailant assumes a passive role, refraining from manipulating communication messages, but instead illicitly capturing sensitive information pertaining to privacy. Should the vehicle openly append its current real-time location to the LBS request and transmit this request to either the RSU or the LBS server, an adversary could exploit the data within the LBS request to promptly trace and monitor the vehicle's movements. For example, Liu [11] proposed a group signature-based pseudonym exchange location privacy protection scheme, even though the vehicle can use signature-based pseudonym exchange and each message is pseudonym exchanged and obfuscated, an attacker can still infer the vehicle's driving route based on its frequently updated location, speed, direction, and road conditions. Consequently, a critical concern addressed in this paper is the formulation of methods to uphold the confidentiality of the vehicle's location, all the while delivering precise LBS services to the vehicle.

In order to solve the location privacy leakage problem of mobile users, Yang et al. [12] proposed a location privacy protection scheme based on elliptic curve cryptosystem for k-anonymity batch authentication, which employs batch processing, fewer number of message interactions, and enhanced the execution efficiency; by combining random location point generation algorithm and location entropy to screen the most k-1 fake locations in the anonymized region, it enhances the user's privacy protection degree. The scheme has certain advantages in terms of completion efficiency and privacy protection degree, and satisfies the security features of anonymity, unforgeability, and resistance to counterfeiting attacks.

3   Preliminaries

Within this segment, we present the fundamental ideas encompassing information entropy and transfer probability.

3.1   Information Entropy

In 1948, Shannon introduced the notion of "information entropy" as a means to quantitatively measure information [13]. In probability theory and statistics, each value of a random variable has its own significance, i.e., each value has its own amount of information. Now there exists a random variable X and the probability space of X satisfies the following mathematical model [14]:

$$\begin{bmatrix} X \\ P \end{bmatrix} = \begin{bmatrix} x_1 x_2 \cdots x_n \\ p_1 p_2 \cdots p_n \end{bmatrix} \tag{1}$$

Within this research article, we employ entropy as a metric to quantify the level of anonymity. This concept can be interpreted as the level of uncertainty associated with pinpointing an individual's present location among all potential alternatives. The increase of information entropy corresponds to the increase of fuzziness, and the uncertainty of physical objects increases, which leads to the decrease of the possibility of successful intrusion by attackers. Conversely, a decrease in information entropy means less ambiguity and less uncertainty about the real thing, which in turn increases the probability of a successful attack. The higher the information entropy, the higher the similarity of the transfer probabilities. In the process of computing entropy, each potential location holds a historical querying probability denoted as P_i. The information entropy H is determined using a formula based on these probabilities

$$H = -\sum_{i=1}^{n} p_i \cdot \log_2 p_i \tag{2}$$

Which

$$\sum_{i=1}^{n} p_i = 1 \tag{3}$$

Our objective revolves around attaining maximum entropy, a state in which all k feasible locations share an equal probability of 1/k. This pinnacle of entropy is reached when uniform probabilities are distributed across all potential locations

$$H_{\max} = \log_2 k \tag{4}$$

3.2 Transfer Probability

The transfer probability aims to guarantee that the chosen group of safeguarded locations exhibits strong resemblance to the user's authentic behavioral pattern. When each location within the safeguarded set closely aligns with the user's genuine transferred location, which is in line with the current user's behavioral pattern, it can avoid the attacker to reason out the user's real location based on the background knowledge such as the geographic location and the user's historical trajectory [15].

The transfer of a mobile subscriber between two consecutive geographic locations can be solved using the transfer probability. Generally, a mobile user will initiate services from consecutive geographic locations within a certain period of time, thus forming a service request trajectory. There are generally correlations between different geographic locations on the mobile service request trajectory [16], i.e., the geographic location of the mobile user at the current moment is correlated with the previous n mobile trajectory points, and thus the mobile trajectory of the mobile user can be regarded as an nth order Markov chain.

$$p_t = M * p_{t-1} \tag{5}$$

$$p_t = \left[p_{t[1]} p_{t[2]} \cdots p_{t[n]} \right] \tag{6}$$

$$0 \leq p_t[i] \leq 1, \sum_{i=1}^{n} p_t[i] = 1 \tag{7}$$

4 System Model

Within this segment, we are poised to introduce a pair of algorithms, namely the information entropy location privacy preserving algorithm and the transfer probability based on location privacy preserving algorithm, Additionally, we will demonstrate, through experimental comparisons, that the proposed algorithms notably enhance the level of privacy preservation.

4.1 Information Entropy Based Location Privacy Preserving Algorithm

Considering the computational cost problem, an appropriate degree of anonymity k is chosen [17]. When using the DLS algorithm [18], The user is required to review all the acquired query probabilities, followed by sorting all the cells the sequence of query probabilities. After sorting, if there are more than one cell with the same query probability as the actual position, the list can be divided into two, half of which is placed before the actual position and the other half is placed after the actual position, which can effectively improve k-anonymity. The k cells before and after the user selects the actual position are used as the 2k candidate cells. The rationale behind opting for a candidate set of 2k locations is to amplify anonymity, and the magnitude of this set can be modified based on the user's preferences. Immediately after that, export m groups of large cells and each cell has k small cells. In every location set, a single cell corresponds to the actual location, while the remaining k-1 cells are selected at random from the pool of 2k candidates [19].

Concretely, select a specific set C and calculate the entropy using the following formula

$$H_j = -\sum_{i=1}^{n} p_{ji} \cdot \log_2 p_{ji} \tag{8}$$

At its peak, maximum entropy is reached when the probabilities of the k submitted locations align with that of the actual location on the server's end. The DLS algorithm identifies and produces the collection exhibiting the greatest entropy value:

$$C = \arg\max H_j \tag{9}$$

Algorithm 1 describes the exact process of the DLS algorithm, which is effective in providing k-anonymity [20], and the algorithm effectively attains a substantial level of privacy in relation to entropy. It becomes evident that, during the process of selecting virtual locations, the DLS algorithm can be further optimized by dispersing them over a larger spatial expanse, leading to enhanced performance, so that the virtual locations can be extended to a larger shaded area.

Algorithm 1: Dummy-Location Selection Algorithm

Input: probability of obtaining a query q, real location, a set of cells m, k

Output: a set of the most accurate virtual locations

1 arrange all cells in order of query probabilities;

2 select 4k virtual objects, split in two, 2k before and 2k after the virtual position;

3 **for** (j=1; j<m; j + +) **do**

4   select a specific set of 2k randomly selected ground truth locations and k-1 other cells;

5   calculate the normalized query probability in each cell;

6      $$H_j = -\sum_{i=1}^{k} p_{ji} \cdot log_2 \, p_{ji}$$

7 **End**

8 output argmax H_{ji}

Algorithm 1 gives the specific procedure of the DLS algorithm, which can effectively corroborate K-anonymity for better location privacy protection, while the DLS algorithm excels in achieving a higher level of privacy in the context of information entropy, it is derived from the analysis that it is desirable to disperse these virtual locations to more distant regions when selecting them. Therefore, research can continue on enhanced DLS algorithms to allow virtual locations to expand to larger and more distant regions.

4.2 Location Privacy Protection Algorithm Based on Transfer Probability

When the user is in a location with fewer routes, the transfer probability is mostly zero, and if the virtual protected location set is constructed directly in this case, it will expose the user's real location to a large extent [21].Therefore, we have to find the set of candidate locations that meet the maximum quality of service loss within the user-defined offset distance, remove the locations with zero transfer probability, and select the offset locations from the remaining valid locations with the following algorithm:

Algorithm 2: User Location Offset Algorithm

Input: Dataset S user location L offset distance D

Output: offset position L'

1 the dataset grid is first divided;

2 calculate the transfer probability for each cell position;

3 Search for locations (excluding the user position) centered on the user position L, with the offset distance D as the radius, and add the searched locations to the set of candidate locations C;

4 filtering the set of candidate locations and removing location units with zero transfer probability;

5 randomly select a position in C as the offset position L';

6 output L'

Algorithm 2 considers the transfer probability of the user's position. The location shift algorithm ensures the anonymity of the position with zero transfer probability, and randomly selects the position to replace the user's position after the shift. This algorithm can protect the maximum entropy of the location group and effectively ensure the location privacy performance.

5 Performance Analysis

Within this segment, we evaluate the performance of location privacy protection in terms of privacy entropy value, running time, and location recognition probability.

5.1 Privacy Entropy

Set the anonymity degree of the location protection group to k. If the attacker can obtain k locations, then the offset probability is $p = 1/k$. From the definition of information entropy, it can be seen that the entropy is relatively large when the transfer probability of the location is closer to the true transfer probability of the user. From the Fig. 3, it can be seen that when the anonymity level k increases, the privacy entropy value of both schemes increases, indicating that the privacy protection ability of the shown schemes are increasing.

5.2 Running Time

In this paper, the algorithm runtime is defined as the time from the beginning of the transfer probability calculation to the generation of a location cluster, and the algorithm runtime is used to reflect the generation efficiency of the algorithm to protect privacy. Figure 4 demonstrates the relationship between anonymity level and running time, as

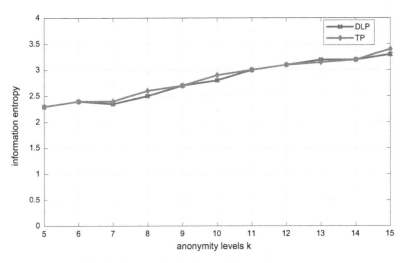

Fig. 3. Anonymity levels and information entropy

shown in the figure, as the anonymity level k increases, the running time increases overall, but the DLS algorithm has a lower running time than the TP algorithm, so the DLS algorithm has a lower time cost and a better algorithmic generation performance while guaranteeing privacy.

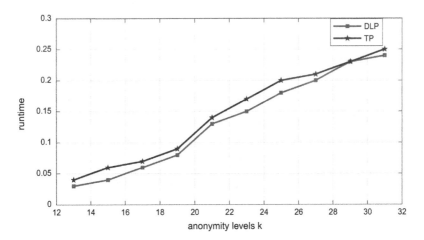

Fig. 4. Anonymity level and runtime

5.3 Location Recognition Probability

In the two algorithms of privacy protection, although it can effectively protect the privacy, but the attacker can still use the effective information to reasoning attack, such as Fig. 5

make the two algorithms recognition probability comparison results graph. From the figure, it can be seen that the recognition rate gradually decreases with the increase of the anonymity level k, which indicates that it is gradually difficult to obtain the user's real location from the location group. This algorithm considers the transfer probability associated with the user's location. Subsequently, an appropriate location is chosen at random, post adjustment, to substitute for the user's original location, and this algorithm can ensure the entropy is maximized, so as to effectively ensure the privacy performance of the location.

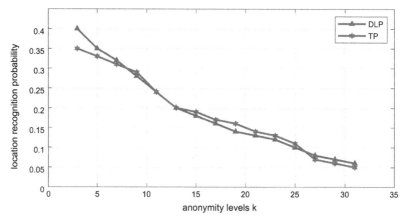

Fig. 5. Anonymity Level and Location Recognition Probability

6 Conclusion

Within this study, we present a virtual location selection approach aimed at shielding the user's location privacy against external information-based attacks and potential threats. Leveraging the derived privacy entropy value, the algorithm adeptly identifies the virtual location that optimally fulfills the desired k-anonymity level. We compare a location privacy preserving algorithm based on transition probability, which considers the case of users with zero transition probability and non-zero transition probability, and guarantees the level of location anonymity with zero transition probability through the location offset algorithm. It can be concluded that the privacy algorithm proposed in this paper can ensure location privacy while having a low time cost, and can effectively defend against external attacks.

In the future development, we need to strengthen the protection of location privacy, propose an enhanced DLP algorithm to cover a larger and farther area, and make greater contributions to the future research on location privacy protection.

References

1. Yu, S., Lee, J., Park, K., et al.: IOV-SMAP: secure and efficient message authentication protocol for IOV in smart city environment. IEEE Access **8**, 167875–167886 (2020)

2. Li, B., Liang, R., Zhou, W., et al.: LBS meets blockchain: an efficient method with security preserving trust in SAGIN. IEEE Internet Things J. **9**(8), 5932–5942 (2022)
3. Ye, J., Kang, X., Liang, Y.-C., et al.: A trust-centric privacy-preserving blockchain for dynamic spectrum management in IoT networks. IEEE Internet Things J. **9**(15), 13263–13278 (2022)
4. Albouq, S., Sen, A., Namoun, A., et al.: A double obfuscation approach for protecting the privacy of IoT location based applications. IEEE Access **8**, 129415–129431 (2020)
5. Yang, M., Ye, B., Chen, Y., et al.: A trusted de-swinging k-anonymity scheme for location privacy protection. J. Cloud Comput. **11**(1), 2 (2022)
6. Xing, L., Jia, X., Gao, J., et al.: A location privacy protection algorithm based on double k-anonymity in the social internet of vehicles. IEEE Commun. Lett. **25**(10), 3199–3203 (2021)
7. Liu, Q., Zhang, J.: Research on Location Privacy Protection Scheme for Internet of Vehicles Information Technology and Informatization, vol. 277, no. 04, pp. 79–82 (2023)
8. Chen, Z., Fu, Y., Zhang, M., et al.: A flexible mix-zone selection scheme towards trajectory privacy protection. In: Proceedings of the 2018 17th IEEE International Conference on Trust, Security and Privacy in Computing and Communications/12th IEEE International Conference on Big Data Science and Engineering (Trust Com/Big Data SE), F 1–3 August 2018 (2018)
9. Liu, S., Liu, A., Yan, Z., et al.: Efficient LBS queries with mutual privacy preservation in IOV. Veh. Commun. **16**, 62–71 (2019)
10. Yang, Z., Ma, H., Ai, M., et al.: A minimal disclosure signature authentication scheme based on consortium blockchain. In: 2022 IEEE International Conference on Blockchain (Blockchain), pp. 516–521 (2022)
11. Liu, S., Cai, Y., Ma, M., et al.: Pseudonymous exchange location privacy protection scheme based on group signatures in VANETs. J. Beijing Inf. Sci. Technol. Univ. (Nat. Sci. Ed.) **37**(03), 68–73 (2022)
12. Nannan, Y., Cheng, S.: K-anonymous bulk authentication for location privacy protection mechanism. J. Chongqing Univ. Posts Telecommun. (Nat. Sci. Ed.) **35**(03), 468–473 (2023)
13. Sessions: information theory applications; proceedings of the 1988 IEEE International Symposium on Information Theory, F 19–24 June 1988 (1988)
14. Jia, J., Zhang, G., Hu, C., et al.: Information hiding method for long distance transmission in multi-channel IOT based on symmetric encryption algorithm. J. Ambient Intell. Humaniz. Comput. (2021)
15. Wang, W., Wang, Y., Duan, P., et al.: A triple real-time trajectory privacy protection mechanism based on edge computing and blockchain in mobile crowdsourcing. IEEE Trans. Mob. Comput. 1–18 (2022)
16. Wang, H., Huang, H., Qin, Y., et al.: Efficient location privacy-preserving k-anonymity method based on the credible chain. ISPRS Int. J. Geo-Inf. **6**(6) (2017)
17. Wang, Y., Cai, Z., Chi, Z., et al.: A differentially k-anonymity-based location privacy-preserving for mobile crowdsourcing systems. Procedia Comput. Sci. **129**, 28–34 (2018)
18. Wang, Y., Wen, J., Zhou, W., et al.: A novel dynamic cloud service trust evaluation model in cloud computing. In: 2018 17th IEEE International Conference on Trust, Security and Privacy in Computing and Communications/12th IEEE International Conference On Big Data Science And Engineering (Trust Com/Big Data SE), pp. 10–15 (2018)
19. Beresford, A.R., Stajano, F.: Location privacy in pervasive computing. IEEE Pervasive Comput. **2**(1), 46–55 (2003)
20. Yang, X., Yu, X., Hou, H., et al.: Efficient asymmetric encryption scheme based on elliptic encryption technology. In: 2023 IEEE 6th Information Technology, Networking, Electronic and Automation Control Conference (ITNEC), pp. 709–714 (2023)
21. Yang, D., Ye, B., Zhang, W., et al.: KLPPS: a k-anonymous location privacy protection scheme via dummies and stackelberg game. Secur. Commun. Netw. **2021**, 1–15 (2021)

Artificial Intelligent Techniques for 6G Networks

Subway Double-Door Anti-pinch Based on RGBD Binary Classification Network

Chunlei Guo[✉], Junjie Yang, Zhicheng Sui, and Nan Dou

Shanghai Dianji University, Shanghai 201306, China
`216003010203@st.sdju.edu.cn`

Abstract. The safe operation of the subway has great significance for the daily order of large cities. Under special circumstances, passengers and various objects may get caught between the subway train doors and the platform screen doors, which poses a safety hazard. We propose a method based on an RGBD dataset and a convolutional neural network binary classification model. By extracting features and using local features for normal and abnormal classification, we can detect in real-time whether there are people or objects caught between the subway double doors. Four sets of experiments were conducted using two classification models, with both RGBD and RGB datasets loaded, to demonstrate the advantages of using an RGBD dataset in improving accuracy. We also found a lightweight and high-accuracy model suitable for this application scenario to be run on edge devices, solving practical problems. By detecting foreign objects in the double-door gap, subway door anti-pinch measures can prevent passengers from getting caught between the subway train doors and the platform screen doors, ensuring passenger safety. This enhances the safety and reliability of the urban subway system, providing assurance for urban development.

Keywords: Subway Double-door Foreign Object Detection · RGBD Dataset · Binary Classification Convolutional Ceural Network

1 Introduction

The subway system in large cities carries the daily commuting needs of residents for work and travel, with thousands of people coming and going every day. Therefore, the safe operation of the subway is of vital importance as it concerns the safety of passengers' lives and property. Once a subway accident occurs, it may not only cause casualties, but also damage to trains and equipment, leading to significant economic losses.

The current subway security measures include the subway door pressure sensor, which sends a signal to the door control system to stop or reverse the door when a foreign object is detected [1]; in addition, platform staff also maintain safety order manually. However, if the sensor accuracy is low or there is a malfunction, or if the staff fails to detect abnormal situations in a timely manner, potential risks may be posed to passengers. There have been a few cases of subway accidents, and there are also

© ICST Institute for Computer Sciences, Social Informatics and Telecommunications Engineering 2024
Published by Springer Nature Switzerland AG 2024. All Rights Reserved
J. Li et al. (Eds.): 6GN 2023, LNICST 553, pp. 155–164, 2024.
https://doi.org/10.1007/978-3-031-53401-0_15

situations where subway doors continue to move despite catching passengers' shoes, backpacks and other items.

To address the safety risks caused by the double doors of subway trains and platform screen doors catching people or objects, this paper introduces a deep learning-based binary classification model using convolutional neural networks. The dataset used to train the neural network was collected from actual subway scenes using an RGBD dataset. Considering that it is a binary classification and needs to be deployed on edge devices for inference detection, there are certain requirements for the model's lightweight and accuracy metrics. Therefore, two convolutional neural network classification models were selected, loaded with RGBD and RGB datasets, respectively, for comparative experiments. The results showed that the classification neural network trained using the RGBD dataset with depth information performed much better in terms of loss and convergence efficiency during the training process than the RGB dataset. It also outperformed the control group in accuracy, with a validation accuracy of up to 99.9%. The experimental results have validated the proposed method in this paper, demonstrating its practicality and reliability in addressing actual subway safety issues.

2 Domestic and International Research Status

Subway train door and platform screen door anti-pinch detection is an important safety measure to prevent passengers from being injured when the door is closed. The current situation of anti-pinch detection of subway train doors and platform screen doors at home and abroad is as follows.

2.1 Foreign Situation of Anti-pinch Detection of Subway Train Doors and Platform Screen Doors

The anti-pinch detection technology of subway train doors and platform screen doors in foreign countries is relatively mature, mainly using sensor technologies such as infrared, laser, and ultrasonic for detection.

Some countries have also adopted new technologies such as capacitive sensors and pressure sensors for detection, improving the accuracy and reliability of detection [2]. In some countries, emergency stopping devices are also installed on subway train doors and platform screen doors. When a pinching incident is detected, the device will automatically stop the train, ensuring passenger safety.

2.2 Domestic Situation of Anti-pinch Detection of subway Train doors and Platform Screen Doors

Currently, the detection of anti-pinch devices for subway train doors and platform screen doors in China mainly relies on sensor technologies such as photoelectric sensors, ultrasonic sensors, and mechanical switches [3].

Due to the inability of sensors to detect the gap between the double doors of the subway in real time, and only opening when an object is caught, it may pose a safety hazard to passengers in certain situations.

Object recognition based on RGBD data has been applied in some industries, such as detecting out-of-stock items on shelves and automatically replenishing them [4]. With the presence of depth information, even small objects like umbrella handles can be well recognized, which is very suitable for detecting foreign objects between subway double doors.

3   Convolutional Neural Network Principles

Convolutional Neural Networks (CNNs) are deep learning models commonly used in image, speech, text, and other fields [5]. Its main feature is the automatic learning of input data features through convolutional layers, achieving tasks such as classification and recognition of input data.

Convolutional neural networks usually consist of multiple convolutional layers, pooling layers, fully connected layers, and so on. Among them, the convolutional layer extracts features of input data by using multiple convolutional kernels to perform convolutional operations on the input data. The pooling layer reduces the size of the feature map by pooling the output of the convolutional layer, thereby reducing computational complexity and the number of parameters. The fully connected layer connects the outputs of the convolutional and pooling layers, achieving tasks such as feature classification or recognition [6].

The core idea of convolutional neural networks is the convolution operation [7]. The convolution operation is a linear filtering operation that applies a convolution kernel to different positions of the input data to obtain a new output. In convolutional neural networks, the convolution operation is used to extract features of input images (Fig. 1).

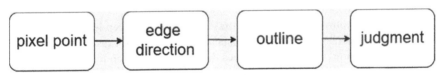

Fig. 1. Convolutional Kernel Filter Principle Flowchart

Pixel points → Edges: Extract the edge information between the pixel points in the image to form an edge image.

Edges → Directions: Filter the edge image to extract features of different directions and form a direction image.

Directions → Contours: Extract connected regions in the image to form a contour image, further reflecting the details of the image.

Details → Judgement: By extracting and classifying features of the contour image, make judgments or recognitions of the image.

The size calculation formula of image matrix after convolution operation is as follows:

$$N = (W - F + 2P)/S + 1 \qquad (1)$$

Input image size: $W \times W$.

Filter size: F × F.

Number of padding pixels: P

Stride: S

Convolutional neural networks are typically trained using backpropagation algorithm to update the model's weights and biases [8]. The backpropagation algorithm calculates the error between the model's output and the actual output, propagates the error back through the network, and updates the weights and biases of each layer to minimize the difference between the model's output and the actual output. The goal is to make the network's output as close as possible to the actual output.

Calculation of error:

Cross Entropy Loss [9]:

$$H = -\sum_{i=1}^{n} o_i^* \log(o_i) \tag{2}$$

Among them o_i^* is the true label value, o_i is the predicted value, by default log is based on e, equivalent to ln.

Calculation of validation set accuracy [10]:

$$\text{Accuracy} = \left(\frac{1}{n}\right) * \sum_{i=1}^{n} 1^{(y_i = y_i^*)} \tag{3}$$

Among them y_i^* is the true label of the i-th sample, y_i is the predicted result of the classification neural network model for the i-th sample. $1^{(y_i = y_i^*)}$ when the true label of the i-th sample equals the predicted result of the classification neural network model, its value is 1, otherwise it is 0.

In general, convolutional neural networks are a powerful deep learning model that automatically learns features from input data and can be used for tasks such as classification and recognition. With the continuous development of deep learning technology, convolutional neural networks have been widely applied in fields such as computer vision and natural language processing.

4   Solution

The following image is an overall process diagram of the solution (Fig. 2).

The RGB sensor and depth sensor of the TOF camera are registered through an algorithm, so that the pixels of the two images correspond one-to-one, and ultimately output RGB and Depth images. The RGBD dataset needs to concatenate RGB images and depth images on the channel dimension to meet the input requirements of the neural network model. Two different types of datasets, RGB dataset and RGBD dataset, were prepared for comparison, with corresponding image preprocessing steps set up.

For this application scenario, a suitable classification neural network model should be selected to output two categories. After loading the preprocessed RGB and RGBD datasets and adjusting the relevant training parameters, the target classification neural network model can be trained.

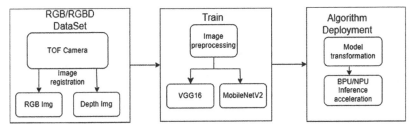

Fig. 2. Overall architecture diagram

The final part of this solution is the actual deployment of the algorithm model. Using the toolchain provided by the target deployment platform, the detection algorithm can be deployed on edge devices through steps such as model conversion and quantization, and accelerated inference can be performed using BPU or NPU.

5 Data Collection and Production

5.1 TOF Camera

The depth camera that uses Time of Flight (TOF) technology for distance measurement can quickly generate 3D images of the measured scene [11], with flexible frame rates. It can effectively recognize small area objects such as cables and cones at close range. It has a working distance of 0.1 m–10 m and is suitable for industries such as robotics, logistics, and security monitoring.

Due to the possibility of objects being trapped between subway doors, in addition to large objects such as passengers and backpacks, there are also small area objects such as umbrella handles, canvas bag straps, and shoes. Therefore, the advantages of TOF cameras are very suitable for our subway security scene (Figs. 3 and 4).

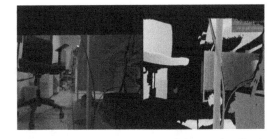

Fig. 3. TOF camera physical diagram

Fig. 4. TOF camera RGB and Depth image visualization

5.2 Making RGBD Dataset

The Advantages of an RGBD Dataset over an RGB Dataset. Depth information: An RGBD dataset not only includes RGB image information but also contains depth information for each pixel. This makes RGBD datasets more suitable for applications that require depth information, such as 3D scene reconstruction, object recognition, and pose estimation.

Robustness: An RGBD dataset is more robust to factors such as lighting, shadows, and occlusion [12]. Since depth information can help distinguish the front and back of objects, RGBD datasets have a greater advantage over RGB datasets when dealing with complex scenes.

Reduced computational complexity: An RGBD dataset can reduce computational complexity. In an RGB dataset, object recognition requires the use of texture, color, and other information in the image, while in an RGBD dataset, depth information can reduce the computational complexity and improve recognition accuracy.

Applicability: RGBD datasets are suitable for a wider range of applications. In addition to traditional computer vision tasks such as object recognition, pose estimation, and scene reconstruction, RGBD datasets can also be applied in fields such as virtual reality, augmented reality, and robot vision.

Hardware Setup. To create an RGBD dataset, an RGB camera and a depth sensor need to be combined and their positions and angles adjusted to align the RGB and depth images. As for the TOF camera mentioned earlier, the corresponding algorithm program has been written to achieve the alignment and output of RGB and depth images.

The following flowchart illustrates the process of creating an RGBD dataset for input into a neural network model (Fig. 5):

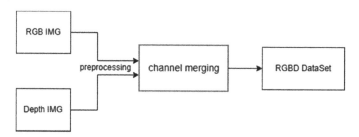

Fig. 5. Production of RGBD dataset

The following Figs. 6 and 7 shows the data collection site:

Fig. 6. Data acquisition site

Fig. 7. Train double-door gap

6 Selection of Experimental Model for Comparative Study

Since our ultimate goal is to successfully deploy the trained binary classification model on related edge devices, the computing power of these devices is limited but timely performance must be ensured. Therefore, in selecting the model, we must choose a lightweight model that ensures recognition accuracy as much as possible. Here, we choose the MobileNetv2 and VGG16 classification models as experimental control models.

VGG16 is a deep convolutional neural network model proposed by the research team at the University of Oxford in 2014 [13]. It is one of a series of convolutional neural network models proposed by the VGG team, and VGG16 is a relatively classic model among them. The structure of the VGG16 model is relatively simple, easy to understand and implement, and has good performance in image classification tasks, thus being widely used in the field of computer vision. At the same time, VGG16 is also used as a basic model for other computer vision tasks.

MobileNetV2 is a lightweight convolutional neural network model proposed by Google's research team in 2018 [14]. It is the second generation of the MobileNet series of models. Compared with the first-generation MobileNet model, MobileNetV2 has improved both in accuracy and speed. The characteristic of the MobileNetV2 model is that it uses a module called an inverted residual structure to build the network. Since the inverted residual structure first increases and then decreases the dimensionality, a large amount of low-dimensional feature information is lost due to the ReLU activation function, so ReLU6 is chosen as the activation function [15]. The MobileNetV2 model performs well in image classification, object detection, and other tasks, and its accuracy and speed are improved compared to other lightweight models (Tables 1 and 2).

Table 1. Comparison of MobileNet and VGG16 models

Model	ImageNet Accuracy	Million Mult-Adds	Million Parameters
1.0MobileNet-224	70.6%	569	4.2
VGG16	71.5%	15300	138

Table 2. Performance comparison of different versions of the MobileNet model

Network	Top 1	Params	MAdds	CPU
MobileNetV1	70.6	4.2M	575M	113 ms
MobileNetV2	72.0	3.4M	300M	75 ms
MobileNetV2(1.4)	74.7	6.9M	585M	143 ms

7 Experimental Strategy and Presentation and Analysis of Results

7.1 Experimental Setting

Configuration file settings: Two output categories of the binary network model are set, one for normal and one for abnormal conditions. In the class_json file, normal conditions are set to 1 and abnormal conditions are set to 0.

Experimental Group 1: The RGBD dataset and RGB dataset are separately input into the VGG16 model, with relevant parameters set within reasonable ranges, trained for 100 epochs, and the training loss and validation accuracy curves are saved separately.

Experimental Group 2: The RGBD dataset and RGB dataset are separately input into the MobileNetV2 model, with relevant parameters set within reasonable ranges, trained for 100 epochs, and the training loss and validation accuracy curves are saved separately.

7.2 Presentation and Analysis of Experimental Results

Loss curve diagrams for different models trained on RGB and RGBD datasets (Figs. 8 and 9).

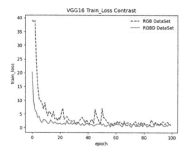

Fig. 8. VGG16 Train Loss   **Fig. 9.** MobileNetV2 Train Loss

Regardless of whether it is the VGG16 or MobileNetV2 model, under the same training of 100 epochs, the RGBD dataset has a faster convergence rate and better convergence effect than the RGB dataset.

Loss curve diagrams for different models trained on RGBD dataset (Fig. 10).

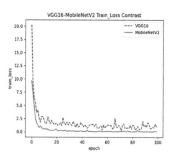

Fig. 10. VGG16&MobileNetV2 Train Loss

Under the same conditions of using the RGBD dataset and training for 100 epochs, the MobileNetV2 model has a faster convergence rate and better convergence effect than the VGG16 model.

Accuracy curve diagrams for validation set of different models trained on RGB and RGBD datasets (Figs. 11 and 12).

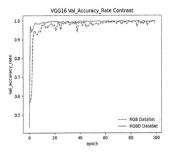
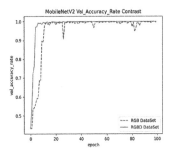

Fig. 11. VGG16 Val Accuracy Rate   **Fig. 12.** MobileNetV2 Val Accuracy Rate

Accuracy curve diagrams for validation set of different models trained on RGBD dataset (Fig. 13).

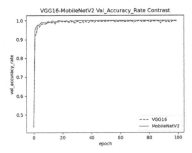

Fig. 13. VGG16&MobileNetV2 Val Accuracy Rate

Regardless of whether it is the VGG16 model or the MobileNetV2 model, the models trained on the RGBD dataset tend to have higher validation accuracy than those trained on the RGB dataset. Moreover, in general, for models trained on the same RGBD dataset, the MobileNetV2 model tends to have slightly better validation accuracy than the VGG16 model.

8 Conclusion

In the narrow gap detection scenario of subway train doors and platform screen doors, using a binary classification model and RGBD dataset collected by TOf cameras, training the MobileNetV2 convolutional neural network can achieve good convergence effect and detection accuracy. Moreover, because the MobileNetV2 model is relatively lightweight, it can be deployed on edge devices and accelerated inference using BPU or NPU, which can accurately and timely identify whether there are people, backpacks, and other abnormal objects between the double doors, and further improve the daily safety operation of the subway security system.

The next focus should be on the noise impact brought by the expansion of the dataset. The larger the dataset, the more noise there will be, and whether the trained network can achieve the best expected effect depends on optimizing the dataset to converge towards the global optimum as much as possible. At the same time, the quantization and calibration of the model during deployment should also be focused on.

Funding. This work was supported by Shanghai Science and Technology Planning Project (Grant NO. 21010501000).

References

1. Hoegg, T., Baiz, C., Kolb, A.: Online improvement of time-of-flight camera accuracy by automatic integration time adaption. In: 2015 IEEE International Symposium on Signal Processing and Information Technology (ISSPIT), pp. 613–618. IEEE (2015)
2. Diana, M., Cordova, E., Juan, R., et al.: A multiple camera calibration and point cloud FusionTool for Kinect V2. Sci. Comput. Program. **143**, 1–8 (2017)
3. David, F., Christian, R., Gernot, R., et al.: Learning depth calibration of time-of-flight cameras. In: British Machine Vision Conference, pp. 102.1–102.12 (2015)
4. Alina, K., Bodo, R.: On calibration of a low-cost time-of-flight camera. In: European Conference on Computer Vision, pp. 415–427 (2014)
5. Haase, S., Forman, C., Kilgus, T., Bammer, R., Maier-Hein, L., Hornegger, J.: ToF/RGB sensor fusion for augmented 3D endoscopy using a fully automatic calibration scheme. In: Tolxdorff, T., Deserno, T.M., Handels, H., Meinzer, H.-P. (eds.) Bildverarbeitung für die Medizin 2012: Algorithmen - Systeme - Anwendungen. Proceedings des Workshops vom 18. bis 20. März 2012 in Berlin, pp. 111–116. Springer, Heidelberg (2012). https://doi.org/10.1007/978-3-642-28502-8_21
6. Krizhevsky, A., Sutskever, I., Hinton, G.E.: ImageNet classification with deep convolutional neural networks. In: International Conference on Neural Information Processing Systems, pp. 1097–1105. Curran Associates Inc. (2012)
7. Xie, S., Girshick, R., Dollar, P., et al.: Aggregated residual transformations for deep neural networks, pp. 5987–5995 (2016)
8. Howard, A.G., Zhu, M., Chen, B., et al.: MobileNets: efficient convolutional neural networks for mobile vision
9. Neerbeky, J., Assentz, I., Dolog, P.: TABOO: Detecting unstructured sensitive information using recursive neural networks (2017)
10. Pham, H., Guan, M.Y., Zoph, B., et al.: Efficient neural architecture search via parameter sharing. arXiv preprint arXiv:1802.03268 (2018)
11. Liu, Z., Sun, M., Zhou, T., et al.: Rethinking the value of network pruning. arXiv preprint arXiv:1810.05270 (2018)
12. Anwar, S., Sung, W.: Coarse pruning of convolutional neural networks with random masks (2016)
13. Maddison, C.J., Mnih, A., The, Y.W.: The concrete distribution: a continuous relaxation of discrete random variables. arXiv preprint arXiv:1611.00712 (2016)
14. Yu, W.: Deep convolutional neural network based on densely connected squeeze-and-excitation blocks. AIP Adv. **9**(6), 065016 (2019). https://doi.org/10.1063/1.5100577

Design and Implementation of Intelligent IoT Paper Pumping System Based on Face Recognition of Loongson Processor

Yichen Hou[✉], Tingjun Wang, Zhenyu Zhang, and Zheyi Chen

Shanghai Dianji University, Shanghai 201306, China
466590278@qq.com, wangtj@sdju.edu.cn

Abstract. In unattended public places, users' excessive paper extraction leads to paper waste, and contact paper extraction also brings the risk of virus transmission. In order to better solve these problems, a paper extraction system based on the Internet of Things (IoT) platform was designed. The structure includes a paper feeding mechanism, a visual sensor and a paper pressing mechanism. This device is equipped with a Loongson chip as the main control device. The UI design uses Qt to complete the writing of the desktop program, and the actual control test of the prototype is carried out. Experimental results show that the intelligent IoT paper extraction system can quickly do face recognition and detection, can control that the same person can only extract a limited number or length of paper towels in a short time, so as to save paper in public places. it meets the demand of non-contact paper extraction and has a good application prospect.

Keywords: Paper extraction system · Intelligent IoT · Loongson processor · Face recognition · UI design

1 Introduction

Every year, the excessive use of paper causes a huge waste of resources and a huge economic loss. The problems of environmental protection and conservation have become increasingly severe in today's society. With the public toilets generally providing self-access toilet paper service, the excessive waste of toilet paper has become one of the burdens of businesses and public toilets [1]. Traditional toilet paper dispensers require staff to always pay attention to the remaining situation of paper, so as to avoid the dilemma that there may be no toilet paper because the toilet paper is not added in time. With the development of technology and the needs of virus prevention, some people's lifestyles have also changed, and a variety of non-contact equipment has gradually become popular, while the traditional paper machine inevitably needs to touch the paper machine, or the web may be touched by the previous person and cause health pollution [2]. With the application of face recognition and detection technology, direct brushing of the face can realize non-contact paper taking [3]. Therefore, in view of the above situation, it is necessary to develop a sophisticated paper extraction equipment that can solve the

J. Li et al. (Eds.): 6GN 2023, LNICST 553, pp. 165–171, 2024.
https://doi.org/10.1007/978-3-031-53401-0_16

above problems. Now researchers studied that automatic-flushing and paper pumping system for squatting toilet based on ARM cortex-M3 [4]. And other researchers studied that automatic flushing paper extraction system for squat toilet based on STC89C52 [5]. Using modern Internet of Things technology, this paper proposes a face recognition technology based on Loongson chip platform to complete non-contact paper extraction equipment.

2 Design Principle

2.1 A Subsection Sample

Loongson Pi and Ubuntu as the server are used to complete the cooperation. The main information exchange is completed through the respective Socket service process, and the effective information is displayed on the screen of the client and server through two QMainWindows processes. The core purpose of the whole system is to use the camera and face recognition program to reasonably start the mechanical part of the device. Take orders - Dc motor - paper transport mechanism - paper pressure mechanism - Paper cutting mechanism.

Loongson 2K1000 chip is selected as the main control device, The user interface (UI) design using Qt to complete the desktop programming, and contains a variety of information content. It's control system is mainly composed of the client and Ubuntu server located in the Loonson Pie, and needs to write the bottom drive in the Loonson pie to control 12V DC motor and push-pull type electromagnet. The paper output per unit time can be set, such as the fixed value of 300 mm, and the error is within 10 mm.The system control flow chart is shown in Fig. 1.

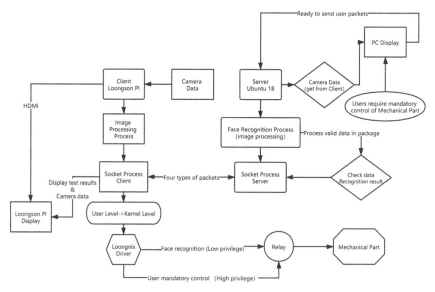

Fig. 1. System control flow chart

Since it is necessary to meet the need for Loongson Pie to complete automatic Boot, the boot scheme is set for it: PMON startup (no display) - Boot kernel (show Boot List) - Call Psplash - Start Xserver and directly start Qt application.

3 Experimental System of Paper Extraction System

3.1 Dragonson Client

After the Client process starts, it initializes the image processing process, socket process, Loongnix underlying driver process, QMainWindow process, and V4L2 driver process. After the V4L2 driver is initialized, it obtains the camera data and sends its memcpy to the socket process and QMainWindow process.

In the QMainWindow thread, the image processing process is called; When there is no face, display the screen directly; When there is a face, the face detection box matrix sent back by the server is used to frame the face on the screen and display it; When the test is successful, it will show a successful detection prompt picture and send it to the underlying driver process.

Socket process, after the completion of initialization, will first connect to wait through the SELECT function, after waiting to the server connection, the camera data packet into V4L2 packet, and sent to the server, and then accept the server back detection data, and put into the global variable, for other processes to call.

When Loongnix's underlying driver process receives incoming data, it uses the incoming parameters to decide whether to turn the machine on or off. In particular, mandatory commands sent back through the server will not pass through the underlying Loongnix driver process, but will instead operate directly on the driver file.

3.2 Ubuntu Server

After the Ubuntu server process starts, it will also carry out a series of initialization operations, including the initialization of image processing process, socket process, OpenCV face recognition process, and QMainWindows process.

In QMainWindows process, users can control Loongson Pi directly through the console on QMainWindows, and its control information is sent by Socket process.

In the Socket process, after waiting for the successful connection to the client, the V4L2 image data will be received back from the client; After the face recognition program is completed, the detection results and recognition results will be sent to the client.

In OpenCV face recognition process, the image processing process is first called to convert V4L2 data into MAT format.

3.3 Face Detection and Recognition

Face detection is mainly carried out through OpenCV cascading classifier. During initialization, Haarcascade_frontalface_alt.xml file is called, data in Mat format is loaded, after grayscale and histogram equalization, detection is carried out, and then face block diagram matrix is returned.

In the OpenCV face recognition process, face recognition will be grayscale image processing, and extract its LBP features, ULBP dimension reduction processing, image segmentation, segmentation into many small cells, and then for each cell histogram processing, update the image model, call Compare Hist prediction, and return similarity.

4 Experimental Results and Analysis

4.1 Experiment System

The paper extraction system experiment system is shown in Fig. 2. The design concept of Server + Client is adopted. Ubuntu 18.04 is run on the Server side and Loongnix system with cross-compiled 3.10.0 kernel is used on the Client side. The communication between them is connected by automatic packet Socket data, and the router is connected by network cable physically.

Fig. 2. The paper extraction system experiment system

The Psplash program in Yocto was transplanted to complete the animation production of the boot. Xinit was used to start the Client when the boot was started to achieve the function of self-starting.

Face recognition part uses OpenCV 3.4.1, through the V4L2 driver to get the camera picture; After loading the OpenCV face detector, the face obtained by the camera is acquired in real time, and the training is completed by the LBPH face recognizer. The training is completed by the Server side, and the collection is completed by the Client side.

Socket transmission contains four kinds of data headers (V4L2 data, detection results, identification results, paper extraction information), in the Socket transmission program through the data header to identify, and decode the effective data respectively.

4.2 Face Recognition Experiment

The system needs to accurately carry out face recognition and detection in order to achieve the purpose of non-contact paper extraction, and it needs to be able to control the underlying driver through the user layer function, otherwise the whole project will not be able to run, and face recognition data needs synchronous transmission between the client and the server. The Qt application interface and synchronized face detection is shown in Fig. 3, and the synchronized face recognition results are shown in Fig. 4.

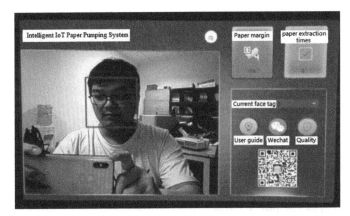

Fig. 3. Qt Application Interface and Synchronous Face Detection (Client)

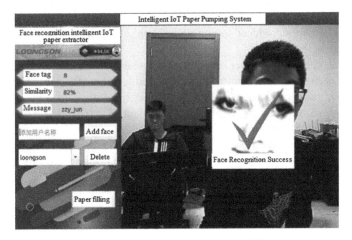

Fig. 4. Synchronous face recognition (Client)

The face recognition part has been verified by several experiments, and the recognition threshold is set to 80%. This threshold can quickly identify the faces in the face library, and only a small number of misjudgments will occur to ensure the accuracy of the system. The simplified experimental data is shown in Table 1. (Misjudgment refers to the recognition of another person's face. In this experiment, two people with similar face shapes were selected for a comparative experiment, 50 times for each person. Failure to detect after 10 s is classified as LOSS). The final realization of accurate face recognition/accurate frame face/no delay and no lag communication.

Table 1. Experimental data of face recognition.

Threshold setting	60%	70%	80%
Number of experiments	100	100	100
Number of successful	64	83	98
Number of mistakes	36	17	2
The success rate	64%	83%	98%
Misjudgment rate	36%	17%	2%

4.3 Analysis of Packet Loss Rate of Socket Communication

By querying the Log run by the program, among the 1423 V4L2 packets counted, the Client side successfully sent 1423 times, and the Server side received 1423 times. The packet loss rate is zero, and neither Client nor Server is stuck.

4.4 Test of Paper Output Accuracy

In the process of paper feeding, the length of each paper should be satisfied to be a fixed value. In order to accurately control the amount of paper produced, the length of paper produced can be calculated by the rotation of the DC motor in one second, which in turn is the amount of paper produced in unit time. Then set the delay time according to the length of each paper.

After measurement, the paper output per unit time is 30 mm, and we need 300 mm paper output each time, then theoretically only need to delay 10 s. However, it is found in practice that there are some errors in each paper quantity, so a general error value should be obtained through experiments, and this method can be used if the error value is within an acceptable range.

The test results of paper output accuracy through 10 experiments are shown in Table 2. It can be seen that the length error of the paper can be controlled within 7 mm, to meet the requirements.

Table 2. Error control experimental data of paper output accuracy.

Number	1	2	3	4	5
Paper length (mm)	295.8	302.6	300.1	297.2	296.3
Number	6	7	8	9	10
Paper length (mm)	304.5	305.2	296.5	298.2	299.4

5 Conclusion

The intelligent IoT paper extraction system based on Loongson has a good application prospect. Loongson send can complete the automatic boot and no operation connected to the server, the preparation of the Dragon son MIPS driver and test, can carry out accurate face recognition and face detection, and accurate control of the driver. It can accurately frame the face between the client and the server, accurately communicate between the client and the server without delay or delay, and automatically complete the process of feeding and pressing paper through the Dragon son self-programmed drive control. If the deep learning face recognition is further used, higher face recognition accuracy can be obtained.

References

1. Cheng, Z., Chen, D.: Research on the potential and benefit of energy saving and emission reduction in Universities. Environ. Sci. Manage. **35**(02), 34–38 (2010)
2. Han, Y.: Free toilet paper needs innovative technology and management. China Tourism News, 29 March 2017
3. Zou, S.: Design and Application of Embedded Linux. Tsinghua University Press, Beijing (2002)
4. Zhou, J., Yang, M.: Design of Automatic-flushing and Paper Pumping System for Squatting Toilet Based on ARM Cortex-M3. J. Anqing Normal Univ. (Nat. Sci.) **26**(01), 84–88 (2020)
5. Liu, L., Liu, J.: Design of automatic flushing paper extraction system for squat toilet based on STC89C52. China Sci. Technol. Inf. **24**, 113–114 (2012)
6. Liang, H., Sun, F., Chang, H., et al.: Design and implementation of face recognition system. Commun. World **23**, 283–284 (2016)
7. Sweet, et al.: MIPS Architecture Perspective. China Machine Press, Beijing (2008)
8. Liu, M., Liu, D., Wang, Y., et al.: On improving real-time interrupt latencies of hybrid operating systems with two-level hardware interrupts. IEEE Trans. Comput. **60**, 978–991 (2011)
9. He, L.: Be alert to the risk of face recognition abuse. Comput. Netw. **46**(16), 12–13 (2020)
10. Dabin, W., Lanlin, L., Maosheng, L., et al.: Scheme of avoiding repeated charges in metro AFC system with face recognition payment. Urban Rail Transit Res. **26**(07), 87–91 (2023)
11. Loongson, Z.: Loongson Group II User Manual. Beijing (2020)
12. Loongson, Z.: Loongson 2K1000 Processor User Manual. Beijing (2020)

Aluminum Defect Detection Based on Weighted Feature Fusion Mechanism

TingTing Sui, JunWen Wang[✉], YuJie Wang, JunJie Yang, ZiHan Xu, YiWen Zou, and YiKai Zhong

School of Electronics and Information, Shanghai Dianji University, Shanghai 201306, China
suitt@sdju.edu.cn, wjw742477658@163.com

Abstract. In the detection of defects on the surface of aluminum materials, such as jet stream and scratch. These defects have the problem of background similarity of different sizes, which brings difficulties and challenges to the detection. Our method introduces a cross-layer linking network to further fuse shallow and deep features. Taking into account more location, semantics, and detailed information, the detection accuracy of the network for aluminum surface defects is improved. Moreover, A weighted feature fusion mechanism is introduced to solve the negative impact of features from different layers and enhance the feature extraction ability of the model. Experimental results show that the improved YOLOv5 network model has good defect detection performance. The mAP on the Tianchi dataset reached 78.4%, which is 2.3% higher than the original YOLOv5 network. The model in this paper can quickly and accurately detect aluminum surface defects while keeping the original detection speed basically unchanged.

Keywords: Defect Detection · YOLOv5 · Surface Defects

1 Introduction

With continuous progress and innovation of science and technology, the application of internet technology, big data and contemporary society has entered the era of intelligent manufacturing. At present, in terms of the detection of aluminum products, there is a situation where intelligent technology gradually replaces manual and automated technology. During the production process of aluminum products, defects such as scratches, dirty spots, and paint bubbles will occur on the surface. These defects destroy the physical structure of aluminum materials, make the corrosion resistance of aluminum materials drop sharply, and seriously affect the quality of aluminum materials. Therefore, how to quickly and effectively detect the surface defects of aluminum materials is an important link to ensure the quality of aluminum materials.

For metal surface defect detection, manual defect detection has always been the mainstream method. However, this method is inefficient, and the detection results are easily affected by human subjective factors, which cannot meet the requirements of real-time detection. Traditional methods of surface defect detection have worked tremendously

© ICST Institute for Computer Sciences, Social Informatics and Telecommunications Engineering 2024
Published by Springer Nature Switzerland AG 2024. All Rights Reserved
J. Li et al. (Eds.): 6GN 2023, LNICST 553, pp. 172–183, 2024.
https://doi.org/10.1007/978-3-031-53401-0_17

for a long time. For example, BP (back propagation) neural network is used to classify defects, but the characteristics of surface defects need to be manually extracted, which is time-consuming, laborious and less accurate. It requires users to have a large amount of knowledge accumulation, and the process is complicated. Nowadays, automated intelligent detection technology is gradually helping the defect detection on the surface of aluminum products, which involves computer vision, deep learning, target detection and other technologies. In computer vision, object detection is an important branch. The traditional target detection algorithm [1, 2] traverses the selected area based on sliding window, and then performs feature extraction and classification. In traditional target detection, region selection algorithms have high computational complexity.

With the rapid development of target detection research and the continuous improvement of computer capabilities, scholars' research on surface defects has rapidly shifted from traditional detection methods to deep learning methods. In deep learning object detection, REN et al. proposed Faster R-CNN [3]. Subsequently, scholars successively proposed algorithms such as the model, you only look once(YOLO)[4–8]. LI Yu et al. used the K-means algorithm in Faster R-CNN to cluster and analyze defect data sets to obtain a better anchor box scheme [9]. This method improves the model's ability to detect small targets. Li Yanzhao and others introduced a lightweight GhostNet network to replace the residual module in the CSP1 module of the YOLOv5 backbone network [10], and used CIOU_loss as the coordinate position loss to improve the convergence speed and accuracy of the algorithm. Ge Zhaoming et al. added the CBAM attention mechanism module to yolov5 to detect lithium battery level slices [11], which improved mAP. However, because the CBAM attention mechanism is a two-channel serial, it has a good improvement effect on some complex and colorful targets such as human faces and animals, but it does not improve the recognition of aluminum materials very well. In 2022, LiuXiaomeng et al. replaced the SPP module with the ASPP module [12] to realize vehicle detection in YOLOv5 traffic monitoring scenarios. Compared with the SPP structure, the improved ASPP structure can obtain a better receptive field and extract more multi-scale information. Xiaoqi Wang et al. added BiFPN to the YOLOv5 network in 2021 [13] to detect PCB boards, which effectively improved the detection ability of the model. Wei Liu et al. added the SimAM non-parameter attention mechanism module to YOLOv5 in 2022 [14], which improved the detection accuracy without changing the memory usage of the model.

This paper proposes an improved aluminum defect detection model based on YOLOv5. In the aluminum defect data set, there are many defects with similar backgrounds and targets. The original YOLOv5 model is not effective in detecting the above defects. We improved the model structure and replaced the previous PANet part with BiFPN. In the feature fusion method, learnable feature fusion is adopted, which improves the ability of the model to detect similar defects in the foreground and background, and reduces the missed detection rate of the model.

2 Proposed Method

2.1 YOLOv5 Algorithm

In recent years, YOLOv5 has been gradually used in defect detection and has good results. The network structure mainly includes the input side, Backbone, Neck network and the Head part. The network structure diagram of YOLOv5 is shown in Fig. 1.

The Backbone architecture is primarily composed of modules such as Focus, Conv, C3, and SPP, all serving the purpose of feature extraction from the input image. The core concept within the Focus structure revolves around the segmentation of the input image. However, YOLOv5 introduces a significant change by replacing the Focus structure with a 6 x 6 convolutional layer. This alteration has proven to be more efficient for certain GPUs and optimization algorithms. Taking YOLOv5m as an example, the initial input is a 640 x 640 x 3 image. Through the Focus structure, this image is divided, resulting in a 320 x 320 x 12 feature image via a slicing operation. Subsequently, these 12 sliced components are concatenated and fed into the Conv module. This module consists of a sequence of layers, including a Convolutional layer, Batch Normalization, and a Leaky ReLU activation function layer. Batch normalization layers contribute to accelerated model training and address gradient vanishing issues. The Leaky ReLU activation variant is employed to rectify non-responsive neurons and prevent gradient-related problems, such as vanishing or exploding gradients, especially as network layers are added.

The Neck section primarily comprises Conv, Upsample, Concat, C3, and Detect modules, serving the purpose of amalgamating the features extracted by the Backbone segment. This integration aims to enrich the target features and facilitate target prediction. The Neck network utilizes the FPN + PAN architecture, inspired by LIN and their well-known Feature Pyramid Networks (FPN). FPN is designed to harmonize feature information from various depths within the network.

The Feature Pyramid Network (FPN) operates by upscaling from the highest to the lowest levels, transmitting robust high-level semantic features while enriching the overall pyramid with semantic information. LIU and their team introduced the Path Aggregation Network (PANet) to enhance information flow in instance segmentation frameworks that rely on proposal-based methods. PANet has also demonstrated impressive performance without the need for extensive batch training. Specifically, PANet follows a bottom-up approach, preserving and conveying the robust positional features from lower layers. It also amalgamates parameters from different detection layers originating in various backbone layers to bolster the entire feature map. This approach effectively shortens the information path between lower-level and topmost features.

The Head segment generates three distinct sets of feature maps when dealing with images of dimensions 640x640. These feature maps come in three sizes: 20 x 20, 40 x 40, and 80 x 80. Each of these size-specific feature sets is employed for predicting three distinct target categories: large, medium, and small.

In YOLOv5, some bounding boxes are considered positive samples in multiple prediction layers, and should be responsible for predicting bounding boxes of 3 different sizes. However, there may be multiple prediction frames for a certain target. In this case, the non-maximum suppression (NMS) method needs to be used to remove the prediction frames with high overlap and low score.

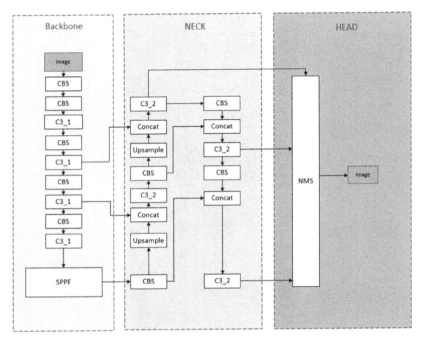

Fig. 1. YOLOv5 network structure diagram

2.2 The Improved Defect Detection Model

This section elaborates the proposed YOLOv5-based aluminum defect detection model. The specific framework can be divided into two parts: Cross-layer linking network (CLLN) and weighted feature fusion (WFU) (Fig. 2).

Cross-Layer Linking Network
Drawing on the idea of the BiFPN structure, a cross-layer linking network structure is added to the model. The core idea of the BiFPN structure is a bidirectional cross-scale connection. The specific operation is to delete some node links that have little impact on the model, so that more effective feature fusion can be obtained while reducing memory consumption. As shown in Fig. 3, Concat2 is an operation of splicing two tensors in a certain dimension, while Concat3 is an operation of splicing three tensors in a certain dimension.

Weighted Feature Fusion. Traditional feature fusion is often just a simple feature map superposition/addition, The previous method is to use concat or shortcut connection without distinguishing the feature maps added at the same time. Different input feature maps have different semantics, and their contribution to feature fusion are also different. Therefore, simply adding or superimposing feature map is not the optimal operation. Here, we propose a weighted special fusion mechanism, as shown in formula (1).

$$0 = \sum_i w_i \cdot I_i \tag{1}$$

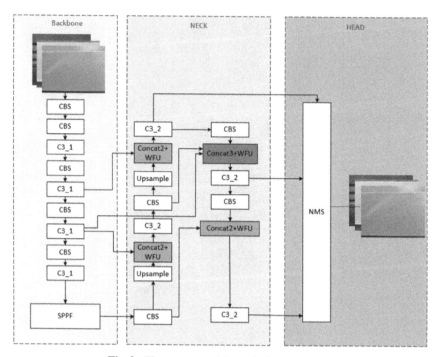

Fig. 2. The structure of the optimized YOLOv5

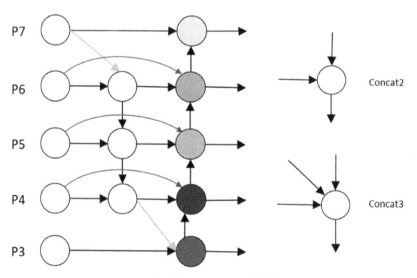

Fig. 3. The structure of CLLN

where w_i is learnable weight. I_i is a scalar (per feature), vector (per channel), or multidimensional tensor (per pixel).

3 Experiment and Analysis

3.1 Experimental Setup

Data Set. In this paper, we use the data from Tianchi aluminum defect data set, and the defects images are collected in real industrial environments. Tianchi Aluminum Defect Dataset contains ten types of aluminum surface defects, leaky bottom (D1), orange peel (D2), pitting (D3), corner leaky bottom (D4), non-conductive (D5), variegated (D6), jet (D7), rubbing (D8), Dirty spots (D9), paint bubbles (D10). Among them, 1893 images are used as the training set, 222 images are used as the verification set, and 890 images are used as the test set. The sample images of the dataset are shown in Fig. 4.

Fig. 4. Aluminum profile sample images

In addition, Mosaic data augmentation is applied to the image. The specific process is as follows: (1) The stitching reference point coordinates of the four images are randomly selected. (2) The four pictures are respectively scaled and placed in the upper left, upper right, lower left, and lower right positions of the large image. (3) According to the size transformation method of each image, the mapping relationship is mapped to the image label. (4) According to the specified horizontal and vertical coordinates, the large image is spliced.

The results of data augmentation for one batch are shown in Fig. 5.

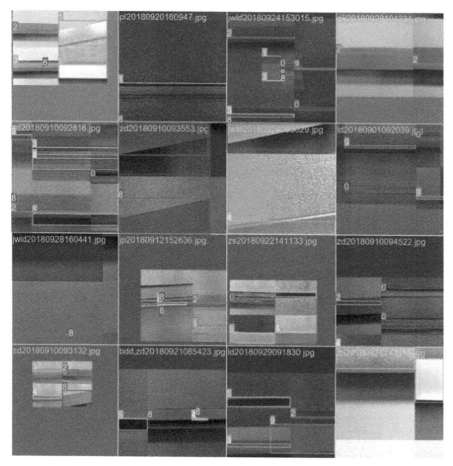

Fig. 5. Aluminum defect images enhanced by Mosaic data augmentation

Model Evaluation Index. This paper delves into an extensive analysis encompassing both detection precision and speed. To evaluate the efficacy at various defect levels, we employ the Average Precision (AP) as our primary metric. The overall network model's performance is then assessed using the mean Average Precision (mAP). Specifically, our evaluation criterion incorporates Map@0.5, signifying that at an Intersection over Union (IoU) of 0.5, we compute the AP for all images within each category and subsequently take the average across all categories. Moreover, to gauge the model's inference speed, we rely on Frames Per Second (FPS).

AP represents the mean value of the precision rate (P) under different recall rates (R). The curve obtained by taking P as the ordinate and R as the abscissa is the PR curve. AP is one of the important indicators used to evaluate the performance of the model. The formula is:

$$P = \frac{TP}{TP + FP} \tag{2}$$

$$R = \frac{TP}{TP + FN} \tag{3}$$

$$AP = \int_0^1 P(R)dR \tag{4}$$

Among them, **TP** represents the number of positive cases judged as positive by the model. **FP** represents the number of negative cases judged as positive cases. **FN** represents the number of positive cases judged as negative cases.

mAP represents the mean value of the average precision of all target detection categories, and its calculation formula is:

$$mAP = \frac{1}{n_j \sum_{j=1}^{n_j} Ap_j} \tag{5}$$

Among them, **n** is the number of a category, Ap_j indicates the detection accuracy of category j.

FPS indicates the number of pictures detected by the model per second. The higher of value displayed by fps indicates that the model works more efficiently. The formula is:

$$FPS = \frac{Frameum}{ElapsedTime} \tag{6}$$

Among them, **Frameum** represents the number of pictures participating in the detection, **ElapsedTime** represents the time it takes for the model to detect all images.

3.2 Experimental Configuration and Training Method

All experiments are implemented under i5-9600KF CPU, 2080ti graphics card, 11G video memory, PyTorch 1.13, CUDA10.1.

The basic parameters are set as follows: epoch is set to 200; batchsize is set to 16; initial learning rate is 0.01. Both preheating and Mosaic data augmentation are used for training.

In order to check whether the model has overfitting phenomenon during training, we analyze the change of loss during training, as shown in Fig. 6. The figure illustrates a notable trend in the training process. Initially, the loss value experiences a sharp decline in the early training stages. As the number of training epochs increases, the training loss undergoes a gradual reduction while exhibiting fluctuations around a particular boundary value. At the 175th epoch, a distinctive observation emerges: the loss reaches a plateau. Notably, the figure provides evidence that overfitting is not observed during the training process.

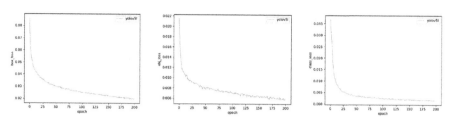

Fig. 6. Three kinds of training loss changes

3.3   The Performance of the Improved Model

The Effect of the Two Sub-models on YOLOv5. We undertake a comprehensive assessment of all the enhancements applied to the model. The outcomes of these enhancements are presented in Table 1, where we employ a "$\checkmark$" to signify the inclusion and usage of a module, and a "$\times$" to indicate its omission and non-utilization.

Performance of the CLLN. Referring to Table 1, we observe that in the YOLOv5_A model, the integration of high-level and low-level feature maps is accomplished through the CLLN technique. This approach yields a noticeable enhancement in *mAP@0.5,* elevating it from 0.761 to 0.771. By doing so, the model gains access to more robust semantic details and enhanced positional information, consequently boosting its overall positioning accuracy.

Performance of the WFU. As can be seen from Table 1, the YOLOv5_B structure adds WFU on the basis of YOLOv5_A. This method improves *mAP@0.5* from 0.771 to 0.784. YOLOv5_B can learn the contribution of each layer to the model while WFU enables YOLOv5 to obtain more layer feature map information. Therefore, this will reduce the occurrence of missed and false detections in the YOLOv5 model.

Table 1.   Comparison of model testing results

Methods	CLLN	WFU	mAP@0.5	FPS
YOLOv5	$\times$	$\times$	0.761	19.7
YOLOv5_A	$\checkmark$	$\times$	0.771	19.6
YOLOv5_B	$\checkmark$	$\checkmark$	0.784	19.1

The Ability of the Model to Detect Different Types of Defects

It can be seen from Fig. 7 that since the defects D1, D2, D3, D4, D5, and D6 are relatively obvious, and the amount of data of this type is abundant, the three methods all have good detection accuracy. However, for defects D7, D8, D9, D10, these defects have similar characteristics of background and target, so the detection effect of the original method is not so good. Because of the addition of CLLN, YOLOv5_A has improved the detection accuracy of D10 defects, but the detection accuracy of defect D8 has decreased. Therefore, we added the WFU mechanism on the basis of YOLOv5_A, which is the YOLOv5_B model. The detection accuracy of this model for various defects is better than that of the YOLOv5_A model.

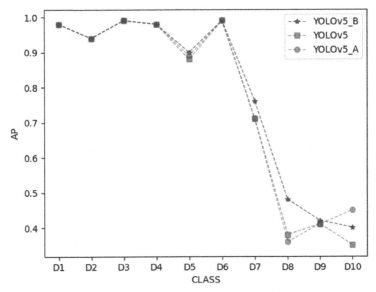

Fig. 7. Line chart of the detection capabilities of the three models for different defects

Performance Comparison of Different Methods in Aluminum Surface Defect Detection

We tested the two methods of CLLN and WFU on the aluminum dataset. The rendering is shown in Fig. 8. Among them, the first column is the original image. The second column is the picture that has been marked. The third column is the detection effect diagram of the YOLOv5. The fourth column is the detection effect diagram of the YOLOv5_A. The fifth column is the detection effect diagram of the YOLOv5_B. There are two defects in Figure (a), and YOLOv5 only detects one defect, and the box is too large to accurately locate the defect target. After using the CLLN, YOLOv5_B can detect all targets and locate them accurately. In Figure (b), YOLOv5 can only detect one target, and can identify two targets after CLLN. However, the positioning is not very accurate. After the WFU method, the model can detect all targets and locate them accurately. It shows that WFU has significantly improved the recognition and positioning of the model. In Figure (c), the defect detection effects of CLLN and WFU are similar. It shows that the feature extraction ability of CLNN to the model is better than that of YOLOv5. In Figure (d) and Figure (e), the YOLOv5 model and the CLLN model have the cases of missed detection. But adding WFU, the model can accurately identify the target and location. This fully demonstrates that our model can have better robustness and can be applied in industrial scenarios.

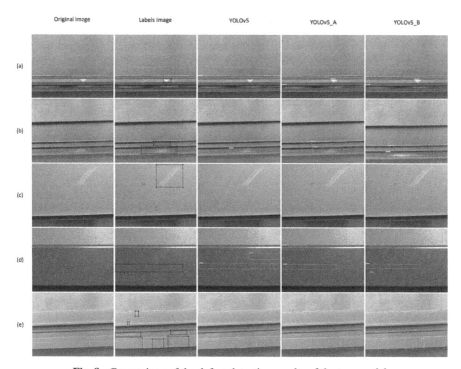

Fig. 8. Comparison of the defect detection results of the two models

4 Conclusion

For the detection of aluminum surface defects, this paper proposes a model YOLOv5 based on WFU and CLLN. We introduce CLLN to enhance feature fusion capabilities. The method make full use of the features extracted by Backbone and achieve better target detection. A weighted fusion mechanism is introduced to fuse the feature maps from different layers according to the way the model learns, which improves the robustness and detection ability of the model. The next step will be to conduct research on the dirty point (D8) defect detection, and further improve the model detection speed and detection speed.

Acknowledgement. This work was supported by National Natural Science Foundation of China under Grant No. 62103256.

References

1. Mao, T.Q., et al.: Defect recognition method based on HOG and SVM for drone inspection images of power transmission line. In: 2019 International Conference on High Performance Big Data and Intelligent Systems, pp. 254–257. IEEE Press, New York (2019)
2. Chu, M.X., et al.: Steel surface defects recognition based on multi-type statistical features and enhanced twin support vector machine. Chemom. Intell. Lab. Syst. **171**, 140–150 (2017)

3. Ren, S.Q., et al.: Faster R-CNN: towards real-time object detection with region proposal networks. IEEE Trans. Pattern Anal. Mach. Intell. **39**(6), 1137–1149 (2017)

4. Redmon, J., et al.: You only look once: unified, real-time object detection. In: 2016 IEEE Conference on Computer Vision and Pattern Recognition, pp. 779–788. IEEE Press, New York (2016)

5. Chang, L., Chen, Y.-T., Hung, M.-H., Wang, J.-H., Chang, Y.-L.:YOLOV3 based ship detection in visible and infrared images. In: 2021 IEEE International Geoscience and Remote Sensing Symposium, IGARSS, Brussels, Belgium, pp. 3549–3552 (2021)

6. Song, D., Yuan, F., Ding, C.: Track foreign object debris detection based on improved YOLOv4 model. In: 2022 IEEE 6th Advanced Information Technology, Electronic and Automation Control Conference (IAEAC), Beijing, China, pp. 1991–1995 (2022)

7. Dong, Z., Chang, H., Pu, X., Luo, P., Weng, J., Liu, Z.:Aircraft segmentation algorithm based on Unet and improved Yolov4. In: 2021 International Conference on Intelligent Transportation, Big Data & Smart City (ICITBS), Xi'an, China, pp. 614–617 (2021)

8. Huang, M., Wang, B., Wan, J., Zhou, C.:Improved blood cell detection method based on YOLOv5 algorithm. In: 2023 IEEE 6th Information Technology, Networking, Electronic and Automation Control Conference (ITNEC), Chongqing, China, pp. 992–996 (2023)

9. Li, Y., Tang, B., Sun, W., Lin, Z., Li, J.: Detection of steel plate surface defects based on improved faster R-CNN. Modular Mach. Tool Autom. Manuf. Tech. (05), 113–115+119 (2022)

10. Li Yanzhao, Y., Jinwen, L.T.: Metal weld defect detection based on improved YOLOv5. Electr. Meas. Technol. **45**(19), 70–75 (2022)

11. Zhao-ming, G.E., Yue-ming, H.U.: Lithium battery electrode defect detection method based on improved YOLOv5. Laser J. **44**(02), 25–29 (2023)

12. Xiaomeng, L., Jun, F., Peng, C.:Vehicle detection in traffic monitoring scenes based on improved YOLOV5s. In: 2022 International Conference on Computer Engineering and Artificial Intelligence (ICCEAI), Shijiazhuang, China, pp. 467–471 (2022)

13. Wang, X., Zhang, X., Zhou, N.:Improved YOLOv5 with BiFPN on PCB defect detection. In: 2021 2nd International Conference on Artificial Intelligence and Computer Engineering (ICAICE), Hangzhou, China, pp. 196–199 (2021)

14. Liu, W., Quijano, K., Crawford, M.M.: YOLOv5-tassel: detecting tassels in RGB UAV imagery with improved YOLOv5 based on transfer learning. IEEE J. Sel. Top. Appl. Earth Observations Remote Sens. **15**, 8085–8094 (2022)

MS TextSpotter: An Intelligent Instance Segmentation Scheme for Semantic Scene Text Recognition in Asian Social Networks

Yikai Zhong(iD), Heng Zhang(iD), Yanli Liu$^{(\boxtimes)}$(iD), Qiang Qian(iD), and Junwen Wang(iD)

School of Electronic Information, Shanghai Dianji University, Shanghai 201306, China
liuyl@sdju.edu.cn

Abstract. Text detection and recognition in natural scenes is an imimportant task in computer vision. However, most of the texts in natural scenes are curved, and the text background is complex and diverse. In recent years, the text detection and text recognition models proposed have inherent defects, especially in applying many false positives, which usually leads to a decline in text detection and text recognition accuracy. To solve this problem, we propose a text detection and recognition model based on instance segmentation: MS TextSpotter (Mask Scoring TextSpotter), which is based on end-to-end training. First, we design a neural network based on Mask R-CNN. The neural network can achieve accurate text detection and recognition through semantic segmentation, especially for multi-directional and curved text in natural scenes. Second, the network block we designed combines text features and predictive masks to learn the quality of text masks and regresses the intersection ratio of text masks to improve the quality of character masks. The model was tested on ICDAR2013, IC- DAR2015, and Total-Text datasets. Experimental results show that the text detection recall rate increases by 1.4% and 0.2% under the first two datasets, and the experimental results under three different vocabularies- is also show that MS TextSpotter has higher accuracy than other text recognition models and is more suitable for curved text recognition in natural scenes.

Keywords: Curved text · Text detection · Text recognition · Semantic segmentation · End-to-end training

This work was supported in part by the National Natural Science Foundation of China under Grant 61963017; in part by Shanghai Educational Science Research Project, China, under Grant C2022056; in part by Shanghai Science and Technology Program, China, under Grant 23010501000; in part by Humanities and Social Sciences of Ministry of Education Planning Fund, China, under Grant 22YJAZHA145.

J. Li et al. (Eds.): 6GN 2023, LNICST 553, pp. 184–198, 2024.
https://doi.org/10.1007/978-3-031-53401-0_18

1   Introduction

Natural scene text detection and recognition provides a fast and automatic way to access text information existing in natural scenes, which is conducive to the realization of various applications in life.e, such as verification code recognition [1], label recognition [17]. However, natural scene text has great differences in font size, arrangement direction, font type, text sparseness, etc.

In recent studies, instance segmentation is often used in an end-to-end text recognition framework, but the direct use of instance segmentation networks to detect text has certain defects: in the character segmentation stage, the score of the text mask is shared with the confidence of the text classification. Mask TextSpotter [2] regards the text classification confidence as the text mask score, but the real text mask quality is quantified as the IoU(Intersection over Union) of the instance mask and its corresponding ground truth value, usually the true mask score does not correlate well with the classification score. As shown in Fig. 1, there is a difference between the text classification confidence Scls and the text mask score Smask. The text instance classification prediction results in accurate box-level positioning results and high classification confidence, but their corresponding character mask scores have certain difference. Thus it is inappropriate to use the confidence of text classification to measure the quality of the mask.

Fig. 1. Classification score cls and comprehensive score mask of test results

To solve this problem, the MS TextSpotter (Mask Scoring TextSpotter) model is proposed in this paper, the model introduces the instance segmentation module and character mask Intersection over the Union module, which not only combines text features with prediction mask, learns the quality of text mask to regress the intersection ratio of text mask, improves the quality of character mask, but also realizes the detection of multi-direction and curved text in the natural scene. Experimental results show that, compared with the previous model, MS TextSpotter has a significant improvement in detection efficiency and recognition accuracy.

In summary, the main contributions of this paper are as follows:

1) Text detection and recognition can be realized by means of semantic segmentation, and curved text in natural scenes can be detected and recognized.
2) The character mask scoring mechanism is used to improve the integrity of character mask, which integrates semantic category and integrity of character mask. The model improves the quality of character mask.
3) It is proposed that the text detection and recognition model is based on end-to-end, which reduces the computational redundancy

The remainder of this paper is organized as follows: In Sect. 3, we describe the proposed method in detail, include character mask scoring mechanism and text recognition model. Experimental results are presented in Sect. 4, Finally, some conclusion remarks and future works are given in Sect. 5.

2   Related Work

As an important research direction of machine vision, text detection of natural scenes has been put into a lot of research in recent years. Before the advent of deep learning, the main trend of scene text detection is bottom-up, mainly using hand-made features.

The end-to-end natural scene text recognition method combines text detection task and text recognition task in a unified network model. This method usually shares the underlying convolution features, detects the text region according to the shared features, and then feeds the shared features of the text area to the recognition module to recognize the text content. For example, Mask TextSpotter [2] proposed a model that can detect and recognize arbitrary shape text instances. Although Mask TextSpotter [2] has achieved good results, but in the character segmentation stage, the score of text mask is shared with the confidence degree of text classification. The confidence degree of text classification is used to measure the quality of mask, which reduces the accuracy of text detection.

The MS TextSpotter model proposed in this paper, which introduces the Mask Head module into the network to realize the detection of multi-directional and curved text in the natural scene, and introduces the MaskIoU Head to learning the quality of the text mask, the cross merge ratio of the text mask is regressed by learning the quality of the text mask. The quality of the character mask is improved, meanwhile the accuracy of the text recognition is improved.

3   Our Proposed MS TextSpotter

The MS TextSpotter model is proposed to detect the text of the input image and convert all the detected text into the corresponding text sequence. The model text detection module uses the target detection model Fast R-CNN [8] to detect the horizontal rectangular area of the text. According to the characteristics of the text area, the text recognition module outputs the text instance probability map, character (English characters and numbers) instance probability map and character background probability map. Finally, the pixel voting algorithm is used to construct the character sequence from left to right.

3.1   Architecture

The overall architecture of the network is shown in Fig. 2, which is improved based on Mask R-CNN [13], It can be seen that the network is composed of the following components: FPN(Feature Pyramid Network) [14] for image feature extraction, RPN(Regional Proposal Network) [9] for generating text region suggestions, R-CNN for boundary box regression, a Mask Head branch for text segmentation and character segmentation, and a MaskIoU Head branch for character mask scoring. The following describes the role of each module.

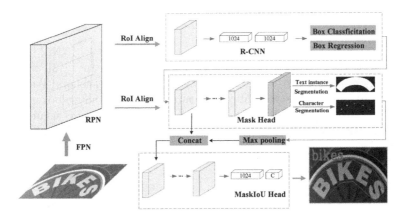

Fig. 2. MS TextSpotter model overview. The solid arrows mean the data flow both in training and inference period, The FPN network is shown in Fig. 3

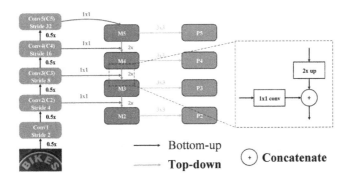

Fig. 3. Feature pyramid network

Backbone. The text in the natural scene is complex and diverse, there are different sizes, and texts of different sizes correspond to different characteristics. The low-level feature semantic information is less, but the object location is accurate. High-level feature semantic information is richer, but the target location is relatively rough. Therefore, text contour, texture and edge can be extracted by low-level features, and text details can be processed by high-level features. As shown in Fig. 3, in order to construct high-level semantic feature maps at various scales, MS TextSpotter adopts a feature pyramid structure with ResNet-50 as the backbone. For single scale image input, FPN uses a top-down architecture to fuse the characteristics of different resolutions, which improves the accuracy at marginal cost.

RPN. Ms TextSpotter generate text suggestion regions for R-CNN, Mask Head and MaskIoU Head branches through RPN. Referring to paper [13], we set different anchors at different stages according to the size of anchors. We use RoI Align [13] to uniformly represent the features of the Bounding box generated by RPN. Compared with RoI pooling, RoI Align retains more accurate position information, which is very important for the segmentation task in Mask branches.

R-CNN. The input of R-CNN branch is generated by RoI Align according to RPN. This branch includes two tasks: Bounding box classification and Bounding box regression. Its main purpose is to provide more accurate location information for the detected text area. In the text detection based on R-CNN, the detection problem is treated as a classification problem and regressed into a more accurate text detection box.

Mask Head. The Mask Head branch is mainly responsible for three tasks: text instance segmentation, character instance segmentation and background instance segmentation, as shown in Fig. 4. After inputting a ROI feature of size, it is passed through three convolutional layers and one deconvolution layer in turn, and the 38-dimensional probability map is output by the last convolutional layerIt includes 36 character instance probability map, one global text instance probability map and one character background probability map. Among them, the text probability map is used to predict the text instance area in the rectangular area; the 36 character probability map includes 26 English letters and 10 numbers to predict different character regions in the rectangular area; the character background probability map is used to predict the non text area in the rectangular area.

MaskIoU Head. MaskIoU Head combines the feature of character instances with their corresponding character masks, regresses the cross merge ratio of character masks, and corrects the deviation between the quality of character masks and the score of character masks by re-scoring strategy.

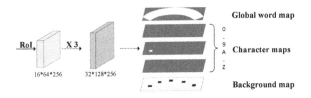

Fig. 4. Illustration of the Mask Head branch. The branch consists of four convolution layers and one de-convolution layer. The last layer produces 38 channels prediction map, including global text map, 0–9 and A-Z character map and background map

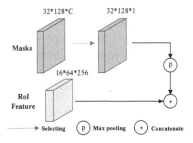

Fig. 5. The input of MaskIoU head, ROI Feature represents the features of the RoI Align layer, and masks represents the features of the prediction mask

The score of character mask prediction as S_{mask}, an ideal S_{mask} value equal to the pixel-level IoU between the predicted character mask and the corresponding ground truth mask. Because each mask belongs to only one class, the ideal one can only have positive values for the category with ground truth value, and the score for other categories is zero. MS TextSpotter divides learning tasks into mask classification and MaskIoU regression. All object categories are represented as follows: $S_{\mathrm{mask}} = S_{\mathrm{cls}} \times S_{\mathrm{iou}}$, Among them, S_{cls} focuses on mask classification, which has been completed in the task of Mask Head branch, so we can get the corresponding classification score directly. The main task of this stage will be the return of S_{iou}.

The input of MaskIoU Head is composed of the features of the RoI Align layer and the features of the prediction mask. As shown in Fig. 5, in order to connect the RoI Feature with the masks, masks need to pass a max pooling layer with a convolution kernel size of 2 and a step size of 2 to obtain the same dimension as the ROI Feature. In the MaskIoU Head branch, there are 4 convolutional layers with a convolution kernel size of 3 and a channel number of 256, and 3 fully connected layers. The 3 fully connected layers follow the Fast R-CNN [8], the output of the first two fully connected layers is 1024, and the output of the last fully connected layer is the number of categories.

3.2   Label Generation

For the image input in training, the general ground truth value includes $P = \{p_1, p_2, \cdots, p_m\}$ and $C = \{C_1, C_2, \cdots, C_n\}$, $C_n = (cc_n, cl_n)$. p_i represents the text instance area and is composed of a polygon box. cc_j and cl_j represent the corresponding location and category of the character pixel. Firstly, cover the polygon with the smallest horizontal rectangle, and then generate objects for RPN network and Fast R-CNN network according to the paper [9], Here, we need to generate two types of targets for mask branch: global map for text instance segmentation and character map for character semantic segmentation using ground truth value P, C and suggested area provided by RPN. Given a suggested region r, we use the matching mechanism of paper [12] to obtain the best matching horizontal rectangle. The corresponding polygons and characters can be further obtained. Next, the matching polygon and character boxes are moved and resized to align the proposed area. $H \times W$ of the object map is calculated according to the following equation.

$$B_x = (B_{x0} - \min(r_x)) \times W/(\max(r_x) - \min(r_x)) \tag{1}$$

$$B_y = (B_{y0} - \min(r_y)) \times H/(\max(r_y) - \min(r_y)) \tag{2}$$

where (B_x, B_y) and $((B_{x0}, B_{y0}))$ respectively represent the updated polygon vertexes and the original polygon vertexes. (r_x, r_y) is the vertexes of the proposal r. After that, just normalize the polygon on the mask initialized to zero and fill the polygon area with a value of 1. The generation of character map is as follows: first, we shrink all character bounding boxes by fixing the center point of the character bounding box and shortening its edge to a quarter of the original edge. Then, the pixel value in the shrunk character bounding boxes is set to its corresponding category index, and the pixel value outside the shrunk character bounding box is set to 0. If there are no character bounding boxes annotations, all values are set to -1.

3.3   Optimization

The MS TextSpotter model we propose is multi-task. Compared with the loss function designed by Mask RCNN, we add a global text instance segmentation loss and character segmentation loss. The loss function is as follows:

$$L = L_{\text{rpn}} + \alpha_1 L_{\text{cls}} + \alpha_2 L_{\text{box}} + \alpha_3 L_{\text{global}} + \alpha_4 L_{\text{char}} \tag{3}$$

where, L_{rpn}, L_{cls} and L_{box} are the loss functions of RPN and Fast RCNN, L_{global} and L_{char} reference paper [2], representing the instance segmentation loss and character segmentation loss.

$$L_{\text{global}} = -\frac{1}{N} \sum_{n=1}^{N} [y_n \times \log(S(x_n)) + (1 - y_n) \times \log(1 - S(x_n))] \tag{4}$$

For L_{global}, N represents the total number of pixels in the global text map, $y_n (y_n \in (0, 1))$ represents the pixel label, and x_n represents the output pixels.

$$L_{\text{char}} = -\frac{1}{N} \sum_{n=1}^{N} W_n \sum_{t=0}^{T-1} Y_{n,t} \log\left(\frac{e^{X_{N,t}}}{\sum_{k=0}^{T-1} e^{X_{n,k}}}\right) \tag{5}$$

$$w_i = \begin{cases} 1 & \text{if} \quad Y_{i,0} = 1 \\ N_{\text{neg}}/(N - N_{\text{neg}}) & \text{otherwise} \end{cases} \tag{6}$$

For L_{char}, T represents the number of categories, N represents the number of pixels in each map, in which the output map X can be seen as a $N \times T$ matrix. Y corresponding to the ground truth value X, the weight W is used to balance the loss value of character class and background class, and N_{neg} represents the number of background pixels, and its weight can be calculated by Eq. (6).

4   Performance Analysis

Our proposed MS TextSpotter model can detect arbitrary shape text in natural scenes. We test its performance on three datasets, ICDAR2013, ICDAR2015, and Total-Text. Here is a brief description of all these related datasets.

Fig. 6. Visual results of MS TextSpotter for text detection and recognition on the ICDAR2013 dataset

Fig. 7. Visual results of MS TextSpotter for text detection and recognition on ICDAR2015 dataset

Fig. 8. Visual results of MS TextSpotter for text detection and recognition on Total-Text dataset

4.1 Horizontal Text

For horizontal text, we evaluate our model on the ICDAR2013 dataset. Firstly, the short edge length of the input image is uniformly set to 1000 pixels. Secondly, our model is compared with six detectors, including FCRNall+multi-filt [15], Textboxes [18], Deep TextSpotter [19], Li et al. [20], Mask TextSpotter [2], Text Perceptron [4], the comparison of the detection results between our model and other scene text recognition models is shown in Tables 1, 3. Strong, Weak and Generic mean a small lexicon containing 100 words for each image, a lexicon containing all words in the whole test set and a large lexicon respectively. Specifically, it can be seen from Table 3 that even if it is only detected on a single scale, MS TextSpotter outperforms some of the previously proposed methods [2,4,20] under the three indicators of Precision, Recall and F-Measure, especially In terms of recall rate, compared with 89.5%of the most advanced detection method Mask TextSpotter [2], it exceeds 1.4 points, and the recall rate reaches 90.9%. As shown in Table 1, in the ICDAR2013 data set test, based on the End-to-End evaluation method, MS TextSpotter outperforms other advanced models in three different constraint vocabularies. The evaluation method based on Word Spotting, It is only slightly lower than the most advanced Text Perceptron in Strong's constraint vocabulary, and it is better than the best method proposed before under Weak and Generic.

4.2 Oriented Text

For multi-directional text in natural scenes, MS TextSpotter evaluates its performance on the ICDAR2015 dataset. First, the short side length of the input image is uniformly set to 1600 pixels, and then MS TextSpotter is compared with 7 detectors, including TextSpotter [21], StradVision, Deep TextSpotter [19], Mask TextSpotter [2], Char- Net [3], Text Perceptron [4], TextDragon [5]. The recognition results of Ms TextSpotter and the comparison results with other text recognition models are shown in Tables 2 and 3. As shown in Table 3, in terms of recall rate, MS TextSpotter once again showed excellent results. Compared

Table 1. MS TextSpotter's text recognition results on the ICDAR2013 dataset and comparison with other text recognition models, the best results are marked in bold.(S for Strong, W for Weak and G for Generic.)

Method	Word Spotting			End-to-End			Speed
	S	W	G	S	W	G	FPS
FCRNall+multi-filt [15]	–	–	84.7	–	–	–	–
Textboxes [18]	93.9	92.0	85.9	91.6	89.7	83.9	1.0
Deep TextSpotter [19]	92	89	81	89	86	77	**9**
Li et al. [20]	94.2	92.4	88.2	91.1	89.8	84.6	1.1
Mask TextSpotter [2]	92.7	91.7	87.7	93.3	91.3	88.2	3.1
Text Perceptron [4]	**94.9**	94.0	88.5	91.4	90.7	85.8	–
MS TextSpotter	94.6	**94.1**	**88.7**	**94.8**	**92.1**	**88.7**	2.9

Table 2. MS TextSpotter's text recognition results on the ICDAR2015 dataset and comparison with other text recognition models, the best results are marked in bold.(S for Strong, W for Weak and G for Generic.)

Method	Word Spotting			End-to-End			Speed
	S	W	G	S	W	G	FPS
TextSpotter [21]	37.0	21.0	16.0	35.0	20.0	16.0	1.0
Stradvision	45.9	–	–	43.7	–	–	–
Deep TextSpotter [19]	58.0	53.0	51.0	54.0	51.0	47.0	**9.0**
Mask TextSpotter [2]	82.4	78.1	73.6	83.0	77.7	73.5	2.0
Char Net [3]	–	–	–	83.1	**79.1**	69.1	–
Text Perceptron [4]	84.1	79.4	67.9	80.5	76.6	65.1	–
TextDragon [5]	**86.2**	**81.6**	68.0	82.5	78.3	65.1	–
MS TextSpotter	85.4	80.0	**75.6**	**84.6**	78.9	**74.6**	1.9

with detectors [2,4,11], it has been significantly improved, and it is better than the most advanced method Char-Net, the recall rate reached 90.6%. As shown in Table 2, in the ICDAR2015 dataset test, based on the End-to-End evaluation method, MS TextSpotter achieved the best results in the two different constraint vocabularies of Strong and Generic. The performance in the table is only lower than Char-Net.

4.3   Curved Text

For curved text, MS TextSpotter evaluates its performance on the Total-Text dataset. First, the short side length of the input image is uniformly set to 1000 pixels, and then MS TextSpotter is compared with 6 detectors, including Ch'Ng et al. [16], Liao et al. [6], Mask TextSpotter [2], Char-Net [3], TextDragon [5],

Table 3. Results of MS TextSpotter text detection on datasets ICDAR2013 and ICDAR2015

Method	ICDAR2013				ICDAR2015			
	P	R	F-M	FPS	P	R	F-M	FPS
CTPN [10]	93.0	83.0	88.0	7.1	74.0	52.0	61.0	–
Seglink [22]	87.7	83.0	85.3	**20.6**	73.1	76.8	75.0	–
EAST [12]	–	–	–	–	83.3	78.3	80.7	–
SSTD [23]	89.0	86.0	88.0	7.7	80.0	73.0	77.0	**7.7**
Wordsup	93.3	87.5	90.3	2.0	79.3	77.0	78.2	2.0
He et al. [11]	92.0	81.0	86.0	1.1	82.0	80.0	81.0	1.1
Mask TextSpotter [2]	94.8	89.5	92.1	3.0	86.6	87.3	87.0	3.1
Char Net [3]	–	–	–	–	**92.6**	90.4	**91.5**	–
Text Perceptron [4]	94.7	88.9	91.7	10.3	92.3	82.5	87.1	**8.8**
TextDragon [5]	–	–	–	–	92.4	83.7	87.8	–
MS TextSpotter	**95.1**	**90.9**	**92.9**	2.9	89.0	**90.6**	89.8	2.8

the comparison of the detection results of MS TextSpotter and other detectors is shown in Table 4.

Compared with the most advanced model, the results show that MS TextSpotter performs better in the detection and recognition of curved text, and the accuracy, recall, and average harmony have been significantly improved. It can be seen from Table 4 that although MS TextSpotter's detection performance is inferior to the most advanced Char-Net (4.7%-6.4%), it is better than other text recognition models, and under the end-to-end evaluation method, MS TextSpotter has improved significantly, intersecting with Char-Net by 8.9%. It can be seen that the improvement of text detection accuracy mainly comes from more accurate positioning output, that is, using polygons instead of horizontal rectangles to detect text areas. The improvement of text detection recall rate

Table 4. Results of MS TextSpotter for text detection and end-to-end recognition on dataset Total-Text

Method	Detection			End-to-End
	P	R	F-M	F-M
Ch'ng et al. [16]	40.0	43.0	36.0	–
Liao et al. [7]	62.1	45.5	52.5	48.9
Mask TextSpotter [2]	69.0	55.0	61.3	71.8
Char Net [3]	88.0	85.0	86.5	69.2
TextDragon [5]	85.6	75.7	80.3	74.8
MS TextSpotter	**72.3**	**64.2**	**68.0**	**75.8**

mainly comes from the scoring of character masks, and correct scoring brings correct text detection and character segmentation.

4.4   Speed

Most existing natural scene text detection and recognition models detect and recognize text in a multi-step manner, which makes them difficult to run efficiently. Compared with these models, Mask TextSpotter has a good compromise between speed and accuracy. On the ICDAR2013 and ICDAR2015 datasets, the model can detect text at 2.9 FPS and 2.8 FPS, respectively. The speed is weaker than Text Perceptron [4], but MS TextSpotter achieves the highest detection accuracy.

4.5   Engineering Applications

With the development of economy, China's demand for railway freight is growing. In order to better match the train number with the train information management system, so as to carry out cargo interaction more quickly and accurately inform the train conductor of the running information in the process of running. The identification of train number is also becoming more and more important. The identification of train number is becoming more and more important.

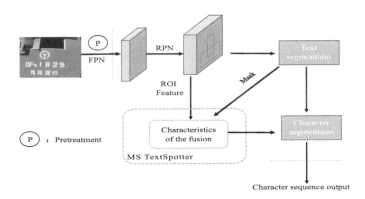

Fig. 9. Overall drawing process

At present, most of the processing steps of this kind of problem can be roughly divided into the following methods, including image pretreatment, location, segmentation and extraction of vehicle number, but the input image requires good lighting conditions, and the vehicle number is relatively easy to locate. In practice, the number pattern of freight train is quite different from the traditional car number pattern.

Therefore, we can integrate the photos of train numbers collected into our MS TextSpotter. Firstly, the train number was preprocessed, including correcting the tilted photos and filtering out the seriously distorted ones, and then the filtered pictures were input into our MS TextSpotter. In the network, RPN and RoIAlign are used to generate a feature map of a specific size. In the Mask Head stage, the input character map is converted into character sequence through pixel processing algorithm. In the MaskIou Head stage, the text Mask and RoI Feature obtained in the Mask Head stage are taken as input, and the final text recognition is obtained through a series of convolutional layers and full-connection layers, as shown in Fig. 9 TextSpotter can effectively improve the accuracy of text detection and greatly reduce computational redundancy.

5   Conclusion and the Future Work

In this paper, we propose MS TextSpotter, an end-to-end network for text detection and recognition in natural scenes. It can effectively detect text and segment characters in complex background. Compared with the previous model, our proposed network is very easy to train and can detect and recognize curved text. In all experiments, MS TextSpotter has achieved excellent performance in horizontal text, multi-directional text, curved text and other datasets, which improves the recognition accuracy and greatly reduces false positives. Our model shows high efficiency and robustness in text detection and end-to-end recognition. In the future work, we will try to optimize our model to improve the speed of text detection in order to realize the application in real life. Secondly, we will explore the recognition of Chinese text.

Author Contributions Conceptualization, Y.L. and Y.Z.; Investigation, Y.Z., Y.L., H.Z.; Software, Q.Q.; Writing-Original Draft Preparation, Y.Z. and Y.L.; writing–review and editing, H.Z.; Funding Acquisition, Y.L. J.W.and H.Z.

Conflicts of Interest The authors declare no conflict of interest.

References

1. Gupta, N., Jalal, A.S.: Traditional to transfer learning progression on scene text detection and recognition: a survey. Artif. Intell. Rev., 1–46 (2022)
2. Lyu, P., et al.: Mask textspotter: an end-to-end trainable neural network for spotting text with arbitrary shapes. In: Ferrari, V., Hebert, M., Sminchisescu, C., Weiss, Y. (eds.) Computer Vision – ECCV 2018. Lecture Notes in Computer Science(), vol. 11218, pp. 67–83. Springer, Cham (2018). https://doi.org/10.1007/978-3-030-01264-9_5
3. Liu, W., Chen, C., Wong, K.K.: Char-Net: a character-aware neural network for distorted scene text recognition. In: Association for the Advancement of Artificial Intelligence, vol. 1, no. 2, pp. 4–12 (2018)

4. Qiao, L., et al.: Text perceptron: towards end-to-end arbitrary-shaped text spotting. In: Proceedings of the AAAI Conference on Artificial Intelligence, vol. 34, no. 7, pp. 11899–11907. arXiv preprint: arXiv:2002.06820 (2020)

5. Feng, W., et al.: TextDragon: an end-to-end framework for arbitrary shaped text spotting. In: 2019 IEEE/CVF International Conference on Computer Vision (ICCV). IEEE (2020)

6. Zheng, Y., Li, Q., Liu, J., Liu, H., Li, G., Zhang, S.: A cascaded method for text detection in natural scene images. Neurocomputing **238**, 307–315 (2017)

7. Liu, W., et al.: TextBoxes: a fast text detector with a single deep neural network. AAAI Press (2017)

8. Girshick, R.: Fast R-CNN. In: IEEE International Conference on Computer Vision (ICCV), pp. 1440–1448 (2015)

9. Ren, S., He, K., Girshick, R., Sun, J.: Faster R-CNN: towards real-time object detection with region proposal networks. In: Advances in Neural Information Processing Systems, pp. 91–99 (2015)

10. Zhi, T., et al.: Detecting text in natural image with connectionist text proposal network. CoRR, abs/1609.03605 (2016)

11. Qin, S., Manduchi, R.: Cascaded segmentation-detection networks for word-level text spotting. In: 2017 14th IAPR International Conference on Document Analysis and Recognition (ICDAR) (2017)

12. Zhou, X., et al.: EAST: an efficient and accurate scene text detector. IEEE (2017)

13. He, K., et al.: Mask R-CNN. In: Proceedings of the IEEE International Conference on Computer Vision, pp. 2961–2969 (2017)

14. Lin, T.Y., et al.: Feature pyramid networks for object detection. In: The IEEE Conference on Computer Vision and Pattern Recognition, Honolulu, HI, USA, pp. 2117–2125 (2017)

15. Gupta, A., Vedaldi, A., Zisserman, A.: Synthetic data for text localisation in natural images. In: Proceedings of the IEEE Conference on Computer Vision and Pattern Recognition (CVPR), pp. 2315–2324 (2016)

16. Karatzas, D., et al.: ICDAR 2015 competition on robust reading. In: 2015 13th International Conference on Document Analysis and Recognition (ICDAR), pp. 1156–1160. IEEE (2015)

17. Ch'Ng, C.K., Chan, C.S.: Total-text: a comprehensive dataset for scene text detection and recognition. In: 2017 14th IAPR International Conference on Document Analysis and Recognition (ICDAR), pp. 935–942. IEEE (2017)

18. Liao, M., et al.: Textboxes: a fast text detector with a single deep neural network. In: Thirty-First AAAI Conference on Artificial Intelligence, San Francisco, California, USA, pp. 4161–4167 (2017)

19. Busta, M., Neumann, L., Matas, J.: Deep textspotter: an end-to-end trainable scene text localization and recognition framework. In: Proceedings of the IEEE International Conference on Computer Vision, pp. 2204–2212 (2017)

20. Li, H., Wang, P., Shen, C.: Towards end-to-end text spotting with convolutional recurrent neural networks. In: Proceedings of the IEEE International Conference on Computer Vision, pp. 5238–5246 (2017)

21. Neumann, L., Matas, J.: Real-time lexicon-free scene text localization and recognition. IEEE Trans. Pattern Anal. Mach. Intell. **38**(9), 1872–1885 (2015)

22. Shi, B., Bai, X., Belongie, S.: Detecting oriented text in natural images by linking segments. In: Proceedings of the IEEE Conference on Computer Vision and Pattern Recognition (CVPR), pp. 2550–2558 (2017)
23. He, P., et al.: Single shot text detector with regional attention. In: International Conference on Computer Vision, pp. 3066–3074 (2017). https://vision.in.tum.de/data/datasets/rgbd-dataset

Improved Solar Photovoltaic Panel Defect Detection Technology Based on YOLOv5

Shangxian Teng, Zhonghua Liu[✉], Yichen Luo, and Pengpeng Zhang

Shanghai Dianji University, Shuihua Road 300, Shanghai, China
liuzh@sdju.edu.cn

Abstract. Nowadays, the photovoltaic industry has developed significantly. Solar photovoltaic panel defect detection is an important part of solar photovoltaic panel quality inspection. Aiming at the problems of chaotic distribution of defect targets on photovoltaic panels, large scale span and blurred features, this paper improves the network structure based on the YOLOv5 model, which can better cope with the defect detection under various conditions. This paper mainly optimizes the following three aspects. Firstly, for the defect targets that are not of the right scale, the SE attention module and GhostBottleneck are introduced on the basis of the YOLOv5 model, which accelerates the extraction of useless features and enhances the precision of small object perception by capturing the features of defects of different scales and fusing semantic features of different depths. Secondly, Ghostconv is introduced to increase the receptive scope of the feature map, which makes the extracted feature discrimination ability stronger, and effectively enhances the defect detection ability of the model. Finally, by utilizing the ELU activation function instead of the Leaky ReLU activation function, the problem of gradient explosion and gradient disappearance is solved, and the model convergence speed is accelerated. Experimental results demonstrate that the improved YOLOv5 model can effectively detect the defects of photovoltaic panels, and the mAP reaches 92.4%, which is 16.2% higher than the original algorithm.

Keywords: Defect detection · Photovoltaic panels · YOLOv5 · Ghostconv · Multiscale training

1 Introduction

With the rapid progress of science and technology, energy has become the main concern of countries around the world today. Countries are striving to find alternative bioenergy, and solar energy has attracted worldwide attention due to its renewable and pollution-free characteristics [1]. The photovoltaic industry that came into being based on solar energy has also become the mainstream of today's world. As of September 2022, China's photovoltaic power generation has reached a staggering 359 million kWh, accounting for about 14% of the total installed power generation, and photovoltaic power plants have become one of the main sources of clean energy in the world. The main component of photovoltaic power station when solar cells are located, its operating conditions are

© ICST Institute for Computer Sciences, Social Informatics and Telecommunications Engineering 2024
Published by Springer Nature Switzerland AG 2024. All Rights Reserved
J. Li et al. (Eds.): 6GN 2023, LNICST 553, pp. 199–213, 2024.
https://doi.org/10.1007/978-3-031-53401-0_19

directly related to the power generation efficiency and stability of the power station, and accurate and efficient monitoring of the status of photovoltaic panels is of great significance to photovoltaic power plants [2].

Therefore, in an effort to ensure the normal operation of the power station, it is particularly important to efficiently detect the defects of photovoltaic panels. Nowadays, methods of photovoltaic panel defect detection are roughly divided into 2 types: one is manual inspection, and the other is machine vision and computer vision inspection. Since manual detection of photovoltaic panel defects is relatively wasteful of time and cost, the current mainstream detection methods are machine vision and computer vision inspection.

With the swift advancement of artificial intelligence technology, detection methods built upon machine vision and computer vision have been continuously produced, such as data augmentation method, robust visual detection method, visual significance detection method and convolutional neural network [3–8]. In addition, domestic and foreign researchers have also proposed some new application methods. Bengio et al. [9] introduced the application of convolutional neural networks (CNNs) to extract features with greater robustness and expressiveness, and CNNs are currently widely used in computer vision. Chen et al. [10] devised a visual defect detection approach utilizing a multispectral depth CNN, in which the CNN model mines visual irregularities observed on the surface in images in multiple spectral regions, which enhances the recognition ability of intricate textured background and defect attributes. Other endeavors have also been described in the literature, such as the introduction of deep learning-based categorization pipeline structures [11] for preprocessing electroluminescent images (such as distortion rectification, image segmentation, and perspective adjustment), deep CNNs designed for the classification of surface defects in solar cells, and studying the effect of a small number of oversamples and data increases on system accuracy [12]. Wang et al. [13] used Fast R-CNN, YOLOv4 and YOLOv5 algorithms to detect surface anomalies of solar cells, among which YOLOv5 algorithm worked best, with a leveling accuracy of 88.2%, which ensured the detection speed while maintaining good accuracy.

The above research has greatly improved the speed and accuracy of solar photovoltaic panel defect detection, but due to the complex background of photovoltaic panel images, variable defect morphology, uneven distribution and other reasons, conventional detection methods will not take care of some special situations. This paper simulates the detection situation in various environments by processing the data picture to ensure the authenticity of the detection. In an effort to improve the precision of defect detection, the following improvements are made based on the YOLOv5s algorithm under the condition that the precision rate and the quantity of model parameters are small.

a) Building upon the original model, the SE attention mechanism module is incorporated to effectively identify the small target defects of photovoltaic panels in special environments.

b) In addition, in an effort to reduce the model parameters while satisfying the precision, this paper improves the YOLOv5 backbone network to the GhostBottleneck module, which reduces the extraction of useless features while satisfying the detection accuracy during feature extraction, thereby improving the model speed.

c) In view of the characteristics of irregular feature size of photovoltaic panels and dense distribution of small targets, Ghostconv is used instead of traditional Conv in the Backbone backbone model of the model, and the C2f module of YOLOv8 is used in the C3 network structure to replace the original C3 module, which improves the detection efficiency and feature extraction ability of the model.

d) In terms of activation function, the exponential linear unit (ELU) activation function is used to replace the Leaky ReLU activation function, which solves the problem of neuronal disappearance while the test output value is closer to 0, accelerating the convergence speed.

The above operation can significantly reduce the quantity of parameters and calculations of the model under the condition of ensuring accuracy, improve the ability of real-time detection of the model, and be more conducive to the deployment of terminals.

2   YOLOv5 Network Introduction

YOLOv5 continues the basic idea of YOLO series, and further improves on the basis of YOLOv4, which not only enhances the detection precision, but also enhances the speed of detection and learning compared to YOLOv4, and the structure is more compact. The YOLOv5 network includes four models, YOLOv5l, YOLOv5x, YOLOv5s, and YOLOv5m, which have the same network structure, and users can adjust the network depth factor and width coefficient according to the task requirements for inspection vision and detection accuracy. Taking YOLOv5s as an example, the network architecture is depicted in Fig. 1, and YOLOv5s starts to adapt image scaling to the original image fed into the network, and obtains an image with a resolution of 680x680. The backbone network then uses a Cross-Stage Partial (CSP) Network architecture for feature selection, the neck network also uses the network architecture of FPN + PAN to integrate the detected features, and finally the detection head generates the type information and location information of the objects.

The detection process of YOLOv5 can be described in the following steps:

1) The input image undergoes preprocessing, such as scaling and normalization, to meet the model's input requirements.

2) The core of the network is a feature extraction network based on a Convolutional Neural Network (CNN). YOLOv5 employs a "backbone" network, typically a lightweight backbone like CSPDarknet53 or EfficientNet. This network is responsible for extracting useful feature information from the input image for subsequent object detection operations.

3) Following the feature extraction network, YOLOv5 introduces a series of "neck" modules that aid in extracting features at different scales. These feature maps have different resolutions and can be used for detecting objects of various sizes.

4) Next comes the "head" section, which includes operations for predicting bounding boxes and class probabilities. YOLOv5 predicts the coordinates of the target boxes and the probabilities of different object classes by applying a series of convolutional and pooling layers on the feature maps at different scales. Each prediction consists of the position of the bounding box (typically represented as the center coordinates along with and width/height) and the class probability.

5) Finally, Non-Maximum Suppression (NMS) is utilized to eliminate overlapping bounding boxes and obtain the final object detection results. NMS removes highly overlapping detection boxes and selects the ones with the highest confidence as the final detections.

In summary, the structure of YOLOv5 consists of preprocessing, a feature extraction network, scale fusion, object prediction, and Non-Maximum Suppression. It enables fast and accurate detection of multiple objects' positions and class information within an image in an end-to-end manner. This makes YOLOv5 one of the widely used algorithms in the field of object detection.

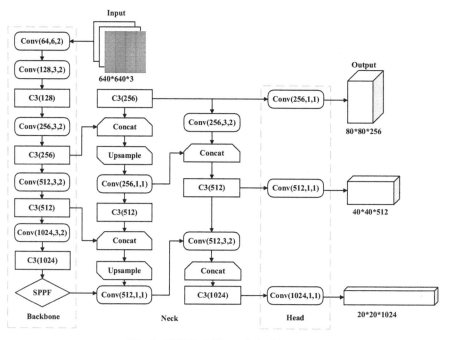

Fig. 1. YOLOv5 Network Architecture.

3 Model Improvement and Optimization

3.1 SENet Attention Mechanism

The SENet attention mechanism establishes certain connections between the features in the feature map channels, and the weights of different channels have different degrees of importance, and iterate on the feature information and weights according to them. SENet improves the importance of key features and suppresses useless features, and reasonably allocates the internal resources of graphics cards according to the weight size of different channels and their importance, which has been well verified in the architecture of various convolutional networks. SENet updates the weights in the channel according to the loss function of training, and obtains a better network model structure for specific task results by increasing the weight values that have a large impact on the results and weakening the weight values with small impact. SENet are removable modules that can be placed between the layers of a network fabric. Although the integration of SENet modules into convolutional neural networks will increase the overall number of parameters, it can improve the ability of the network to extract and integrate feature information. Figure 2 is a schematic illustration of the SENet attention mechanism module.

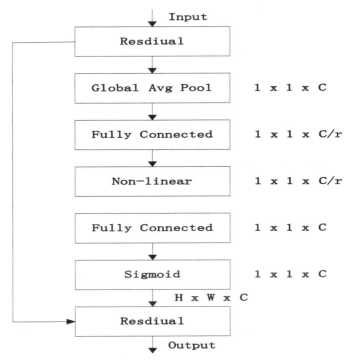

Fig. 2. Schematic visualization of SENet attention mechanism module.

The SENet attention mechanism module first passes the input image or feature map through the global flat pool (GAP) module according to the pixel scale size to obtain a feature map with a scale compressed to 1x1, and Eq. 1 represents the process.

$$z_c = F_{sq}(U_c) = \frac{1}{H \times W} \sum_{i=1}^{H} \sum_{j=1}^{W} u_c(i,j) \tag{1}$$

where z_c represents the weight of the c channel, which F_{sq} is the compression objective function, H and W are the height and width dimensions of the pixel scale, and the feature map obtained by convolution is u_c, and i and j are the positions of pixels on the high and wide boundaries of the feature map.

Then, in an effort to obtain the connection between each channel, the compressed feature map pooled by the global average of the previous layer is passed through the first fully connected layer, the Rectified Linear Unit (ReLU) activation function and the subsequent fully connected layer, and then the weight coefficient of each individual channel is computed by the Sigmoid activation function, and Eq. 2 represents the process.

$$\hat{z} = T_2(ReLU(T_2(z))) \qquad (2)$$

$\hat{z}$ is the weight coefficient of the channel, which T_2 represents the fully connected layer, the ReLU activation function,and the compression weight z.

Finally, each weight is applied to the original channel, reclassifying the importance of different information on the initial feature map, that is, the correlation between the channel dimensions is obtained, and Eq. 3 expresses the process.

$$\widehat{X} = X \cdot \sigma\left(\hat{z}\right) \qquad (3)$$

where $\widehat{X}$ is the channel after the product, X is the initial channel, and σ is the Sigmoid activation function.

It should be noted that the purpose of global average pooling is to enable the layers in the SENet module to share the global range of sensing fields. After global average pooling, the fully connected layer performs dimensionality reduction operations uniformly, which optimizes computing resources to a certain extent. The purpose of the final fully connected layer is to adjust the channel dimensionality, aligning it with the number of channels in the subsequent layer.

3.2 GhostBottleneck

GhostNet [14] is a lightweight feature extraction network structure proposed by Huawei, the core concept behind this approach is to devise a convolutional computational module that operates in a sequential manner, first perform a certain nonlinear convolution on the feature map, and then perform a linear convolution operation on this basis, and the updated feature map obtained is called the "ghost" of the previous feature map. In this way, useless features are reduced and the parameter count of the network model is adjusted to obtain a lightweight network structure. The nonlinear convolution operation referred to here is a complete combination of (Conv-BN-Leaky ReLU), while linear convolution refers to ordinary convolution, without batch normalization and nonlinear activation functions. The structure of the GhostNet module and its improved Ghost Bottleneck module is shown in Fig. 3, where DW Conv is deep convolution, and during the course of YOLOv5 network lightweighting, this paper uses GB module to replace the original CSP module.

3.3 Ghostconv

Traditional convolutional feature extraction has a lot of redundancy, as shown in Fig. 4. The operation of a traditional convolutional network, which generates n feature maps for the input data, can be mathematically represented by Eq. 4:

$$Y = X \times f + b \left(X \in R^{c \times h \times w}\right) \qquad (4)$$

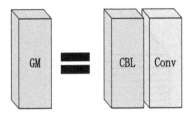

Ghost Model

Ghost Bottleneck(stride=1)

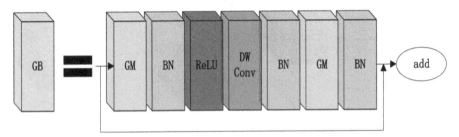

Ghost Bottleneck(stride=2)

Fig. 3. GhostNet and GhostBottleneck modules.

where c represents the number of input channels, while h and w correspond to the height and width of the input data, respectively; b is a bias constant; $f \in R^{c \times h \times w}$ represents a convolution kernel; Y represents the output feature map.

Compared to traditional convolutional networks, Ghostconv has two steps, as shown in Fig. 5. First, Ghostconv obtains a partial feature map through conventional convolution methods, then obtains more feature maps through cheap linear operations, and then combines each feature map into a new output result. It is expressed by the formula as follows:

$$Y' = X * f' \tag{5}$$

$$y_{i,j} = \varphi_{i,j}(y'_i), \forall i = 1, \ldots, m, \forall j = 1, \ldots, s \tag{6}$$

whereY' $\in R^{h' \times w' \times m}$;$*$ represents a convolution operation; m is the number of channels in the convolution operation; $\varphi_{i,j}$ is a convolutional operation of a linear transformation.

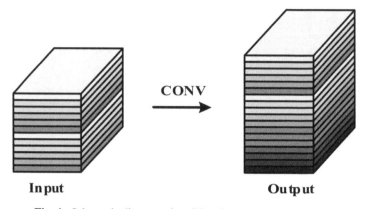

Fig. 4. Schematic diagram of traditional convolution processing

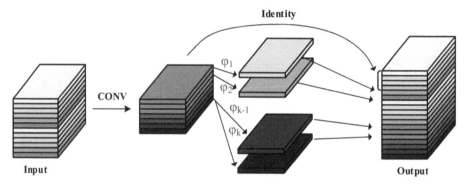

Fig. 5. Schematic diagram of GhostConv processing process

3.4 ELU Activation Function

When solar photovoltaic panel surface defect detection is applied to industrial inspection, the primary focus lies in achieving a highly accurate and precise model with exceptional localization capabilities, and the training model will basically not affect the detection speed. In this study, the ELU activation function is employed as a replacement for the Leaky ReLU activation function within the network.

The Leaky ReLU activation function mainly solves the problem of neuronal "death" and ensures that the gradient disappears during the backpropagation process. The ELU activation function has all the advantages of the Leaky ReLU activation function, as shown in Eq. 7:

$$f(x) = \begin{cases} x, & x > 0 \\ a(e^x - 1), & x \le 0 \end{cases} \tag{7}$$

where α is a hyperparameter and $\alpha = 1$.

The ELU activation function, being continuously differentiable across all points of the image, effectively mitigates the issues of gradient vanishing and gradient explosion.

This characteristic of the ELU activation function facilitates faster training and improved accuracy in neural network training when compared to the Leaky ReLU activation function. Although ELUs are exponential linear transformations, unlike Leaky ReLU, ELUs also have a negative value, which causes their mean to shift towards 0. Because of this shift, the model will be trained faster than the Leaky ReLU activation function converges. Although the operation speed of the ELU activation function is slower, due to its faster convergence, the model training speed is faster than that of the Leaky ReLU activation function, and it has better generalization, so the accuracy is also higher.

4   Experimental Results and Discussion

4.1   Experiment Configuration

The training platform of this experiment uses the RDX3090 graphics card cloud server, Pytorch and CUDA versions 1.11.0 and CUDA versions 11.3, respectively. During training, the pre-trained model "yolov5s.pt" is loaded. Training initializes relevant parameters such as categories, class names, and training paths on COCO datasets and VOC datasets. The image input size is 640 x 640, and the Adam algorithm is used for training and optimization, which solves the problem of insufficient data, epoch = 400, batch size = 16, and the learning rate is adjusted to 0.001.

4.2   Datasets

This paper uses the publicly available dataset (which can be downloaded from the website: https://github.com/CCNUZFW/PV-Multi-Defect) of solar photovoltaic panel defects annotated by Li Longlong et al. [22] from Central China Normal University, which has 1106 images. In order to detect photovoltaic panels in some special environments, a part of the dataset is selected for image processing, and the photovoltaic panel scene in some special scenarios is simulated by adding noise, rotation transformation, contrast transformation, color enhancement and other methods. In training, the dataset is randomly partitioned into a training set comprising 1000 images and a separate test set.

The annotation software LabelImg is used to annotate the defect location and category of the dataset in YOLO format, and the annotation categories are broken, scratch, black_border, no_electricity, and hot_spot, as shown in Table 1 below. During the marking process, defects in the PV panel image are surrounded by rectangular boxes, which provides information regarding the category and precise location of each defect. The annotations are stored as XML files following the PASCAL VOC format.

4.3   Evaluation Indicators

In this experiment, four metrics were utilized to evaluate the model's performance, including Recall (R), Average Precision (P), mean Average Precision (mAP), and Frames per Second (FPS). The calculation method is as follows.

$$R = \frac{TP}{TP + FP} \tag{8}$$

Table 1. Sample diagram of battery panel defects.

Name of Defect	Description	Image
broken	Photovoltaic panels with damaged regions	
scratch	Photovoltaic panels in areas with scratches	
black_border	Photovoltaic panels exhibiting black or gray edges	
no_electricity	The photovoltaic panel is not powered on and displays a black area	
hot_spot	Photovoltaic panels with prominent bright regions	

$$P = \frac{TP}{TP + FN} \tag{9}$$

where TP (True Positives) represents the number of positive samples correctly identified as positive samples; Where FP represent the number of negative samples incorrectly identified as positive samples; While FN denotes the number of positive samples incorrectly detected as negative samples. The PR curve can be plotted with different accuracy and recall, and the area under the PR curve is defined as AP. The mean of all APs of the detection class is mAP. The performance evaluation index AP (P_{AP}) and mAP (P_{mAP}) calculation methods are as follows.

$$P_{AP} = \int_0^1 p(r)dr \tag{10}$$

$$P_{mAP} = \frac{1}{n} \sum_{i=1}^{n} P_{AP} \tag{11}$$

4.4 Experimental Results and Analysis

In the identification of PV panel defects, in an effort to reflect the influence of different improvement strategies on the accuracy of detection of surface defects on PV panels, an ablation experiment was carried out, and the experimental results are presented in Table 2. Compared with the YOLOv5s model, the optimization model has the following indicators: the accuracy of YOLOv5s-1 is increased by 5.4%; The accuracy of YOLOv5s-2 increased by 9.1%; The accuracy of YOLOv5s-3 increased by 13.9%; The accuracy of YOLOv5s-4 improved by 16.2%. From the experimental results, it is evident that each step of improvement enhances the performance metrics of the original model, which proves that each step of improving the model is meaningful.

Table 2. Comparison Table of Ablation Experiment Results.

Models	Description	FPS(f/s)	mAP(%)	Precision (%)	Recall (%)
YOLOv5s	YOLOv5s	38.4	76.2	73.4	75.6
YOLOv5s-1	YOLOv5s + SE	42.3	81.6	75.6	81.4
YOLOv5s-2	YOLOv5s + SE + GConv	43.5	85.3	79.3	80.7
YOLOv5s-3	YOLOv5s + SE + GB + GConv	45.1	90.1	85.6	85.7
YOLOv5s-4	YOLOv5s + SE + GB + GConv + ELU	45.6	92.4	87.4	89.7

This can be found from the data in Table 2. After adding the SE attention mechanism module and Ghostconv, YOLOv5s has improved mAP value, accuracy and recall compared with the original YOLOv5. According to the above results, it can be found that Ghostconv reduces the learning cost of non-critical features, which can help the model extract features from the target more accurately and quickly, thereby greatly improving the training efficiency of the model. SE attention mechanism also has a significant effect on the defect detection of photovoltaic panels, mainly because the effect of the original algorithm on feature extraction is not very stable, it will be affected by environmental noise, resulting in missed detection and wrong detection, and the model after adding SE attention mechanism will be more accurate in feature extraction of photovoltaic targets.

On the basis of comparison with the original network model, in order to verify the effectiveness of the improved network, the same photovoltaic panel dataset is trained using SSD, YOLOv4, YOLOv7 and other networks while keeping the configuration information of the training platform unchanged, and finally the best weights of each network model are selected for comparative analysis on the test set, and the comparison results are shown in Table 3 below.

Table 3. Table of Comparison Results between Improved Algorithms and Other Algorithms.

Models	Backbone	FPS(f/s)	mAP(%)
YOLOv5s	Focus + CSPNet	38.4	76.2
SSD	VGG16	25.8	68.1
YOLOv4	CSPDarknet53	27.3	70.3
YOLOv7	E-ELAN	29	83.4
Ours	SE + GhostNet	45.6	92.4

It can be seen from Table 3 that compared with SSD, YOLOv4, YOLOv7 and YOLOv5s models, the mAP values of the improved YOLOv5s are increased by 24.3%, 22.1%, 9% and 16.2%, respectively, and the detection speed is increased by 7.2f/s compared with YOLOv5s, which further verifies that the improved model has stronger performance for PV panel defect detection. The mAP curve of the improved model for detecting each defect on the surface of the photovoltaic panel is shown in Fig. 6 below, and the detection results are shown in Fig. 7.

It can be seen from Fig. 7 that the improved YOLOv5s can not only accurately identify large defects on the surface of solar photovoltaic panels, but also accurately locate defects with smaller scales, and have high recognition results.

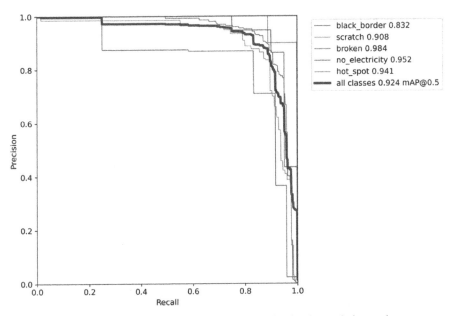

Fig. 6. MAP values for various defects detection in photovoltaic panels.

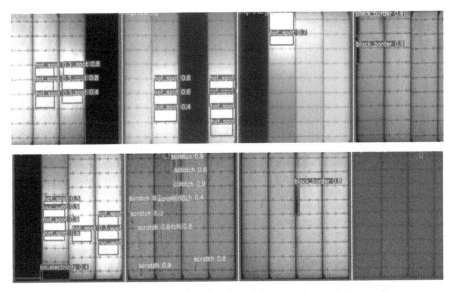

Fig. 7. Identification results of surface defects on photovoltaic panels.

5 Conclusion

Aiming at the defect characteristics of solar photovoltaic panels, this paper comprehensives an improved model based on YOLOv5 object detection, introduces the Ghostconv module, SE attention mechanism, and uses GhostBottleneck to replace the CSP module of the original model, which enhances the ability of feature extraction and realizes defect detection at different scales.

At the same time, in order to optimize the model, the activation function is also improved to avoid gradient explosion and gradient disappearance, so that the model can converge more quickly, with higher robustness and generalization ability. Comparative experiments and ablation experiments show that the mAP of the improved model reaches 92.4%, which is 16.2% higher than that of the original detection model, and the speed reaches 45.6FPS, which has a significant enhancement effect. The next work is to reduce the complexity of the model by building and distilling the model, and achieve faster detection speed while ensuring a high detection rate, so as to achieve lightweight improvement of the model.

References

1. Romero-Cadaval, E., et al.: Grid-connected photovoltaic generation plants: components and operation. Ind. Electr. Mag. IEEE **7**(3), 6–20 (2013)
2. Bin, S.: Research progress and development prospect of solar photovoltaic power generation materials. China Powder Ind. (1), 22–24 (2020)
3. Wang, Y., Sun, Z., Zhao, B.: Zhao does not bribe. Cracking detection of silicon wafers of solar cells based on machine vision. Comb. Mach. Tool Autom. Process. Technol. (12), 95–97 (2019)
4. Balzateguo, J., Eciolaza, L., Arexolaleiba, A.: Defect detection on polycrystalline solar cells using electroluminescence and fully convolutional nerural networks. In 2020IEEE/SICE International Symposium on System Integration(SH), pp. 949–953 (2020)
5. Chen, H.Y., et al.: Accurate and robust crack detection using steerable evidence filtering in electro-luminescence images of solar cells. Opt. Lasers Eng. **118**, 22–33 (2019)
6. Xiaoliang, Q., et al.: Surface defect detection of solar cells based on machine vision saliency. J. Instrum. **38**(7), 1570–1578 (2017)
7. Ying, Z., et al.: Application of improved CNN in defect detection of solar panels. Comput. Simul. **37**(3), 458–463 (2020)
8. Yunyan, W., Zhigang, Z., Shuai, L.: Data enhanced defect detection of solar cells. J. Electr. Measur. Instrum. **35**(1), 26–32 (2021)
9. Hinton, G.E., Salakhutdinov, R.: Reducing the dimensionality of data with neural networks. Science **313**(5786), 504–507 (2006)
10. Chen, H., et al.: Solar cell surface defect inspection based on multispectral convolutional neural network. J. Intell. Manuf. **31**, 453–468 (2020)
11. Shanableh, T.: Saliency detection in MPEG and HEVC video using intra-frame and inter-frame distances. SIViP **10**, 703–709 (2016)
12. Akram, M.W., et al.: CNN based automatic detection of photovaoltaic cell defects in electroluminescence images. Engergy **189**, 116319 (2019)
13. Shuqing, W., et al.: Surface defect detection of solar cells based on improved YOLOv5s. Instrum. Technol. Sens. **5**, 111–116 (2022)

14. Han, K., et al.: GhostNet: more feature feom cheap operations. In: Proceedings of the IEEE/CVF Conference on Computer Vision and Patten Recognition, pp. 1580–1589 (2020)
15. Zhou, T., Jing, X.: Surface-based detection and 6-Dof pose estimation of 3-D objects in cluttered scenes. IEEE Trans. Robot. (99), 1–15 (2016)
16. Janssen, P., Swanepeol, J.: Efficiency behaviour of kernel-smoothed kernel distribution function estimators. South Afr. Stat. J. **54**(1), 15–23 (2020)
17. Yang, L., et al.: Research on fault diagnosis method of photovoltaic module. Mech. Des. Manuf. (12), 82–87 (2021)
18. Liu, W., et al.: SSD: single shot MultiBox de-tector[EB/OL] (2015). arXiv:1512.02325. https://arxiv.org/abs/1512.02325
19. Redmon, J., et al.: You only look once:unified,real-time object detection. In: 2016 IEEE conference on Computer Vision and Pattern Recognition, June 27–30, 2016, Las Vegas, NV, USA, pp. 779–788. IEEE Press, New York (2016)
20. Lin, T.Y., et al.: Focal loss for dense object detection. IEEE Trans. Pattern Anal. Mach. Intell. **42**(2), 318–327 (2020)
21. Zhao, Q.J., et al.: M2Det: a single-shot object detector based on multi-level feature pyramid network. In: Proceedings of the AAAI Conference on Artificial Intelligence, vol. 33, pp. 9259–9266 (2019)
22. Li, L., Wang, Z., Zhang, T.: GBH-YOLOv5: ghost convolution with BottleneckCSP and tiny target prediction head incorporating YOLOv5 for PV panel defect detection. Electronic **12**, 561 (2023)

YOLO-L: A YOLO-Based Algorithm for Remote Sensing Image Target Detection

Wang Yinghe$^{(\boxtimes)}$, Liu Wenjun, and Wu Jiangbo

Shanghai Dianji University, Shanghai, China
wangyinghe@sdju.edu.cn, liuwenjun@email.cn

Abstract. In response to the poor performance of object detection in remote sensing images under complex backgrounds, this paper proposes a new object detection algorithm model called YOLO-L based on the YOLOv5 algorithm. This algorithm model constructs a novel backbone network called BottleCSP, which refreshs the C3 module and SPP module in the original YOLOv5 backbone network. It integrates feature maps of different sizes to beef up contextual information, thereby better extracting deep linguistic information from images. The BottleCSP backbone network solves the problem of poor performance in handling images with small target recognition by customizing and adjusting the original Backbone backbone network. Additionally, the CoordAttention attention mechanism is inserted in both the Backbone backbone network and the Head prediction head, which helps the network to more accurately locate the objects of interest and enhance the model's perception ability in different regions of the image. This paper conducts experiments on the RSOD public dataset, and compared with the original YOLOv5 algorithm and newer detection methods such as DC-SPP-YOLO, MRFF-YOLO, and YOLOv5-slight, the mAP (mean average precision) is improved by 12.3%, 16.22%, 4.47%, and 7.67%, respectively. This effectively improves the recognition performance of images with small targets.

Keywords: Remote sensing image target detection · BottleCSP module · CoordAttention mechanism

1 Introduction

With the rapid evolution of satellite and imaging technology, the resolution and pixels clarity of optical remote sensing images have been improving, and the research on remote sensing image analysis and understanding has attracted widespread attention [1]. Target detection techniques in remote sensing images have wide application value in environmental monitoring, urban planning and traffic management, military prevention and so on. Unlike conventional images, targets in remotely sensed images are often small in scale, densely distributed

J. Li et al. (Eds.): 6GN 2023, LNICST 553, pp. 214–225, 2024.
https://doi.org/10.1007/978-3-031-53401-0_20

and multi-directional [2], which displays a great challenge to general detection networks. Traditional target detection algorithms are mostly based on manual extraction of features, which has the disadvantages of complicated operation, having singularity and poor applicability [3]. With the suddenly appear of deep learning technology, deep learning detection methods are gradually applied to remote sensing image target detection. The algorithms based on deep learning target detection are currently split into two main categories: region-based two-stage target detection and single-stage target detection based on regression analysis [3]. The principle of region-based two-stage target detection algorithms focuses on region proposal followed by refined classification and position regression on candidate frames. The two-stage approach thus has higher detection accuracy, but also results in slower detection [4]. The principle of the single-stage target detection algorithm based on regression analysis is relatively simple; it performs classification and position regression prediction of targets directly at each position in the image by using the entire image as input to the network [5]. it is faster on speed, but the relatively poor detection accuracy results from the fact that the single-stage method makes predictions at each location [6].

In order to promote the detection accuracy and detection speed of targets in remote sensing images, this paper proposes a new YOLOV5-L detection model based on the YOLOV5 algorithm, which mainly makes the following contributions: on the basis of the original YOLOV5 backbone network, a new backbone network of BottleCSP is constructed, which replaces the infant C3 and the SPP module, and insert the BottleCSP feature extraction module into different stages of the backbone network to fuse deep and shallow contextual information, so as to effectively retrain deep contextual information from pixels. In addition, the CoordAttention attention mechanism is integrated in the BackBone backbone network and Head prediction head to enable the model to discover the field of interest more accurately, enhance the model's ability to perceive different regions in the image, and reduce the complexity of the model.

2 Introduction of YOLOV5 Detection Network

YOLO (you only look once) was came up by Redmon et al. in 2015, and is the earliest single-stage detector in the field of deep learning [7]. The name "You Only Look Once" suggests a completely different approach to detection, and the biggest feature is speed of recognition especially quick. [8]. Then for each grid cell, YOLO predicts whether that grid cell contains a target object and the location and size of its bounding box [9]. Each bounding box is typically represented by five attributes: the (x, y) coordinates of the bounding box, (w,h) as bounding box's width and height, and the confidence score c [10]. In addition, each box also predicts the category of the target [11]. The output of the YOLO model is a tensor containing all predicted bounding boxes, each with a confidence score indicating the confidence that the box contains target objects, and a category label indicating the predicted category of the target. To eliminate overlapping boxes, non-maximal suppression is adopted to filter the final detection results.

Where non-maximum suppression [12] calculates the IoU (intersection-to-merge ratio) between different bounding boxes [13] and removes redundant bounding boxes with an IoU above a certain threshold. Finally, after non-maximum suppression, the YOLO algorithm outputs the final detection results, including the location, size, category labels and confidence scores of the target bounding boxes. The YOLOV5 structure primarily consists of four parts: the input, BackBone backbone network, Neck connection layer and prediction head [14], YOLOv5 extracts image features through the backbone network, fusing features at different scales through the connection layer, and finally predicting the location and class of the target through the detection head. In addition, the YOLOV5 model uses the CIoU loss function as the boundary loss, which can better take into account the relationship between the location, size and shape of the bounding box, and thus more accurately measure the similarity of the bounding box.

3   Improvement of YOLOV5

As remote sensing images have many problems such as large size variations, richer inclusion of feature and background information, uneven distribution and easy occlusion of target objects [15], it makes it difficult to extract specific information about the target using commonly used convolutional networks in natural scenes, and as the depth of the network deepens, the location information extracted from the features becomes more and more blurred, making it difficult to extract detailed information about the detected objects, thus reducing the impact of the detection of small The detection effect of small target objects is reduced. In this paper, based on the YOLOV5 algorithm model, a new remote sensing image target detection algorithm model is constructed, which is suitable for the recognition of small target images. The overall framework of the YOLO-L algorithm model is shown in Fig. 1. The network first uses the Focus feature extraction module for processing the input feature map, which is shortened to half of the original size. By reducing the size of the feature map, the complexity and computation of the subsequent computation can be significantly reduced and the speed of the model can be increased. The Convolution module is then adopted to extract local feature information using convolution operations, followed by the arrangement of BottleCSP features to stitch the output features of the backbone network with the output features of the branches by channel dimension to form a richer feature representation. This cross-stage concatenation helps information to pass better within the module, improves the expressiveness of the model, helps to reduce the sums of model parameters, and improves computational efficiency while maintaining model performance (Fig. 2).

3.1   Building the BottleCSP Backbone Network

Drawing on the ideas of the CSP bi-directional feature extraction network model, this paper proposes a new BottleCSP backbone network, which replaces the C3 and SPP modules, and inserts the BottleCSP feature extraction module into different stages of the backbone network to better recognize deep semantic information from the images. The BottleCSP backbone network is suitably customised

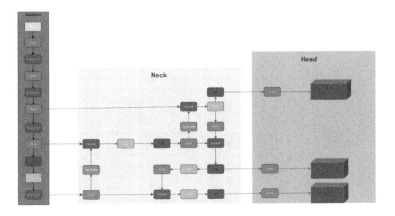

Fig. 1. Network structure of the YOLO-L algorithm

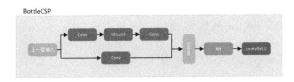

Fig. 2. Structure of BottleCSP feature extraction module

and adapted from the BackBone backbone network to solve the problem of poor performance of the original BackBone backbone network in processing images with small target recognition, and to improve the performance and generalisation capacity of the deep learning model in various tasks.

The BottleCSP feature extraction module achieves the main function of fusing and splicing the output features of the backbone and branch networks in the channel dimension, resulting in a richer feature representation. This reduces the computational load and the number of model parameters, while keeping the key feature information intact. Secondly, the feature extraction module makes use of a larger hybrid combination module, named Mixunit, which captures the relationships between pixels at different locations, and smoothes and abstracts the input image through convolution operations, helping to extract more global and semantic features. Finally, another 1×1 convolutional layer is used to restore the feature dimension to its original size.

For MixunitMixunit's output draws on the idea of residual learning [16], as shown in Fig. 3. This output module obtains a more comprehensive output feature mainly through the superposition of two convolutional networks, allowing the network to learn deeper feature representations more easily, using the operation of superimposing the extracted features with the original features for accumulation at the feature output port, and passing the input directly This effectively solves the gradient disappearance problem and improves the expressiveness and fitting ability of the network.

Mixunit

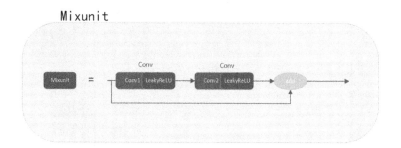

Fig. 3. Structure of the Mixunit module

The output process formula for Mixunit is derived as follows:

$$C1 = W1 \cdot X + b1 \tag{1}$$

$$Conv1 = LeakyReLU(C1) \tag{2}$$

$$C2 = w2 \cdot Conv1 + b2 \tag{3}$$

$$Conv2 = LeakyReLU(C2) \tag{4}$$

In the above equation, X represents the input layer, C1 represents the value obtained by simple linear combination of the input layer after Conv1 convolution, Conv1 represents the feature obtained by non-linear transformation function after Conv1 convolution, and C2 and Conv2 repeat the above operation to extract more global and semantic features. Get the final result:

$$Mixunit = Conv + X \tag{5}$$

3.2 Injecting the CoordAttention Attention Mechanism

CoordAttention (Coordinate Attention) is an attention mechanism for computer vision tasks that aims to improve a model's ability to model spatial information. Traditional attention mechanisms typically focus on the channel dimension on the feature map, weighting features by calculating the importance weight of each channel, thus enabling adaptive feature fusion. However, they typically ignore site information, which is significant for generating spatial selectivity attention graphs. Using embedding site feature information into attention module, this can also be called "coordinate attention" [17]. The structure of the CoordAttention attention mechanism is shown in Fig. 4, where a feature map is first taken as input, usually the output of one of the convolutional layers in a convolutional neural network. This feature map can be viewed as a three-dimensional tensor with height, width and number of channels. The input pattern is globally averaged to give a tensor with the equivalent amount of channels but with a height and width of 1. The globally averaged pooled tensor is then import io a fully connected layer for learning the relationships between the channels. Finally, after processing by the attention mechanism, a character map with enhanced representation is obtained as an output.

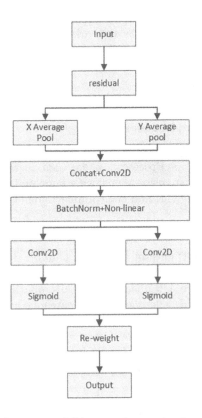

Fig. 4. Structure of CA attention mechanism model

To help the attention module capture long-range interactions spatially using accurate location information, this paper embeds the CoordAttention attention mechanism in the BackBone backbone network and Head prediction head, the main principle of which is to decompose the global pool into two spatially oriented 1D feature encoding operations to enable the model to more accurately locate objects of interest and enhance the model's ability to different regions of the image. Specifically, the two spatial ranges (H, 1) or (1, W) of the pooling kernel are respectively used to encode each channel along the horizontal and vertical coordinates, given the input X. Thus, the output of the cth channel at height h can be formulated as:

$$z_c^h(h) = \frac{1}{W} \sum_{0 \le i < W} x_c(h, i) \tag{6}$$

$$z_c^w(w) = \frac{1}{H} \sum_{0 \le j < H} x_c(j, w) \tag{7}$$

Both transformations allow the constructed attention blocks to capture long-range dependencies along one spatial direction and retain precise location information along the other, helping the network to locate objects of interest more accurately, enhancing the model's ability to focus on different regions in the image and improving the perception of spatial structure and local detail [18].

4   Results and Analysis

4.1   Experimental Environment

This thesis uses an RTX A5000 graphics card, 24G of running memory, an AMD EPYC 7371 16-Core Processor, a data set increased to 1152 by using data enhancement and expansion, cuda with version 11.5, cudnn with version 8.3, pytorch with version 1.11, and the compiler language is python 3.8.

4.2   Introduction to the RSOD Dataset

This thesis uses the RSOD dataset [19] as the training dataset, which is an open dataset contains four main categories, which are aircraft, oil tanks, playgrounds and overpasses. There are 446 images in the aircraft category file, with a total of 4993 aircraft annotated. The playgrounds category has 189 images, with 191 playgrounds annotated; the overpasses category has 176 images, with 180 overpasses annotated; and the fuel tanks category has 165 images, with 1586 fuel tanks annotated. The dataset was formatted as PASCAL VOC and randomly splited into a training datas and a validation datas in the ratio of 8:2.

4.3   Dataset Pre-processing

Due to the relatively large size of the original images of the RSOD dataset. If the initial size is applied as the training dataset, it will result in too many parameters. Therefore, in order to enhance the recognition effect of the model and enhance the computing speed of the model, the pixel size of the initial aerial image is reduced to 608 pixels × 608 pixels by pixel transformation in this paper. And based on the initial image, adopted flipped, rotated, colour enhanced and noise added in various ways to make the remote sensing image have different presentation and scale, which helps to avoid the occurrence of overfitting and thus promote the generalisation capacity of the training network. The Fig. 5 shows the aircraft category dataset as an example for the dataset expansion operation.

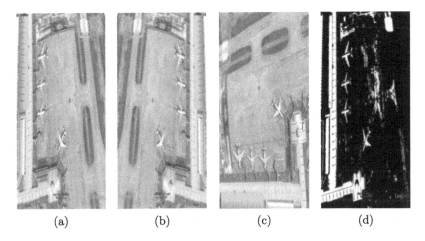

Fig. 5. Schematic diagram of the pre-processing of the RSOD dataset, where (a) is the original image, (b) is the flipped image, (c) is the rotated image and (d) is the grayscale converted image.

4.4 Assessment Indicators

Precision. Precision is the proportion of objects detected by the algorithm that are true targets. It indicates how many of all detections classified as targets are correct. Precision can be collected by calculating ratio of the number of correct targets detected to the total number of targets detected. High accuracy reveals that the algorithm has a low false detection rate, i.e. the algorithm rarely misclassifies non-target areas as targets.

Recall. Recall is the proportion of true targets successfully detected by the algorithm to the all amount of true targets [20]. It indicates how many of all real targets are successfully detected. The recall can be expressed by calculating ratio of the number of correctly detected targets to the total number of real targets. A high recall indicates that the algorithm has a low rate of missed detections, i.e. the algorithm is able to capture the targets in the image better.

mAP. Mean Average Precision (mAP) is a composite metric commonly used in target detection tasks, which calculates the mean value of the average precision for different target classes [21]. In a multi-category target detection task, the AP is estimated for each category and these AP values are averaged to obtain mAP. mAP provides a more comprehensive measure of the algorithm's performance on different target categories, rather than being limited to a single category.

4.5 Analysis and Comparison of Experimental Results

The partial detection effect of the improved algorithm in this paper is shown in Fig. 6. It can be clearly display that the prediction framework is more accurate in recognising targets such as densely arranged aircraft, playgrounds and oil drums, and the new YOLO-L algorithm model arises the recognition accuracy of small target images and effectively decreases the influence of background on target classification. At the same time, the overlap problem caused by the horizontal boundary prediction frames of dense targets is reduced, and the location information of the region of interest is better captured. 300 rounds were selected as the best training rounds for this experiment after analysis, with an average target detection accuracy of 1.00, a target detection tie regression rate of 0.98, and an average accuracy mean mAP of 0.928.

Fig. 6. Identification of various types of targets in the RSOD dataset using the YOLO-L algorithm model

The performance of the proposed detector was compared with four newer general-purpose detectors, including YOLOV5, DC-SPP-YOLO [22], MRFF-YOLO [23] and YOLOV5-slight [24], by conducting remote sensing image detection experiments on the RSOD dataset. To ensure the reliability of the experiments, all training and test data were trained in the same experiment environment with a guaranteed number of 300 rounds per training. After the optimal best model weights were used for testing. As shown in the table below, this paper tests the performance of different networks on RSOD in terms of four evaluation metrics, P, R, mAP and FPS respectively. The testing speed is trained on the same server. Compared to YOLOv5, the proposed scheme can achieve a considerable improvement of 12.3% in mAP and a decrease in FPS speed cost of about 20.3. Compared to DC-SPP-YOLO, the proposed scheme can achieve an improvement of 16.22% in mAP and a decrease in FPS speed cost of 13.0. Compared to MRFF-YOLO, the proposed scheme can Compared to YOLOV5-slight, the proposed scheme can improve mAP by 7.67% and reduce FPS speed cost by 7.8. The new YOLO-L algorithm model is experimentally proven to be more effective in detecting small target recognition (Table 1).

Table 1. Results of 3D target detection based on SUN RGB-D. The evaluation metric is the average accuracy of the proposed 3D IoU threshold of 0.25. The best results are bold.

methods	P	R	mAP	FPS
YOLOV5	80.00	78.00	80.50	40.8
DC-SPP-YOLO	78.00	75.50	76.58	33.5
MRFF-YOLO	90.50	88.00	88.33	25.1
YOLOV5-slight	88.78	83.33	85.13	37.7
YOLO-L	100.00	98.00	92.80	20.5

5 Conclusion

algorithm for detecting common target objects in remote sensing images in complex backgrounds. The algorithm mainly optimises the YOLOv5 backbone network model by replacing the C3 module and SPP and inserting the BottleCSP feature extraction module into different stages of the backbone network, so as to better capture deep linguistic information processing from the images and solve the problem of poor performance when recognising images with small targets. Meanwhile, the CoordAttention attention mechanism is embedded in the Back-Bone backbone network and Head prediction head to enhance the perception of spatial structure and local details. This paper performs target detection on the remote sensing image dataset of RSOD. Experiments show that by comparing with YOLOV5, DC-SPP-YOLO, MRFF-YOLO and YOLOV5-slight detection methods respectively, it is concluded through comparison that the improved YOLOV5-based method proposed in this paper improves the mAP evaluation metrics by 12.3%, 16.22%, 4.47% and 7.67% respectively, and the detection speed FPS is also improved with guaranteed accuracy. However, as remote sensing satellite images are easily affected by weather factors, the recognition effect is relatively poor for remote sensing images with large scale variations. In subsequent experiments, relatively clear multi-scale images can be selected to learn different types of dataset features.

References

1. Cheng, Y., et al.: A multi-feature fusion and attention network for multi-scale object detection in remote sensing images. Remote Sens. **15**(8), 2096 (2023)
2. Zhong, Y., Wang, J., Zhao, J.: Adaptive conditional random field classification framework based on spatial homogeneity for high-resolution remote sensing imagery. Remote Sens. Lett. **11**(6), 515–524 (2020)
3. Girshick, R., Donahue, J., Darrell, T., Malik, J.: Rich feature hierarchies for accurate object detection and semantic segmentation. In: Proceedings of the IEEE Conference on Computer Vision and Pattern Recognition, pp. 580–587 (2014)
4. Girshick, R.: Fast R-CNN. In: Proceedings of the IEEE International Conference on Computer Vision, pp. 1440–1448 (2015)

5. SunX, X., et al.: Fair1m: a benchmark dataset for fine-grained object recognition in high-resolution remote sensing imagery. ISPRS J. Photogramm. Remote. Sens. **184**, 116–130 (2022)

6. Liu, W., Anguelov, D., Erhan, D., Szegedy, C., Reed, S., Fu, C.-Y., Berg, A.C.: SSD: single shot multibox detector. In: Leibe, B., Matas, J., Sebe, N., Welling, M. (eds.) ECCV 2016. LNCS, vol. 9905, pp. 21–37. Springer, Cham (2016). https://doi.org/10.1007/978-3-319-46448-0_2

7. Redmon, J., Divvala, S., Girshick, R., Farhadi, A.: You only look once: unified, real-time object detection. In: Proceedings of the IEEE Conference on Computer Vision and Pattern Recognition, pp. 779–788 (2016)

8. Qu, Z., Zhu, F., Qi, C.: Remote sensing image target detection: improvement of the YOLOV3 model with auxiliary networks. Remote Sens. **13**(19), 3908 (2021)

9. Zheng, Z., Liu, Y., Pan, C., Li, G.: Application of improved YOLOv3 in aircraft recognition of remote sensing images. Electron. Opt. Control. **26**(4), 28–32 (2019)

10. Wu, D., Lv, S., Jiang, M., Song, H.: Using channel pruning-based yolo v4 deep learning algorithm for the real-time and accurate detection of apple flowers in natural environments. Comput. Electron. Agric. **178**, 105742 (2020)

11. Tan, S., Bie, X., Lu, G., Tan, X.: Real-time detection of personnel mask placement based on the YOLOv5 network model. Laser J. 147–150 (2021)

12. Ren, S., He, K., Girshick, R., Sun, J.: Faster R-CNN: towards real-time object detection with region proposal networks. In: Advances in Neural Information Processing Systems, vol. 28 (2015)

13. Hou, T., Jiang, Y.: Application of improved YOLOv4 in remote sensing aircraft target detection. Comput. Eng. Appl. **12**(57), 224–230 (2021)

14. Yasir, M., et al.: Multi-scale ship target detection using SAR images based on improved YOLOv5. Front. Mar. Sci. **9**, 1086140 (2023)

15. Wu, Z., Su, L., Huang, Q.: Decomposition and completion network for salient object detection. IEEE Trans. Image Process. **30**, 6226–6239 (2021)

16. He, K., Zhang, X., Ren, S., Sun, J.: Deep residual learning for image recognition. In: Proceedings of the IEEE Conference on Computer Vision and Pattern Recognition, pp. 770–778 (2016)

17. Hou, Q., Zhou, D., Feng, J.: Coordinate attention for efficient mobile network design. In: Proceedings of the IEEE/CVF Conference on Computer Vision and Pattern Recognition, pp. 13713–13722 (2021)

18. Wen, G., Li, S., Liu, F., Luo, X., Er, M.J., Mahmud, M., Wu, T.: YOLOV5s-CA: a modified yolov5 network with coordinate attention for underwater target detection. Sensors **23**(7), 3367 (2023)

19. Körez, A., Barışçı, N., Çetin, A., Ergün, U.: Weighted ensemble object detection with optimized coefficients for remote sensing images. ISPRS Int. J. Geo Inf. **9**(6), 370 (2020)

20. Xia, G.S., et al.: DOTA: a large-scale dataset for object detection in aerial images. In: Proceedings of the IEEE Conference on Computer Vision and Pattern Recognition, pp. 3974–3983 (2018)

21. Lin, T.Y., Goyal, P., Girshick, R., He, K., Dollár, P.: Focal loss for dense object detection. In: Proceedings of the IEEE International Conference on Computer Vision, pp. 2980–2988 (2017)

22. Huang, Z., Wang, J., Fu, X., Yu, T., Guo, Y., Wang, R.: DC-SPP-YOLO: dense connection and spatial pyramid pooling based YOLO for object detection. Inf. Sci. **522**, 241–258 (2020)

23. Xu, D., Wu, Y.: MRFF-YOLO: a multi-receptive fields fusion network for remote sensing target detection. Remote Sens. **12**(19), 3118 (2020)
24. Lang, L., Xu, K., Zhang, Q., Wang, D.: Fast and accurate object detection in remote sensing images based on lightweight deep neural network. Sensors **21**(16), 5460 (2021)

Power and Energy Systems

An Effective N-BEATS Network Model for Short Term Load Forecasting

Chang Tan$^{(\boxtimes)}$ 🆔, Xiang Yu 🆔, Lihua Lu 🆔, and Lisen Zhao 🆔

School of Electronics and Information, Shanghai Dianji University, Shanghai 201306, China
tanchang317@gmail.com, {yux,lulihua}@sdju.edu.cn

Abstract. This paper addresses the issue of short-term load forecasting in the power system, a domain where risks are liable to take place during the peak period of power consumption. To prevent such risks, it is crucial to have a precise load forecasting that can be carried out beforehand, allowing for the arrangement of the power peak period in advance, and greatly minimize the occurrence of power accidents. To achieve this goal, a power load forecasting method of high reliability and precision is indispensable. In this paper, an effective approach is proposed that is capable of effectively resolving short- and medium-term power forecasting issues. Prior to the forecasting, the Seasonal-Trend decomposition procedure based on Loess (The full term for Loess is locally weighted scatterplot smoothing, LOWESS or LOESS) (STL) time series decomposition is conducted on the data set, enabling the observation of the trend, periodicity, and corresponding residual items. Based on the STL decomposition results, the prediction lengths of the previous and subsequent items in the N-BEATS (Neural basis expansion analysis for interpretable time series forecasting) network model are adjusted, and the corresponding network structure is altered to identify the optimal model through continuous experiments. The method's efficacy was evaluated on datasets obtained from two distinct regions, and the results indicated a better performance compare with other similar forecasting methods in terms of accuracy and forecasting bias.

Keywords: Short term load forecasting · N-BEATS · STL · Data analysis · Loss optimization

1 Introduction

In recent years, the world has been experiencing a decrease in non-renewable energy sources, highlighting the need for the development and reasonable utilization of renewable resources. Electric power, in particular, has become a pressing concern for everyone, especially in terms of storing large amounts of power effectively. Balancing the continuous equilibrium between power consumption and production is already a challenge for the current power system, but this challenge is intensified as the share of renewable energies continue to increase. The key to balancing the power system lies in reliable forecasting of demand and power generation at any point in time, which can help reduce and avoid energy shortage or oversupply [1, 2].

© ICST Institute for Computer Sciences, Social Informatics and Telecommunications Engineering 2024
Published by Springer Nature Switzerland AG 2024. All Rights Reserved
J. Li et al. (Eds.): 6GN 2023, LNICST 553, pp. 229–243, 2024.
https://doi.org/10.1007/978-3-031-53401-0_21

Forecasting methods for electric load can be divided into two categories: statistical and machine learning-based methods. Statistical methods, such as those based on the Kalman filter [3] and Box-Jenkins model [4], typically use hand-crafted features to obtain forecasting results and do not need to learn in the process. Time series is a representative statistical method used in power load forecasting. For instance, Mbamalu et al. used autoregressive (AR) to forecasting the power load and estimated the parameters in the forecasting using the least squares method [5–8].

On the other hand, machine learning-based methods involve training models that can find complex relationships between the input and output. One such method is Chen et al., which clusters inputs into distinct groups and uses the vector machine (SVM) to predict the result [9]. The long short-term memory (LSTM) network model is another machine learning-based method that is particularly effective for long-term electric load forecasting [10]. Additionally, the Neural basis expansion analysis for interpretable time series forecasting (N-BEAT) network model is a deep neural network and a representative of data-driven models for time series data analysis. It has demonstrated advanced forecasting capabilities in the analysis of large time series data in different fields, such as the M3, M4 and TOURISM competition datasets [11].

In this paper, the Seasonal-Trend decomposition procedure based on Loess (STL) is used to preprocess the data set [12], followed by an N-BEATS network to carry out power load forecasting. Our N-BEATS model is particularly useful in automatic analysis of time series data, offering advantages such as high precision, high reliability, and a certain degree of tolerance. This research focuses on the short-term power load data set, using the N-BEATS model to solve the problem of short-term power load forecasting.

The contributions of our model are:

(1) Applying Seasonal-Trend decomposition procedure based on Loess (STL) on data set to optimize N-BEATS parameters. The authors remove the trend items and residual items decomposed from the middle area, and then pre-determine the front and rear items of the N-BEATS model by obtaining the periodic trend Screening, so that the N-BEATS model can obtain data that is updated faster and contains more information as much as possible in each learning.
(2) Designing an effective method for our modeling parameter selection: choosing relatively small value of backcast_length helps to improve model efficiency, as setting too large a value will cause overlapping training samples to become more and more relevant. A small backward prediction value will allow the data entering the model to update quickly, and avoiding the situation of less data entering the model training.

The power load forecasting task based on deep learning and time series decomposition STL is mainly divided into three steps: data organization, time series analysis before and after item forecasting, and building a model for forecasting. Firstly, the data is screened to ensure the same seasonality in the entire data set is included to form a new data set for convenient learning of trend and seasonality by the subsequent N-BEATS network. Then, time series decomposition is performed on the screened one-dimensional time series data to analyze change characteristics and determine the input and output window setting strategy of. This model was tested on the Australian power load data set and the American data set. The average MAPE is around 3.5, which is 17% more accurate than the CNN-LSTM model proposed by Liu et al. [13].

2 Our Method

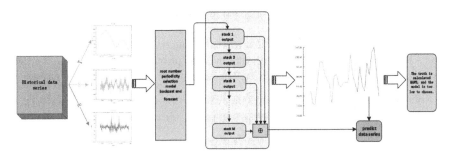

Fig. 1. Structure diagram of method

For the optimal performance of deep learning, a comprehensive approach is essential. It is not enough to rely solely on human optimization and hyper parameter debugging, as the organizational structure of data must also be carefully regulated. Furthermore, the performance of the computer poses a significant constraint on the amount of data that can be imported, making it necessary to manually select valuable data for input training. Various experiments have demonstrated that the composition of data has a profound impact on the prediction effect and training ability of the model (see Fig. 1).

In this paper, a new dataset was created by selecting data from the same season in the Australia load dataset [14], which was used to train an N-BEATS model [15, 16]. This model was capable of more effectively extracting trend and cycle information from the data set, resulting in a significant improvement in the model's prediction effectiveness.

When setting the input and output window of the model, it is crucial to consider the characteristics of the time series itself. For example, for time series with periodic characteristics, the input window's length should reflect the real period to avoid losing periodic information. For time series with absurd changes, the input window's length should cover the entire period. Similarly, for time series that include other nonlinear features, the input window's length should be representative of the data.

The output window's size must also be set according to the real prediction time frequency and time extension requirements to ensure that the deep learning prediction results are accurate. There are several ways to analyze the regularity of time series data, but this study adopted a strategy that decomposed the data set into trend factors, seasonal factors, and random factors using the seasonal trend decomposition method based on Loss (STL).

$$Y_V = T_V + S_V + R_V \tag{1}$$

where $(Y_V, T_V, S_V, R_V$ represent data, trend item, seasonal item and remainder item respectively, and the range of v is 0 to N) It is a very general and robust method for decomposing time series, where Loess is a method for estimating nonlinear relationships. The STL decomposition method was proposed by R. B. Cleveland et al. [17, 18]. This method is ideal for power load forecasting, and its results are interpretable, making it suitable for this study's purposes.

2.1 N-BEATS Model

The primary objective of N-BEATS is to address univariate time series forecasting problems using deep learning. It adopts a deep neural network structure that comprises backward and forward residual links and very deep fully connected layers. This structure boasts several desirable characteristics, including interpretability, among others. The N-BEATS network model consists of various perceptron units and represents a deep network created using a specific strategy. The network structure serves two purposes:

(1) Simulation, it extracts and emulates the features of the original time series data;
(2) Prediction, it extends the simulated sequence data characteristics along the time dimension to complete the prediction.

The design strategy of the N-BEATS network model [19] comes from time series decomposition. The model also incorporates trend and seasonal compositions, making the stack output easier to interpret. Hence, the model's advantage is not just that it can achieve high accuracy in predicting results, but it can also interpret time series changes like trends and seasonality, similar to the STL method. Figure 2 illustrates the model's framework diagram. The network training input is a historical data sequence with a predetermined window size. The sequence data passes through M stacks to identify different data characteristics, with each stack comprising several blocks that undergo processing through the residual module. Each block features several computational neurons.

The N-BEATS model showed good learning, interpretability and easy adjustment in this experiment. First of all, it can learn and capture the periodicity and trend of time series data well, and it can also share the information of each stack in the model, which is helpful for prediction accuracy and training efficiency. Second, the N-BEATS model is interpretable because each block contains a forecasting sublayer and a processing sublayer, which is used to capture long-term dependencies such as periodicity and trends in time series data, while forecasting Sublayers are used to generate future forecasts of the time series. The functions and roles of these sublayers are very clear, which can help users understand the internal mechanism of the model. Finally, since the architecture of the N-BEATS model is composed of multiple blocks, each of which can contain multiple sublayers, it is easily adjusted. The complexity of the model can be varied by increasing or decreasing the number of blocks or changing the parameters of the blocks to suit different datasets and tasks.

The primary focus of this model is to solve univariate time series point forecasting problems using deep learning. The structure of the model is a deep neural network that incorporates backward and forward residual links and very deep fully connected layers, which have desirable properties, including interpretability. The N-BEATS network is composed of several perceptron units and built using a specific strategy that has two functions: 1. Simulation, which extracts and simulates the characteristics of the original time series data; and 2. Prediction, which continues the simulated sequence data characteristics from the time dimension to complete the prediction.

The construction strategy of the N-BEATS network model is inspired by time series decomposition, which designs trend and seasonal decomposition in the model. This design makes the stack output easier to explain, and the advantage of this model is not only that it can obtain high accuracy but also that its prediction results have the ability to

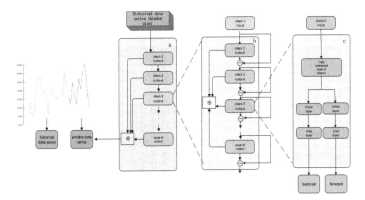

Fig. 2. Structure diagram of N-BEATS model [11]

explain time series changes such as trends and seasonality, similar to the STL method. The input of a network training is the historical data sequence with a set window size, and the sequence data passes through M stacks to identify different characteristics of the data. Each stack has several blocks, and the blocks are processed through the residual module. Each block contains several computational neurons.

The block layer learns the time series evolution mode. The original time series data goes through four layers of fully connected layers. The forward forecasting result predicts the direction of the future sequence along the development direction of the time series, while the backward forecasting result recovers the historical time series for signal separation. Therefore, the input in the block area is the original sequence data, and the output is the forward and backward prediction results.

The stack layer decomposes the remaining time series features. The output of each block is its forward and backward prediction results, as shown in the box in Fig. 2. The forward prediction results are integrated to become the output of the stack, and the backward prediction results exist between the blocks in the form of residuals and connect the blocks, which is more conducive to the backpropagation of the gradient and makes the next block's forecasting easier. Therefore, the input in this stack is the forward and backward prediction results of each block, and the output is the forward prediction result obtained through integration in this part.

The time series feature prediction components are integrated. The output of each part of the stack is finally integrated, and the feature components obtained by each stack are superimposed to obtain the overall prediction result, which is the sequence of prediction data, as shown in the box in Fig. 2.

Advantages of N-BEATS Network Architecture. The neural architecture of this model can achieve the explication of time series variations such as trends and seasonality. Its fundamental mechanism is to provide diverse linear layers and bias layers to the neural components in different stacks. Its key strength lies in its interpretability, which is manifested in the fact that the trend and cycle components are allocated into two separate stacks.

In the trend model, its characteristics are typically represented by linearly increasing or decreasing functions that change slowly over time. Therefore, the linear layer and bias layer in part (c) of the trend stack are configured as

$$y = T \cdot x \tag{2}$$

where $T = [1, t^1, t^2, ..., t^p]$, with p being a small value that regulates the speed of linearly increasing or decreasing behavior, resulting in gradual trend changes; t is a time vector; x is the input to the linear layer and bias layer in the block; and y is the output of the linear layer and bias layer in the block.

For the cycle model, periodic patterns are typically viewed as cyclic and fluctuating. Therefore, the linear layer and bias layer in part (c) of the periodic stack are set to the Fourier series form

$$y = S \cdot x \tag{3}$$

$$S = \left[1, \cos(2\pi t), \cdots, \cos\left(2\pi\left[\tfrac{H}{2} - 1\right]t\right), \sin(2\pi t), \cdots, \sin\left(2\pi\left[\tfrac{H}{2} - 1\right]t\right)\right] \tag{4}$$

In the equation, t represents the time series vector; H represents the forward prediction window length; x is the input of the linear layer and bias layer in the block; and y is the output of the linear layer and bias layer in the block.

3 Experiment

3.1 Dataset

In the experiment, the electrical data sets from two different regions and different years were utilized, namely the Australian data set and the Midwest data set in the United States. The diversity in the geographical locations where the data sets are collected ensures that the model can be adapted to various training environments. The collection methods of the two data sets are also distinct. In the Australian data set, data was collected from 2006 to 2010 with 48 sampling points per day. On the other hand, the US data set was collected from 2015 to 2022 with 24 sampling points per day, where samples were taken hourly. To make it easier for the N-BEATS model to learn and since N-BEATS is more suitable for smaller data sets, the daily average value of the two data sets was used to create a new data set for prediction (Fig. 3).

To conduct a more comprehensive analysis, the experiment focused on three sets of data, specifically February and August in the Australian data set, and February in the US data set, which were selected from a vast number of experiments. Because February is at the end of winter in the United States, the temperature gradually rises at this time, the temperature difference is large, and the fluctuation of electricity consumption also becomes larger. This increases the difficulty of model prediction.

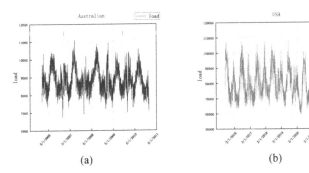

Fig. 3. (a) Australia dataset (b) USA dataset

3.2 Our Model Settings

The N-BEATS model is a structure that determines the prediction data based on the window size. To predict the output results of m future moments, it uses k historical moments as the input sequence, where k is determined by the window size. In order to deal with seasonality, trend, and other deformation signals, the experiment set up 3 stacks, each consisting of 3 blocks, and 4 fully connected layers were set in each block, with 64 hidden layer units. The parameter dimensions of the linear layer and the bias layer after the connection layer are set to [3, 8], representing other deformation stack models, seasonal stack models, and trend stack models. The structure of the N-BEATS landslide prediction network model is shown in the figure below.

During the training process, the mini-batch gradient descent method was used, with a small batch size of 16, and all samples were trained for a total of 1200 times. To ensure the adaptability of the model to different training environments, electricity data sets from two different regions and different years were used, namely the Australian data set and the Midwest data set in the United States. The daily average value of the two datasets was used to form a new data set prediction, which was convenient for the N-BEATS model to learn, as it is more suitable for smaller data sets. In the experiments, February and August in the Australian data set, and February in the US data set were selected for further analysis. The Australian dataset spanned from 2006 to 2010 and had 48 sampling points in one day, while the US dataset spanned from 2015 to 2022 and had 24 sampling points in one day, with samples taken every hour (Fig. 4).

The aim of deep learning is to optimize the difference between the model predicted value and label value by minimizing the value of the loss function using the optimizer. This optimization process leads to the attainment of a highly efficient and accurate model. The loss function used in this research paper for the N-BEATS network model is the mean absolute error loss function Mean Absolute Percentage Error (MAPE).

MAPE is a statistical measure that determines the absolute difference between the model predicted value and the label value, expressed as a percentage of the label value's average. It uses absolute values to prevent the cancellation of positive and negative errors and relative errors for comparing the accuracy of forecasts from different time series models.

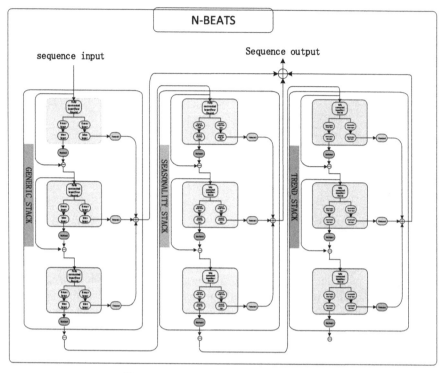

Fig. 4. N-BEATS network model settings

The optimizer used in this research paper is Adam, and the learning rate is set to 0.0001. Adam is an optimization algorithm that independently designs adaptive learning rates for different parameters through the calculation of the first-order and second-order moment estimations of the gradient. Adam is known to perform better than other stochastic optimization algorithms, hence its selection.

The hyper parameter settings used in this research paper are shown in the table below. The network hyper parameters used for the N-BEATS network model were obtained after several experiments and applying the recommended parameters for empirical debugging (Table 1).

Table 1. Model hyper parameter setting table

hyper parameter	set value
Stacks	3
Blocks per stack	2
Parameter dimension	[8, 8, 3]

(continued)

Table 1. (*continued*)

hyper parameter	set value
Hidden layer unit	64
Loss function	MAPE
Optimizer	Adam, learning rate 1e-4
Iterations	1200
Batch	16

3.3 Prediction Task Dataset Settings

This paper aims to investigate the effects of input and output window settings on forecasting accuracy using a simple time series decomposition strategy and the STL algorithm to decompose the research target time series into trend factors, seasonal factors [20], and random factors. The characteristics of the data components of the power load sequence are roughly 84% trend items, 8% cycle items, and 8% uncertainty factors. The trend component is stable and needs no consideration in setting the input and output window size. The uncertain component presents a form of irregular development and cannot be adjusted in the window setting, while the periodic component is an important basis for setting the input and output window size (Fig. 5).

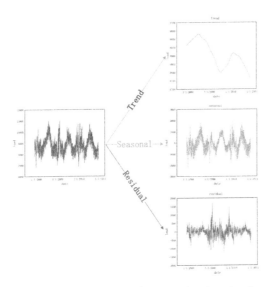

Fig. 5. Seasonal-trend decomposition procedure based on loess STL

To ensure the sample set of training data is rich while considering the periodic characteristics of the sequence, this paper sets backcast_length to {12, 24, 36, 48, 60, 72}, and forecast_length to {3, 7, 12, 24, 28, 31}, resulting in 36 combinations of trained and tested data. The accuracy of the prediction results decreases with the increase of the input and output windows, and the larger the input window, the more complete the periodic components considered during model training, but the learning data of the model is reduced, and the training effect is decreased. The larger the output window, the more "greedy" it becomes to predict future unknown time series data, and the corresponding price will be paid for the decrease of prediction accuracy.

4 Results and Discussion

The error detection technique utilized in this manuscript is none other than the Mean Absolute Percentage Error (MAPE), which is a statistic that measures the discrepancy between the predicted and actual values in percentage. The employment of the absolute value is beneficial in negating the possibility of positive errors and negative errors getting lower. MAPE is widely employed in determining the forecasting accuracy of diverse time series models. The outcomes of the experiments conducted in this paper will be juxtaposed against the CNN-LSTM model. The CNN-LSTM model was formulated by Liu et al. [12], who applied the data subjected to the similarity filter to CNN-LSTM for training, thus making it a successful short-term power load forecasting method.

In numerous experiments, we selected three data sets for comparing SFCL with the technique employed in this paper, which are February and August in Australia, and the prediction of February in the central US data set (Table 2).

Table 2. Forecasting errors on midwest U.S. in february

Forecast	Backcast					
	12	24	36	48	60	71
3	6.30	5.95	7.12	6.86	6.74	8.04
7	6.06	7.23	5.51	4.87	7.37	8.39
12	4.91	4.04	4.40	4.43	5.86	6.12
24	4.06	4.16	4.50	7.17	9.01	6.8
28	5.25	4.33	4.88	5.95	7.29	6.07
30	**3.75**	4.73	7.32	6.29	4.95	4.78

In contradistinction, this manuscript opts for a composition model featuring a backcast_length of 12 and a forecast_length of 30, which satisfies the prerequisite for achieving the utmost precision in short- to medium-term power load forecasting. This composition model yields an average MAPE of 3.75 for February in the central United States (Fig. 6) (Table 3).

In this experiment, the author chose a composition mode with a forecasting of 28 and a backcast length of 48. In this composition mode, the average MAPE of Australia in February was 3.91 (Fig. 7) (see Table 4).

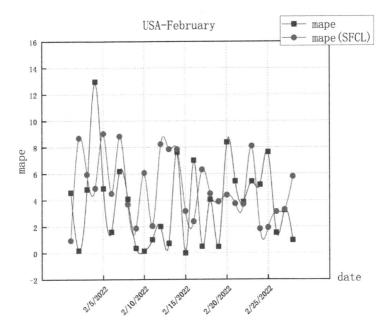

Fig. 6. U.S. february forecasting mape comparison chart

Table 3. U.S. february forecasting average error table

	The method in this article	SFCL
MAPE	3.75	4.87

Table 4. Forecasting errors: australia data set february

Forecast	Backcast					
	12	24	36	48	60	71
1	7.15	6.51	7.01	6.9	6.28	7.06
3	6.07	6.41	6.35	6.13	7.04	6.61
7	6.68	5.78	6.46	5.58	5.57	5.68
12	7.19	6.54	5.72	5.37	5.69	6.21
24	6.38	5.31	4.64	5.33	4.66	5.22
28	5.94	5.35	5.24	**3.91**	5.27	6.04
30	6.29	6.06	5.50	4.87	4.98	5.93

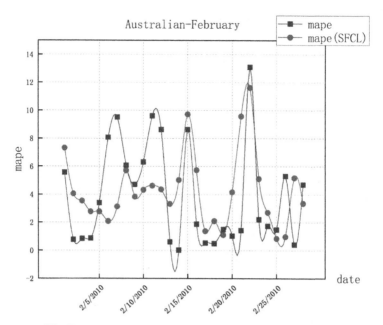

Fig. 7. Australian february forecasting mape comparison chart

Table 5. Australian february forecasting average error table

	The method in this article	SFCL
MAPE	3.91	4.31

Table 6. Australian february forecasting average error table

Forecast	Backcast					
	12	24	36	48	60	71
3	6.20	6.37	6.15	5.64	6.64	6.89
7	6.08	5.97	6.72	6.52	6.70	6.55
12	5.59	5.67	6.17	5.90	6.57	5.89
24	5.94	5.32	5.88	5.48	5.23	5.43
28	3.84	4.67	5.46	4.63	5.01	4.62
30	**2.85**	4.24	4.43	4.26	4.40	4.95

In Australia's August forecasting, this paper chooses the composition model with a forecasting length of 30 and a backcast length of 12. This model's output error on Australia data set against real data is 2.85%, better than CNN-LSTM (Fig. 8) (Table 5).

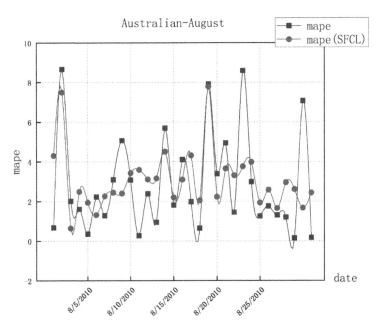

Fig. 8. Australian august forecasting mape comparison chart

Table 7. Australian february forecasting average error table

	The method in this article	SFCL
MAPE	2.85	3.07

Upon conducting the experiments on predicting short and medium-term power load for various regions, this paper has identified a critical observation. It was discovered that as the value of the backcast increases, the corresponding MAPE value will also surge up to a certain threshold. This can be attributed to the fact that, in the event of selecting a more significant value for the backcast in the model, the repetition of training samples will become more and more correlated, leading to a drastic decrease in the number of effective samples that may further aggravate overfitting issues (see Table 6).

The February forecasting for Australia brought another interesting revelation. It was observed that while ensuring the data periodicity, the sample set should also be abundant enough. With our task of predicting the rapid fluctuations of power loads in the short and medium term, we believe that the backcast value must be kept minimal (Table 7).

As illustrated in the chart above, our method has outperformed the SFCL model in terms of prediction error. Nonetheless, certain days had a considerable forecasting error, which could have been influenced by holidays.

5 Conclusions

This paper introduces an effective approach based on N-BEATS model for Australia and U.S. monthly load predictions. Lab results demonstrate that the performance of this method is better compared with existing methods such as CNN-LSTM. Experimental outputs from Australia and the United States show the following key findings:

An N-Beat model forecasting input and output window size of {(12,30), (48,28), (12,30)} can meet the needs of short-term power load forecasting for the data structure of Australia and the United States. This achieves a better forecasting effect under the premise of forecasting. The N-BEATS network model yields high forecasting accuracy of about 3.5 MAPE in power load forecasting, confirming the validity and reliability of the N-BEATS network model in time series forecasting tasks and engineering practice. This paper proved that with the increase of the backcast value, although the data injected into the model at a single time will increase, the overlapping training samples in the subsequent training will continue to increase, which greatly reduces the effective samples and causes the model to over fit. To avoid this issue, the author selects a small backcast value to input the model while ensuring the periodicity of the data, which solves the problem of overfitting.

While the use of the N-BEATS network model for short-term power load forecasting has achieved good results, there are still many problems to be addressed, such as the ability of the model to cope with holidays or sudden data growth, and the effect of this model on mid-term and long-term forecasting. Additionally, other features could be added to the network to improve power load forecasting.

References

1. Oreshkin, B.N., et al.: N-BEATS neural network for mid-term electricity load forecasting. Appl. Energy **293**, 116918 (2021)
2. Hadjout, D., Torres, J.F., Troncoso, A., Sebaa, A., Martínez-Álvarez, F.: Electricity consumption forecasting based on ensemble deep learning with application to the algerian market. Energy **243**, 123060 (2022)
3. Santhosh, M., Venkaiah, C., Vinod Kumar, D.M.: Current advances and approaches in wind speed and wind power forecasting for improved renewable energy integration: a review. Eng. Rep. **2**(6), e12178 (2020)
4. Donald, I.I., Cios, K.J.: Time series forecasting by combining RBF networks, certainty factors, and the box-Jenkins model. Neurocomputing **10**(2), 149–168 (1996)
5. Mbamalu, G., El-Hawary, M.E.: Load forecasting via suboptimal seasonal autoregressive models and iteratively reweighted least squares estimation. IEEE Trans. Power Syst. **8**(1), 343–348 (1993)
6. Hu, Y., et al.: Industrial artificial intelligence based energy management system: Integrated framework for electricity load forecasting and fault prediction. Energy **244**, 123195 (2022)
7. Wan, C., et al.: An adaptive ensemble data driven approach for nonparametric probabilistic forecasting of electricity load[J]. IEEE Trans. Smart Grid **12**(6), 5396–5408 (2021)
8. Razavian, A.S., et al. CNN Features off-the-shelf: an astounding baseline for recognition. In: 2014 IEEE Conference on Computer Vision and Pattern Recognition Workshops, pp. 806–813. IEEE (2014)

9. Tjhi, W.C., Chen, L.: Adapting SVM classifiers to data with shifted distributions. In: Data Mining Workshops, 2007. ICDM Workshops 2007. Seventh IEEE International Conference on, pp. 69–76. IEEE (2008)

10. Hochreiter, S., Schmidhuber, J.: Long short-term memory. Neural Comput. **9**, 1735–1780 (1997)

11. Oreshkin, B.N., et al.: N-BEATS: neural basis expansion analysis for interpretable time series forecasting. arXiv preprint: arXiv:1905.10437 (2019)

12. Cleveland, R.B.: STL: a seasonal-trend decomposition procedure based on loess. J. Off. Stat. **6**, 3–73 (1990)

13. Liu, J., Lu, L., Yu, X., Wang, X.: SFCL: electricity consumption forecasting of CNN-LSTM based on similar filter. In: 2022 China Automation Congress (CAC), pp. 4171–4176. IEEE (2022)

14. https://github.com/Tanchang777/An-effective-N-BEATS-network-model-for-short-term-load-forecasting. Accessed 17 Apr 2023

15. Puszkarski, B., Hryniów, K., Sarwas, G.: Comparison of neural basis expansion analysis for interpretable time series (N-BEATS) and recurrent neural networks for heart dysfunction classification. Physiol. Meas. **43**(6), 064006 (2022)

16. Sbrana, A., Rossi, A.L.D., Naldi, M.C.: N-BEATS-RNN: deep learning for time series forecasting. In: 2020 19th IEEE International Conference on Machine Learning and Applications (ICMLA), pp. 765–768. IEEE (2020)

17. Arneric, J.: Multiple STL decomposition in discovering a multi-seasonality of intraday trading volume. Croatian Oper. Res. Rev. **2021**(1), 61–74 (2021)

18. Sanchez-Vazquez, M.J., et al.: Using seasonal-trend decomposition based on loess (STL) to explore temporal patterns of pneumonic lesions in finishing pigs slaughtered in England, 2005–2011. Prev. Vet. Med. **104**(1–2), 65–73 (2012)

19. Olivares, K.G., et al.: Neural basis expansion analysis with exogenous variables: forecasting electricity prices with NBEATSx. Int. J. Forecast. **39**(2), 884–900 (2023)

20. Mishra, A., Sriharsha, R., Zhong, S.: OnlineSTL: scaling time series decomposition by 100x. arXiv preprint: arXiv:2107.09110 (2021)

A Novel Ultra Short-Term Load Forecasting Algorithm of a Small Microgrid Based on Support Vector Regression

Lin Liu[✉]

School of Electronic Information Engineering, Shanghai Dianji University, Shanghai 201306, People's Republic of China
liul@sdju.edu.cn

Abstract. Based on the support vector regression (SVR), a novel ultra short-term load of forecasting algorithm is proposed in this paper. The algorithm is used to realize the online cycle prediction of the load. The historical load data is divided into three categories: working days, weekends and holidays. The similarity method is used to compute the correlation between the training dataset and the prediction load at the current time. The correlation values are sorted in descending order. The first kth load data are used for training to obtain three types of online forecasting models based on SVR. A novel particle swarm optimization (PSO) algorithm is designed to get the optimal parameters of SVR. The experiment on a commercial building of a small mirogrid system shows the designed algorithm works efficiently and stably.

Keywords: Support vector regression · Online cycle prediction · Particle swarm optimization

1 Introduction

Many urban commercial buildings are equipped with small microgrid systems consisting photovoltaic or wind turbine power generation. The accuracy of load forecasting directly affects the utility of microgrid energy management system. It is very important for the optimal scheduling of controllable micro sources such as wind power, photovoltaic and micro gas turbine. The load forecasting results will affect the operating strategy of the microgrid. Due to the strong randomness of the load and the large fluctuations of the load, it is much more difficult in short-term and ultra short-term load forecasting of the microgrid compared with the large grid. It is necessary to design a load forecasting algorithm considering ultra short-term time series for better performance.

In last several decades, a lot of research work is mostly focused on short-term load forecasting of microgrid. Time series analysis and machine learning are two commonly used in short-term power load forecasting methods with high accuracy [1–17]. Time series analysis includes the simplest linear regression [1], multiple regression [2, 3],

J. Li et al. (Eds.): 6GN 2023, LNICST 553, pp. 244–256, 2024.
https://doi.org/10.1007/978-3-031-53401-0_22

autoregressive moving average (ARMA) [4] and autoregressive integrated moving average (ARIMA) [5, 6], exponential smoothing (Holt–Winters) [7] and Kalman filter [8]. The disadvantage of time series analysis is that it cannot be used efficiently in considering irregular changes in the selected time series. The hybrid methods are designed to solve this problem [9, 10]. A hybrid method for short-term load forecasting is proposed with time series, regression analysis and wavelet decomposition [9]. The time series and regression analysis are used to select the best set of inputs for the load forecasting of the special days. Zhai M-Y. [10] proposes a method combined with self-similarity and fractal interpolation. The algorithm analyzes the self-similarity of load historical data using multi-resolution wavelet.

Machine learning includes several methods, such as support vector machine [11, 12], random forest [13, 14], generalized regression neural network and integration algorithm [15–17]. Li S. et al. [11] propose a novel short-term load forecasting method based on wavelet transform, extreme learning machine (ELM) and modified artificial bee colony (MABC) algorithm. The wavelet transform is used to decompose the load series for capturing the complicated features at different frequencies. The global searching technique MABC is used to find the best parameters of input weights and hidden biases for ELM. In [12], a new combined forecasting method (ESPLSSVM) based on empirical mode decomposition, particle swarm optimization (PSO) and least squares support vector machine (LSSVM) model is proposed.

The ultra short-term load forecasting is to predict the load demand from the current time to the next 4 h [18, 19], which is updated every constant interval. The sampling period is 15 min, thus there are 16 sampling points in the next 4 h. Ultra short-term load forecasting requires shorter computing time and higher online performance compared with short-term load forecasting. It depends on the forecasting model with high accuracy and calculation efficiency.

In this paper, a novel ultra short-term load forecasting algorithm based on SVR is designed for a small microgrid system. The influence of weather factors has been reflected in the historical load data. Then weather factors are no longer considered separately as input to train the forecasting model. The historical load data is divided into three categories: working days, weekends and holidays. The similarity method is used to compute the correlation between the training dataset and the output at the current time. The correlation values are sorted in descending order. The first kth load data are used for training to obtain three types of online forecasting models based on SVR. The testing results on cases show the effectiveness of the algorithm, which can be applied to a larger microgrid energy management system in the future.

This paper is organized as follows. Section 2 introduces the detail procedure of the proposed algorithm which including the data preprocessing, composition of input dataset, similarity computation for SVR, and online cycle forecasting strategy. Section 3 presents and discusses the forecasting results of the algorithm for a commercial building with a small microgrid system. Section 4 draws the conclusions.

2 Materials and Methods

2.1 Establish Input Dataset

Data Preprocessing. Failures of supervisory control and data acquisition (SCADA) system cause bad data points. Under normal conditions, SCADA system probably records distorted data due to special events or major errors. The value of load data at a certain time in the data table is null. Under abnormal conditions, the maximum or minimum value is not the actual peak value. The load value at the time is far greater than or less than that of adjacent days [20].

There are probably some noise data in the historical dataset. The data preprocessing is used to find abnormal values of the original dataset and fill missing values. In this paper, the principle named 3σ is adopted to process the load time series. The principle calculates quartiles of the load data respectively, and sets the rational range in acceptable conditions. Equation (1) gives how to set an acceptable value.

$$Q_1 - \beta(Q_3 - Q_1) \sim Q_3 + \beta(Q_3 - Q_1) \tag{1}$$

where Q_1 is the 1st quartile which is the 25th% of the number in the sequence, Q_3 is the 3rd quartile which is the 75th% of the number in the sequence.

Experiments show that if there are more missing values in the dataset, it will lead to lower forecasting accuracy. Therefore, the linear interpolation method and mean interpolation method are used to calculate the average value which is taken to fill in the corresponding missing data. Equation (2) shows the computing procedure.

$$F = g(y, a_1, a_2, \ldots a_m) = a_1 g_1(y) + a_2 g_2(y) + \ldots + a_m g_m(y) \tag{2}$$

where a_i is the weight, y is the load value of adjacent time points, m is the number of adjacent load values.

Wavelet analysis is a signal analysis method based on time domain and frequency domain for non-stationary conditions [21]. It shows great advantages in signal decomposition, reconstruction, and noise separation, etc. It has been widely used in signal processing and image processing. Equation (3) shows the wavelet transformer.

$$WT(\alpha, \tau) = \frac{1}{\sqrt{\alpha}} \int_{-\infty}^{\infty} f(t) * \psi\left(\frac{t - \tau}{\alpha}\right) dt \tag{3}$$

where α is the scale factor which controls the expansion and contraction of the wavelet function, τ is the shift factor which controls the shifting of the wavelet function.

Composition of Input Dataset. The proposed ultra short term load forecasting algorithm based on SVR is to train a model to map related factors to the load with good generalization ability. The performance of the load forecasting model trained by the algorithm fundamentally depends on the quantity and quality of historical data. There are several factors related to the load, such as historical data of load, temperature and weather. The ultra short-term load forecasting takes 15 min as the sample period of load forecasting, therefore the temperature changes slowly, and the weather will not change suddenly in most cases. The load at the prediction time is greatly affected by the load in the adjacent interval. Therefore, the load data at the corresponding time is selected as the input sample. Because the ultra short-term load forecasting is to realize online forecasting, it needs a simple forecasting model. In this paper, only the historical load data is taken as the characteristic input.

The training dataset and testing dataset of three types are constructed respectively for the ultra short-term load forecasting. The three types includes working days, weekends and holidays. The ultra short-term load forecasting predicts the next 4 h load values. The sampling interval is 15 min, so there are 16 elements in each input sample to train the online forecasting model. Each vector of input sample includes the load from 1st to 16th sampling times backtracking from the current time. Equation (4) shows the input vector as follows.

$$x_{i,t} = \left(L_{i,t-16}, L_{i,t-15}, L_{i,t-14}, \cdots, L_{i,t-1}, L_{i,t}\right) \tag{4}$$

where $x_{i,t}$ is input vector which has 16 elements, $L_{i,t}$ is the load at the time t on the day i, $t = 1, 2, ..., 96$.

The Min-Max scaler is used to linearly transform the input vector into the interval 0 to 1. The input vector x is transformed to $\bar{x}$ as shown in Eq. (5).

$$\bar{x} = \frac{x - x_{min}}{x_{max} - x_{min}} \tag{5}$$

where x_{max} and x_{min} are the maximum and minimum values of the input vector respectively.

Similarity Algorithm. Microgrid load is affected by many factors, such as weather, power market, load structure, user consumption level and etc. The load fluctuation and the correlation among load data are not same in different regions and different periods. Therefore, the similarity degree is generally adopted considering the similarity of load sequence and influencing factors. In this paper, the similarity algorithm is used to compute the correlation coefficients between the training sample data set at the current prediction time. Then the correlation coefficients are sorted in descending order. The first k sample data are used for model training to obtain three types of online SVR models for weekdays, weekends and holidays.

Equation (6) computes the correlation coefficient between the sample data set and the load data at the current prediction time.

$$\varepsilon_i(j) = \frac{\min\limits_{i}\min\limits_{j}|x(j) - x_i(j)| + \rho\max\limits_{i}\min\limits_{j}|x(j) - x_i(j)|}{|x(j) - x_i(j)| + \rho\max\limits_{i}\max\limits_{j}|x(j) - x_i(j)|} \tag{6}$$

where $x(j)$ is the jth element of the input sample at the current prediction time t, $x_i(j)$ is the jth element of the last ith sample at the historical same time, $\varepsilon_i(j)$ is the correlation coefficient between $x(j)$ and $x_i(j)$, ρ is the resolution coefficient.

The total correlation coefficient η_i between the input vector x and x_i is as shown in Eq. (7).

$$\eta_i = \prod_{j=1}^{n} \varepsilon_i(j) \qquad (7)$$

where n is the dimension of the input vector, here $n = 16$.

2.2 Ultra Short-Term Load Online Forecasting

Online Cycle Forecasting Strategy. The procedure of online cycle forecasting strategy is shown in Fig. 1.

Step 1 Input historical load data.

Step 2 Divide historical load data into three types: weekdays, weekends and holidays.

Firstly, the historical load data is divided into three types: weekdays, weekends and holidays. Secondly, the missing data is filled by Eq. (2). Thirdly, both the training sample dataset and test sample dataset are processed by the wavelet transformer as shown in Eq. (3).

Step 3 Min-Max normalization.

The 16 time-adjacent elements are selected as the input sample vector, and then the load data in both two sample datasets are normalized by the Min-Max scaler as shown in Eq. (5).

Step 4 Establish the SVR model for ultra short-term load online forecasting.

Step 5 Train SVR model for ultra short-term load online forecasting.

As shown in Eqs. (6) and (7), the correlation coefficients between the training sample dataset are computed at the current prediction time. The first k sample data are used for model training to obtain three types of online SVR models on weekdays, weekends and holidays.

Step 6 Optimize kernel function parameters of SVR model using PSO.

The kernel function of SVR model is radial basis function (RBF), and an improved particle swarm optimization algorithm is used to optimize the width parameters and penalty coefficients of RBF.

Step 7 Predict load values in the next 4 h using cycling strategy.

The cyclic prediction method is used to calculate the load values of next 16 sampling points from the current prediction time, and then the predicted load values are continuously taken into the input sample set. At the current prediction time t, the input sample is as shown in Eq. (8).

$$\begin{aligned}
x_t &= (L_{t-16}, L_{t-15}, L_{t-14}, \cdots, L_{t-1}, L_t) \\
x_{t+1} &= (L_{t-15}, L_{t-14}, L_{t-13}, \cdots, L_t, L_{t+1}) \\
&\vdots \\
x_{t+15} &= (L_{t-1}, L_t, L_{t+1}, \cdots, L_{t+14}, L_{t+15})
\end{aligned} \qquad (8)$$

where $x_t, x_{t+1}, \ldots x_{t+15}$ is the input sample at the current prediction time t, L_t is the load value at time t.

In this paper, if the load samples span days, the following processing method will be adopted. When $t < 17$, the required sample data includes the historical data of the previous day with same day type. When $t > 81$, the required sample data includes the historical data of the next day with same day type. Then load values of the next 16 sampling points including the current prediction time are computed step by step using the method above.

Step 8 Output forecasting load values in the next 4 h.

Optimization of the SVR Kernel Function Based on Improved PSO. In this paper, RBF is the SVR kernel function, and an improved PSO is used to optimize the width parameters and penalty coefficient of RBF. PSO is used widely in nonlinear optimization such as multi-objective optimization. The flight speed and position of each particle are randomly distributed. The particles dynamically adjust their flight velocities and positions according to the global optimum and individual optimum [22–25].

Assume in a D-dimensional search space, there are U particles. The position of particle j is shown in Eq. (9).

$$x_j = [x_{j1}, x_{j2}, \cdots x_{jD}], j = 1, 2, \cdots U \tag{9}$$

where D is the dimension, U is the number of particles.

The velocity of particle j is shown in Eq. (10).

$$v_j = [v_{j1}, v_{j2}, \cdots v_{jD}], j = 1, 2, \cdots U \tag{10}$$

The individual optimum of particle j is shown in Eq. (11).

$$s_j^b = [s_{j1}, s_{j2}, \cdots s_{jD}], j = 1, 2, \cdots U \tag{11}$$

The global optimum of particle j is shown in Eq. (12).

$$s_g^b = [s_{g1}, s_{g2}, \cdots s_{gD}], j = 1, 2, \cdots U \tag{12}$$

After the kth flight of particle j, the velocity and position are updated according to Eq. (13) and Eq. (14) respectively.

$$v_j^{k+1} = \omega v_j^k + c_1 r_1 \left(s_j^b - x_j^k \right) + c_2 r_2 \left(s_g^b - x_j^k \right) \tag{13}$$

$$x_j^{k+1} = x_j^k + v_j^{k+1} \tag{14}$$

where v_j^k an x_j^k are the velocity and position of particle j at the kth flight respectively, c_1 and c_2 are learning factors, ω is inertia weight, r_1 and r_2 are random number between 0 and 1.

The performance of PSO largely depends on the control parameters of the algorithm. This paper designs an improved PSO, which dynamically optimizes three important control parameters including bounds of velocity, learning factor and inertia weight [23–25].

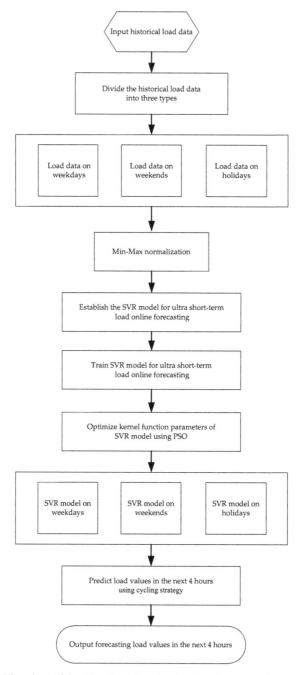

Fig. 1. Flowchart of the ultra short-term load online forecasting based on SVR.

Equation (15) shows that the velocity of particle j is limited in upper and lower bounds.

$$\begin{cases} v_j^{k+1} = v_j^{max}, v_j^{k+1} \geq v_j^{max} \\ v_j^{k+1} = v_j^{max}, v_j^{k+1} \leq -v_j^{max} \end{cases} \quad (15)$$

where $\pm v_j^{max}$ are upper and lower bounds of velocity.

Learning factors are updated linearly by Eq. (16).

$$\begin{aligned} c_1 &= c_1^{max} - \frac{c_1^{max} - c_1^{min}}{k_{max}} \cdot k \\ c_2 &= c_2^{max} - \frac{c_2^{max} - c_2^{min}}{k_{max}} \cdot k \end{aligned} \quad (16)$$

where $c_1{}^{max}$ and $c_1{}^{min}$ are upper and lower bounds of c_1, $c_2{}^{max}$ and $c_2{}^{min}$ are upper and lower bounds of c_2, k_{max} is maximum number of iterations, here $c_1{}^{max} = c_2{}^{max} = 2.5$ and $c_1{}^{min} = c_2{}^{min} = 0.5$.

The inertia weight ω is updated dynamically by Eq. (17) [23, 24].

$$\begin{aligned} d_j &= \frac{1}{U} \sum_{l=1}^{U} \sqrt{\sum_{q=1}^{D} (x_j(q) - x_l(q))^2} \\ E_f &= \frac{d_g - d_{min}}{d_{max} - d_{min}} \\ \omega &= 0.5 E_f + 0.4 \in [0.4, 0.9], \forall E_f \in [0, 1] \end{aligned} \quad (17)$$

where d_j is average distance between particle j and other particles, d_g is average distance between the global optimal particle and other particles, d_{max} and d_{min} are respectively maximum and minimum of average distances among the whole particles, E_f is dynamic factor.

The detail procedure of the improved PSO is as follows.

Step 1 Initialize the population of particles. The population size is U. The current iteration is $k(k = 0)$. The flight velocity and position of each particle are randomly generated within a certain range.

Step 2 Calculate the fitness of all initial particles.

Step 3 Update the individual optima of all particles and the corresponding optimal position.

Step 4 Update the global optima of all particles and the corresponding global optimal position.

Step 5 Update the flight velocity and position of all particles.

Step 6 According to the fitness of the updated particles, update the global optima and individual optimal position of the particle.

Step 7 If $k > k_{max}$, then stop, otherwise return to step 5.

3   Experiments and Results Analysis

3.1   Experimental Datasets and Criteria in the Proposed Ultra Short-Term Load Forecasting

In this paper, the experimental dataset is the load of a commercial building in a small microgrid system from March to May 2019. The historical data from March to April is selected as the training data and that of May is selected as the testing data. The sampling period is 15 min. The prediction interval is next 4 h. The input sample data is the load dada backtracking 4 h from the prediction time. Taking 4 h as the sliding time window, the historical load data corresponding to the prediction interval is selected as the training samples.

The ultra short-term load forecasting is evaluated by two criteria including the mean Absolute Percentage error (MAPE) and the root mean square error (RMSE).

The MAPE is shown in Eq. (18).

$$MAPE(y, \hat{y}) = \frac{1}{n} \sum_{i=1}^{n} \left| \frac{y_i - \hat{y}_i}{y_i} \right| \times 100\% \tag{18}$$

where y_i an $\hat{y}_i$ are real value and predicted value respectively,n is sampling number.

The RMSE is shown in Eq. (19).

$$RMSE(y, \hat{y}) = \sqrt{\frac{1}{n} \sum_{i=1}^{n} (y_i - \hat{y}_i)^2} \tag{19}$$

3.2   Experimental Results

The performance of the designed algorithm is verified by the ultra short-term forecast load from May 1 to May 12, 2019. May 1 to May 4 are legal holidays (see Fig. 4), May 5 to May 10 are working days (see Fig. 2), and May 11 to May 12 are weekends (see Fig. 3). The interval of updating the proposed algorithm adopts two strategies. Strategy 1 computes the prediction results every 15 min at the current time. Strategy 2 computes the prediction results every 1 h at the current time.

3.3   Results Analysis

The results given in Table 1, 2 and 3 show that strategy 1 is better than strategy 2 for the prediction of three types of working days, weekends and holidays. On working days, as shown as Table1, the MAPE equals 6.14% and the RMSE equals 1.94 Kw using strategy 1, the MAPE equals 9.13% and the RMSE equals 2.93 Kw using strategy 2. The MAPE values of working days are significantly better than those on weekends and holidays. One reason is that there are many working day data in the training data set, and the accuracy of the trained model is higher. The other reason is that the power consumption is relatively stable and the value is large, therefore the relative deviation of the prediction results is small (see Fig. 4).

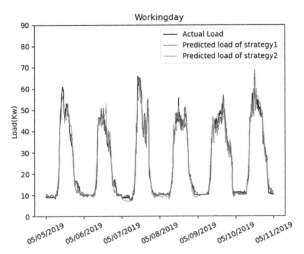

Fig. 2. Results of working days using the proposed algorithm with strategy 1 and strategy 2

Table 1. MAPE and RMSE of working days using the proposed algorithm with strategy 1 and strategy 2.

Criteria	MAPE(%)	RMSE(Kw)
Strategy 1	6.14	1.94
Strategy 2	9.13	2.93

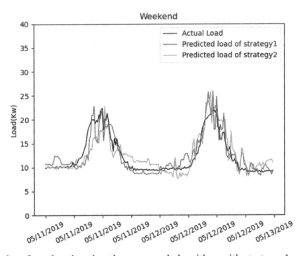

Fig. 3. Results of weekends using the proposed algorithm with strategy 1 and strategy 2

In the case of weekends, as shown as Table 2, the MAPE equals 10.66% and the RMSE equals 1.76 Kw using strategy 1, the MAPE equals 13.56% and the RMSE equals

Table 2. MAPE and RMSE of weekends using the proposed algorithm with strategy 1 and strategy 2.

Criteria	MAPE(%)	RMSE(Kw)
Strategy 1	10.66	1.76
Strategy 2	13.56	2.17

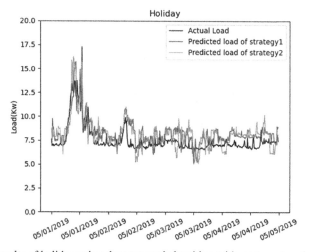

Fig. 4. Results of holiday using the proposed algorithm with strategy 1 and strategy 2

Table 3. MAPE and RMSE of holiday using the proposed algorithm with strategy 1 and strategy 2.

Criteria	MAPE(%)	RMSE(Kw)
Strategy 1	12.78	1.12
Strategy 2	14.61	1.26

2.17 Kw using strategy 2. The MAPE increases compared with the working days. The reason is that some employees may work overtime on weekends, and the overtime is temporary, so that the randomness of the historical load data is very high. The trained model cannot fully map the characteristics of weekends, and the algorithm will continue to be improved.

In the case of holidays, as shown as Table 3, the MAPE equals 12.78% and the RMSE equals 1.12 Kw using strategy 1, the MAPE equals 14.61% and the RMSE equals 1.26 Kw using strategy 2. Compared with working days and weekends, the MAPE is worse. Because there are few historical data of holidays to be trained into a prediction model that accurately maps the characteristics of holidays. In the future, training data sets of holidays will be increased to improve the accuracy of prediction.

4   Discussion

A novel ultra short-term load forecasting algorithm of a small commercial microgrid based on SVR is proposed and discussed in this paper. Weather factors are no longer considered separately as input to train the forecasting model because the influence of them has been reflected in the historical data of load. The power user is a commercial building, and the power consumption is significantly different between workdays and weekends or holidays. Therefore, the historical load data is divided into three categories including working days, weekends and holidays, and the three types of prediction model based on SVR are trained respectively. The proposed algorithm is updating load forecasting every 15 min or 1 h using the online cyclic trained model based on SVR. The correlation coefficients between the training dataset and the output are computed according to the similarity method at current prediction time. Then the correlation values are sorted in descending order. The first kth samples are selected to train the online forecasting models for three types of working days, weekends and holidays. The prediction accuracy is evaluated by MAPE and RMSE. The testing results on cases show that the performance of updating every 15 min is better than that of updating every 1 h. The results of working days are apparently better than that of weekends and holidays. The randomness of the historical load data is high in case of weekends, so the proposed algorithm will be improved in this situation. In the case of holidays, lack of historical load data causes that the trained model cannot map the characteristics of holidays accurately. The performance of holidays trained model will be enhanced by collecting more historical load data. In the future, the proposed algorithm will be applied to a large microgrid energy management system.

References

1. Goia, A., May, C., Fusai, G.: Functional clustering and linear regression for peak load forecasting. Int. J. Forecast. **26**(4), 700–711 (2010)
2. Ramanathan, R., Engle, R., Granger, C.W., Vahid-Araghi, F., Brace, C.: Short-run forecasts of electricity loads and peaks. Int. J. Forecast. **13**(2), 161–174 (1997)
3. Amral, N., Ozveren, C., King, D.: Short term load forecasting using multiple linear regression. In: UPEC 2007. 42nd International Universities Power Engineering Conference, Brighton, UK, pp.4–6 (2007)
4. Pappas, S., Ekonomou, L., Karamousantas, D., Chatzarakis, G., Katsikas, S., Liatsis, P.: Electricity demand loads modeling using autoregressive moving average(ARMA) models. Energy **33**(9), 1353–1360 (2008)
5. Lee, C.M., Ko, C.N.: Short-term load forecasting using lifting scheme and ARIMA models. Expert Syst. Appl. **38**(5), 5902–5911 (2011)
6. Cho, M., Hwang, J., Chen, C.S.: Customer short term load forecasting by using ARIMA transfer function model. In: Proceedings of 1995 International Conference on Energy Management and Power Delivery EMPD '95, Singapore, pp. 21–23 (1995)
7. Taylor, J.W.: An evaluation of methods for very short-term load forecasting using minute-by-minute British data. Int. J. Forecast. **24**(4), 645–658 (2008)
8. Al-Hamadi, H., Soliman, S.: Short-term electric load forecasting based on Kalman filtering algorithm with moving window weather and load model. Electr. Pow. Syst. Res. **68**(1), 47–59 (2004)

9. Ghayekhlooa, M., Menhaj, M.B., Ghofrani, M.: A hybrid short-term load forecasting with a new data preprocessing framework. Electr. Power Syst. Res. **119**, 138–148 (2015)
10. Zhai, M.-Y.: A new method for short-term load forecasting based on fractal interpretation and wavelet analysis. Int. J. Electr. Power Energy Syst. **69**, 241–245 (2015)
11. Li, S., Wang, P., Goel, L.: Short-term load forecasting by wavelet transform and evolutionary extreme learning machine. Electr. Power Syst. Res. **122**, 96–103 (2015)
12. Chen, Y.H., Yang, Y., Liu, C.Q., Li, C.H., Li, L.: A hybrid application algorithm based on the support vector machine and artificial intelligence: an example of electric load forecasting. Appl. Math. Model. **39**, 2617–2632 (2015)
13. Moon, J., Kim, Y., Son, M., Hwang, E.: Hybrid short-term load forecasting scheme using random forest and multilayer perceptron. Energies **11**, 3283 (2018). https://doi.org/10.3390/en11123283
14. Lahouar, A., Slama, J.B.H.: Day-ahead load forecast using random forest and expert input selection. Energy Convers. Manage. **103**, 1040–1051 (2015)
15. Arvanitidis, A.I., Bargiotas, D., Daskalopulu, A., Laitsos, V.M., Tsoukalas, L.H.: Enhanced short-term load forecasting using artificial neural networks. Energies **14**, 7788 (2021). https://doi.org/10.3390/en14227788
16. Wang, Y.Z., Zhang, N.Q., Chen, X.: A short-term residential load forecasting model based on LSTM recurrent neural network considering weather features. Energies **14**, 2737 (2021). https://doi.org/10.3390/en14102737
17. Machado, E., Pinto, T., Guedes, V., Morais, H.: Electrical load demand forecasting using feed-forward neural networks. Energies **14**, 7644 (2021). https://doi.org/10.3390/en14227644
18. Zhang, D.Y., Tong, H.X., Li, F., Xiang, L.Y., Ding, X.L.: An ultra-short-term electrical load forecasting method based on temperature-factor-weight and LSTM model. Energies **13**, 4875 (2020). https://doi.org/10.3390/en13184875
19. Tan, M., Yuan, S.P., Li, S.H., Su, Y.X., Li, H., He, F.: Ultra-short-term industrial power demand forecasting using LSTM based hybrid ensemble learning. IEEE Trans. Power Syst. **35**(4), 2937–2948 (2020)
20. Guan, C., Luh, P.B., Michel, L.D., Wang, Y., Friedland, P.B.: Very short-term load forecasting: wavelet neural networks with data pre-filtering. IEEE Trans. Power Syst. **28**(1), 30–41 (2013)
21. Morlet J.: Wave propagation and sampling theory. Geophysics (1982)
22. Kennedy, J., Eberhart, R.: Particle swarm optimization. In: Proceedings of ICNN'95 - International Conference on Neural Networks, Perth, WA, Australia, vol. 4, pp. 1942–1948 (1995)
23. Zhan, Z., Zhang, J., Li, Y., Chung, H.: Adaptive particle swarm optimization. IEEE Trans. Syst. Man. Cybern. B **39**(6), 1362–1381 (2018)
24. Tang, Y., Wang, Z., Fang, J.: Parameters identification of unknown delayed genetic regulatory networks by a switching particles swarm optimization algorithm. Expert Syst. Appl. **38**(6), 2523–2535 (2011)
25. Clerc, M., Kennedy, J.: The particle swarm: explosion, stability, and convergence in a multi-dimensional complex space. IEEE Trans. Evol. Comput. **6**(1), 58–73 (2002)

Laser Welding Process of Lithium Battery Lugs Based on Finite Element Simulation

Tianpeng Ren[1(✉)] and Yanbing Guo[2]

[1] School of Mechanical Engineering, Shanghai Dianji University, Shanghai 201306, China
1282288154@qq.com
[2] College of Ocean Science and Engineering, Institute of Marine Materials Science and Engineering, Shanghai Maritime University, Shanghai 201306, China

Abstract. To investigate the application of laser welding in the production of lithium battery modules for electric vehicles, this study employs the finite element method to simulate the welding process of lugs and busbars in lithium batteries under different parameters. The objective is to analyze the temperature change in the lugs, and its influence on the depth and width of the welds. The temperature field is simulated with varying welding heat inputs to examine the distribution of stress during welding and residual stress patterns in the weldments, as well as the deflection of the weldments under different welding parameters. The analysis of stress field simulations allows for the assessment of residual stress distribution and weldment deflection. The findings indicate that the width of the molten pool is significantly affected by the laser power and welding speed. Moreover, the laser power has a greater influence on the residual stress of the weldment compared to the welding speed.

Keywords: laser welding · finite element method · temperature field · stress field

1 Introduction

Power battery is one of the core components of electric vehicles and the power source of electric vehicles. Lithium-ion batteries are applied in the field of electric vehicles with the advantages of high single voltage and high energy density, and have gradually gained market recognition [1]. Currently, the battery module is composed of a large number of single batteries through series and parallel connection, heat insulation plate and shell and other parts [2], single batteries in series and parallel connection need to be welded to the pole lugs and busbar. In the traditional welding method, it will produce welding defects such as false welding, welding through, excessive deflection of the welded parts, etc. [3, 4], once the above defects occur, the whole battery pack will fail, which will cause huge economic losses, so the quality of lithium battery lug welding directly affects the use of the whole battery pack.

National Natural Science Foundation of China (51975346)

J. Li et al. (Eds.): 6GN 2023, LNICST 553, pp. 257–266, 2024.
https://doi.org/10.1007/978-3-031-53401-0_23

The materials used for lithium battery lugs are generally ultra-thin and easy to deform, and the laser welding method is very suitable for welding these materials. Laser welding has the advantages of small heat-affected zone and deflection, high energy density, high welding accuracy, etc., and good weld seam can be obtained by modifying the welding process parameters of laser power or welding speed [5–11]. Therefore, the paper adopts a reasonable heat source model, through the simulation results of the temperature field and residual stress field under the laser heat source analysis, on the one hand, the study of different laser welding process parameters on the electrode welding weld depth and width of the effect, and on the other hand, the study of different laser welding process parameters on the electrode welding residual stress distribution and weldment deflection, to provide a basis for the application of laser welding in the actual production of lithium batteries. The second is to study the effect of different laser welding process parameters on the residual stress distribution of lug welding and the deflection of the weldment, in order to provide a basis for the application of laser welding in the actual production of lithium battery.

2   Laser Welding Materials

The laser welding of the lugs refers to the laser overlay welding of the lugs and the busbar, in which there are two types of materials for the lugs, namely 1060 aluminium alloy and TU1 oxygen-free copper, and the material for the busbar is 6061 aluminium alloy. The main chemical compositions of the materials are shown in Table 1. From the chemical composition of the materials, the corresponding alloy type is selected in Jmatpro software, the content of each of its chemical compositions is entered, and then the thermal-physical property parameter calculation module is selected, so that the required thermal-physical property parameters can be obtained. As shown in Fig. 1, the thermal conductivity and specific heat capacity of the three materials are plotted against temperature.

Table 1. Chemical composition of 1060 aluminium alloy, 6061 aluminium alloy and tu1 oxygen-free copper (mass fraction, %)

	Al	Si	Cu	Mg	Zn	Mn	Ti	V	Fe			
1060 aluminium alloy	margin	≤ 0.25	≤ 0.05	≤ 0.05	≤ 0.05	≤ 0.05	≤ 0.03	≤ 0.05	0~0.4			
	Cu	As	Sb	Fe	Ni	Pb	Sn	S	P	Bi	Zn	O
Tu1 oxygen-free copper	margin	0.002	0.002	0.004	0.002	0.003	0.002	0.004	0.002	0.001	0.003	0.002

(*continued*)

Table 1. (*continued*)

	Al	Si	Cu	Mg	Zn	Mn	Ti	V	Fe			
	Al	Si	Cu	Mg	Zn	Mn	Ti	Cr	Fe			
6061 aluminium alloy	margin	0.4~0.8	0.15~0.4	0.8~1.2	0.25	0.15	0.15	0.04~0.35	≤0.7			

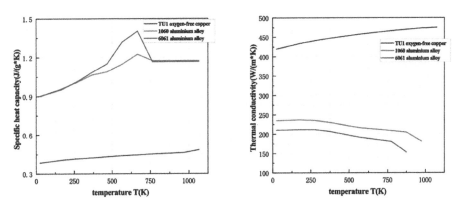

Fig. 1. Thermo-physical parameters of the three weldment materials

3 Auricle Laser Welding Temperature Field Simulation

3.1 Boundary Conditions and Heat Source Selection for Temperature Field Simulation

The welding temperature analysis of lithium battery electrode lugs for electric vehicles is a nonlinear transient thermal analysis, and the initial conditions and boundary conditions need to be set before solving the temperature field results [12]. The initial condition is the initial temperature of the lithium battery lugs and busbar, set to room temperature 22 °C. The boundary condition is the convective heat transfer between the surface of the lugs and the busbar in the welding process, and the surface convective heat transfer coefficients between the lugs and the busbar and the surrounding environment are calculated as shown in Table 2.

It is crucial to choose a suitable heat source model to simulate the heating effect of the laser on the weldment. Considering that the welding mode of the lugs and the busbar is stacked welding and the thickness of the lugs is only 0.3 mm, which is small, as well as the selected heat source model should be able to simulate the heating effect of the surface molten pool on the weldment, the final choice of the planar Gaussian heat source is shown in Fig. 2. The calculation formula of the planar Gaussian heat source model is:

$$q(r) = \frac{3Q\eta}{\pi R^2}exp\left(-\frac{3r^2}{R^2}\right) \tag{1}$$

Table 2. Convective heat transfer coefficients for different temperatures

Temperature T/°C	Convective heat transfer coefficient h (W/m2·°C)
100	5.76
200	7.25
300	8.3
500	9.84
700	11.01
1000	12.4

In the equation:

$q(r)$- the heat flux at the center r of the heat source; Q-the laser energy at the laser emission; η-the laser absorption by the material of the weldment; R-the radius of the heat source; r-distance from the centre of the heat source.

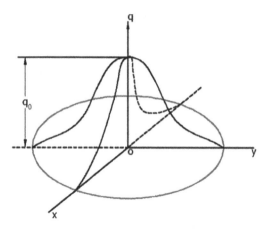

Fig. 2. Red laser heat source model

The actual structure of EV lithium battery lugs and busbar is relatively simple, so 1:1 to establish the welding geometry model of lugs and busbar as shown in Fig. 3. Then according to the distance from the weld position of the mesh divided into different density, the establishment of the lug welding finite element model, as shown in Fig. 4.

When performing laser welding temperature field simulations, the following assumptions are made to simplify the calculations within the allowable error range [13]:

(1) Compensation of the effect of convection in the molten pool on the temperature field by an equivalent heat transfer coefficient;
(2) Absorption of laser light by lithium battery lugs and sinks does not vary with time;
(3) Laser welding is performed at a constant speed, not counting the acceleration and deceleration phases.

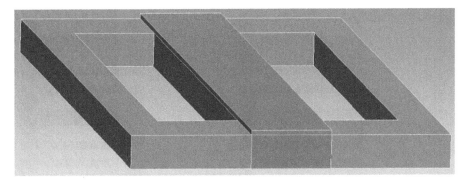

Fig. 3. Geometric model of laser welding of lugs and busbar

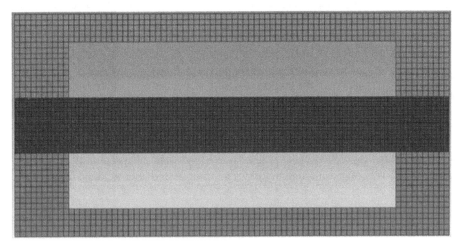

Fig. 4. Laser welded finite element model of pole lugs and busbar

3.2 Temperature Field Cloud Diagram Analysis

The temperature field of the pole lug welding is depicted in Fig. 5 when the welding power is 1600W and the welding speed is 40 mm/s. At t = 0.001 s, the spot welding of the starting point of welding is completed. The spot position reaches the center point of the weld at t = 0.375 s, followed by the completion of laser continuous welding for the entire weldment at t = 0.75 s, with the maximum temperature reaching 2743.9 °C. After 5 s of welding, the weldment cools rapidly, causing the the maximum temperature to decrease to 110.2 °C. After 50 s of welding, the maximum temperature further reduces to 101.71 °C, indicating a significant decrease in the rate of temperature drop. Finally, after 100 s of welding, the maximum temperature drops to 98.378 °C, resulting in a more uniform temperature distribution throughout the entire weldment.

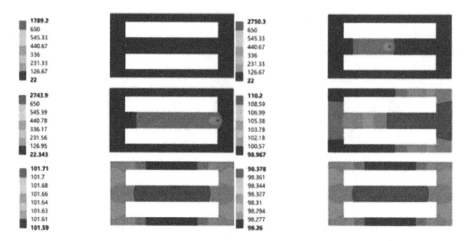

Fig. 5. Cloud view of the welding temperature field of the pole lug at different moments

3.3 Effect of Laser Welding Parameters on Weld Depth and Width

Weld depth and width are one of the important standards to test the welding quality, and they also have a large impact on weld shaping and its mechanical properties. Figure 6(a) and (b) show the effects of laser power and welding speed on the depth and width of fusion of aluminium-aluminium welded joints under red laser welding, respectively; Fig. 7(a) and (b) show the effects of laser power and welding speed on the depth and width of fusion of copper-aluminium welded joints under red laser welding, respectively.

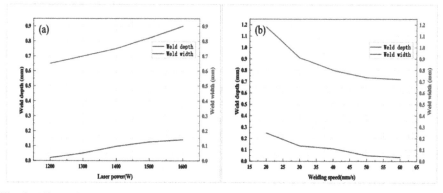

Fig. 6. Effect of (a) laser power and (b) welding speed on weld depth and width of welds in aluminium-aluminium welded joints

It is important to note that the welding parameters were designed using a single-variable method. Therefore, the effect of laser power on both the weld depth and weld width of the weld was investigated with a welding power of 40 mm/s. Similarly, the impact of welding speed on the weld depth and weld width of the weld was examined by controlling the laser power at 1450 W and 2750 W, respectively.

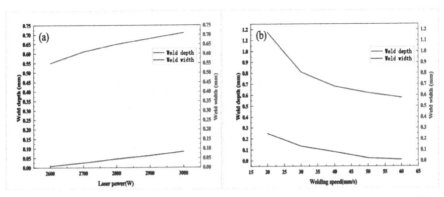

Fig. 7. Effect of (a) laser power and (b) welding speed on weld depth and width of welds in copper-aluminium welded joints

From Fig. 6(a) and 7(a) demonstrate that at a constant welding speed of 40 mm/s, increasing the laser power results in an increasing trend for both the weld width and weld depth of the weld seam in both aluminium-aluminium and copper-aluminium welded joints. Additionally, the increase in weld width is significantly greater than the increase in melt depth. From Fig. 6(b) and 7(b) reveal a decreasing trend in the weld width and weld depth of the weld seam in both aluminium-aluminium and copper-aluminium welded joints when the laser power is held constant. This trend occurs because as the welding speed increases, the time spent in the laser unit area decreases, reducing the heat absorbed by the weld and resulting in a decrease in both the weld depth and weld width of the weld seam.

4 Stress Field Simulation for Laser Welding of Pole Lugs

4.1 Stress Distribution During the Welding Process

Figure 8 shows the distribution of longitudinal stress along the welding direction at the points on the centre line of the weld when the welding time is 0.1875 s, 0.375 s, 0.5625 s, and 0.75 s, respectively, and the longitudinal stress are the stress along the welding direction. From the figure, it can be seen that the front of the molten pool is subjected to a large compressive stress, and the compressive stress in the compressive stress region close to the molten pool is gradually converted to tensile stress, while the tail of the molten pool is a tensile stress region.

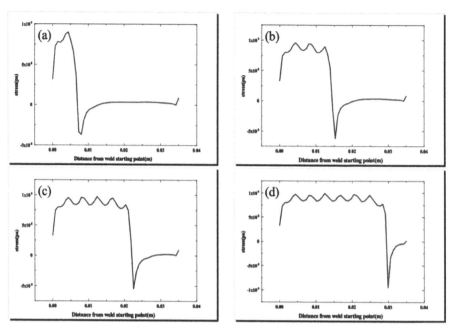

Fig. 8. (a) Welding 0.1875s (b) Welding 0.375s (c) Welding 0.5625s (d) Welding 0. 75s
Longitudinal stress variation curve

4.2 Influence of Laser Welding Parameters on the Residual Stress in the Weldment and the Amount of Deflection of the Weldment

From Figs. 9(a)(b) and 10(a)(b), it is evident that as the control welding speed increases to 40 mm/s, the aluminum-aluminum and copper-aluminum weldments show an increasing trend in residual stress and deflection with the rise in laser power. This is due to the fact that higher laser power leads to increased heat input during welding, resulting in a concentration of stress in the weld. Moreover, laser welding is a rapid cooling and heating process, which prevents the release of stress in the weld region. Conversely, in Figs. 9(c)(d) and 10(c)(d), with a constant laser power, an incremental increase in welding speed shows a downward trend in residual stress and deflection in the aluminum-aluminum and copper-aluminum weldments. This is because higher welding speeds reduce the amount of energy applied to each unit area, thereby reducing overall thermal stress in the weldments and resulting in a decrease in residual stress. Comparing the effect of changes in laser power and welding speed on the magnitude of residual stress and deflection, it is evident that laser power has a greater impact. Therefore, in the welding process, it is crucial to select an appropriate welding power to mitigate stress concentration and excessive deflection.

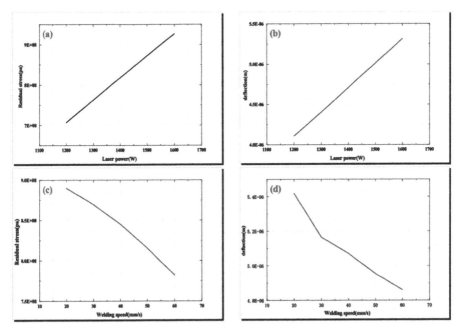

Fig. 9. Effect of laser power (a), (b) and welding speed (c), (d) on residual stress as well as deflection in aluminium-aluminium welded joints

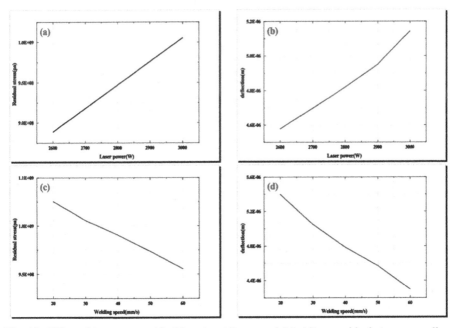

Fig. 10. Effect of laser power (a), (b) and welding speed (c), (d) on residual stress as well as deflection in copper-aluminium welded joints

5 Conclusion

1. The heat during the laser welding of lithium battery lugs is distributed centrally within the weld region, resulting in a significant temperature gradient in front of the molten pool and a smaller gradient at the rear. During the cooling process after welding, the temperature decreases rapidly within 5 s. Subsequently, the heat is gradually conducted away from the weld region towards the lower sink, causing an inconspicuous increase in sink temperature.
2. The laser power and welding speed significantly affect the weld depth and width of the weld during the process of laser welding lithium battery lugs. Therefore, it is crucial to select appropriate welding parameters to prevent over-welding or under-welding.
3. The selection of suitable welding parameters, such as laser power and welding speed, is crucial in minimizing residual stress in the weldment.

References

1. Xu, D., Zhao, X.: An analysis of lithium-ion power battery module design. Power Supply World (2), 26–29 (2017)
2. Xu, D.: Design and Mechanical Characterisation of Lithium-ion Battery Module. Hunan University, Changsha (2018)
3. Zeng, T., Chao, Y., Zou, L., et al.: Study on the organisation and properties of 5A06 aluminium alloy active TIG welded head. Weapon Mater. Sci. Eng. **41**(04), 82–86 (2018)
4. Ning, J., Zhang, X., Zhang, L.: Research progress of laser application in purple copper welding. Laser J. **36**(03),1–6 (2015)
5. Bai, J.: Research on laser welding process of new energy vehicle battery busbar. Changchun University of Science and Technology (2017)
6. Zhang, Y.: Discussion on the development and prospect of laser welding technology. Sci. Technol. Innov. **22**, 180–181 (2019)
7. Zhang, Z., Wang, R., Gou, G., Chen, H., et al.: Droplet transfer behaviour of narrow gap laser wire filling welding. Int. J. Mod. Phys. B **33**(1n03), 1940045 (2019)
8. Yuan, R., Deng, S., Cui, H., et al.: Interface characterization and mechanical properties of dual beam laser welding-brazing Al/steel dissimilar metals. J. Manuf. Process. **40**, 37–45 (2019)
9. Mei, J.G.: Research on pulsed laser welding of T2 copper with 1060 aluminium. Zhongyuan Institute of Technology (2018)
10. Chen, C.: Aluminium/steel laser welded head organisation, mechanical properties and calculation of interfacial compounds. Nanchang University (2018)
11. Zou, Y.: Research on pulsed laser welding process of ultra-thin T2 copper/6061 aluminium alloy. Huazhong University of Science and Technology (2019)
12. Bachmann, M., Avilov, V., Gumenyuk, A., et al.: Numerical simulation of full-penetration laser beam welding of thick aluminium plates with inductive support. J. Phys. D Appl. Phys. **45**(3), 035201 (2012)
13. Geng, L., Yang, Y., Fu, H.: Laser welding process of explosion-proof valve based on temperature field simulation. Welding **531**(09), 18–21+68–69 (2017)

Failure and Stress Analysis of Cylindrical Springs

Wei Liao[1], Mingze Sun[2(✉)], and Yanbing Guo[3(✉)]

[1] State Grid Shanghai Electric Power Research Institute, Shanghai 200437, China
liaowei@sh.sgcc.com.cn

[2] School of Mechanical Engineering, Shanghai Dianji University, Shanghai 200437, China
1421500755@qq.com

[3] College of Ocean Science and Engineering, Institute of Marine Materials Science and Engineering, Shanghai Maritime University, Shanghai 201306, China
yanbingg1984@126.com

Abstract. This paper presents the causes of fractures in the column head springs of switchgear. By using fracture spectral material detection, macro-morphological analysis, electron microscope scanning analysis, fracture energy spectrum elemental analysis, the fracture characteristics of the spring and the chemical composition of its fracture material were determined, and it was found that the material of the spring at the fracture was chromium-manganese-type austenitic steel, and that the content of manganese at the fracture was excessive, which reduced the plasticity of the steel and ultimately led to the occurrence of fatigue fracture. In addition, in the above environment, the finite element model of the spring was established, and the stress analysis of the spring was carried out to provide a basis for further optimization of the spring design.

Keywords: Spring · Finite Element Method · Fracture

1 Introduction

Cylindrical springs are an indispensable component of substation systems, playing a crucial role in ensuring the safety, efficient operation, and stability of power systems [1]. With their exceptional attributes, such as high reliability, quick responsiveness, high load-carrying capacity, and precise control capability, cylindrical springs are frequently utilized for the opening and closing of circuit breakers and isolation devices. They are employed to isolate faults or overloaded portions, mitigate circuit fluctuations, and protect equipment from damage [2]. Cylindrical springs also find applications in motor drive devices, serving as auxiliary forces to provide the necessary energy for starting or operating motors. In load switch systems, cylindrical springs offer reliable driving forces, enabling rapid switching when required. Moreover, cylindrical springs play significant roles in power control, elastic support, and mechanical interlocks [3].

Technology Projects of State Grid Shanghai Municipal Electric Power Company (SGTYHT/21-JS-223).

J. Li et al. (Eds.): 6GN 2023, LNICST 553, pp. 267–276, 2024.
https://doi.org/10.1007/978-3-031-53401-0_24

The fatigue level and durability of springs determine the operating lifespan of sub-station systems. Most springs are prone to fracture due to fatigue, but the reasons for their failures vary [4]. The quality of raw materials and irregular designs are the primary factors influencing their service life [5]. Internal micro-cracks and inclusions in the raw materials tend to cause stress concentration [6]. Surface indentations and machine processing can increase surface roughness, potentially leading to crack initiation [7]. Additionally, improper heat treatment is a significant cause of premature spring failure [8].

Therefore, in order to ensure the durability of springs and reduce the cost of wear and tear, this paper takes the cylindrical springs that have been broken as the target of research, analyzes the causes of their breakage and stress tests, and identifies the main factors affecting the plasticity and hardness of the springs, which provides a basis for future research discussions.

2   Spectral Material Testing

To determine the material of the spring, a portable X-ray fluorescence spectrometer was used for semi-quantitative spectral material analysis of the 4th spring on the upper column of Fe12 C phase and 19 other springs. The detection results are shown in Table 1.

Table 1. Spring Spectral Material Testing Results (Mass Fraction: %)

Converting station	Cr	Mn	Ni	Fe	Cu	Mo	V	Deformation situation
Hua Tie station	10.44	18.85	0.126	70.22	0.214	/	0.059	
	17.86	9.44	5.15	65.54	0.939	0.175	/	
	18.05	9.42	5.21	66.03	0.589	0.142	0.071	
	17.92	9.62	5.26	65.57	0.941	0.136	0.09	
	17.82	9.99	5.38	64.9	1.11	0.146	/	
Yin Ze station	17.54	9.58	4.3	67.51	0.442	0.133	0.073	
	17.30	9.46	5.34	66.23	0.276	0.247	/	
Feng Shun station	17.41	9.67	5.31	66.58	0.388	0.098	0.087	deformity
	17.55	10.04	5.36	66.18	0.398	0.1	0.108	deformity
Guan Sheng Yuan station	17.51	9.51	5.52	66.32	0.426	0.16	0.104	
	17.26	9.59	5.35	66.71	0.385	0.108	0.077	deformity
Jin Ge station	17.88	9.68	5.36	66.27	0.423	0.1	0.094	
	17.81	9.58	5.3	66.06	0.418	0.106	/	

(*continued*)

Table 1. (*continued*)

Converting station	Cr	Mn	Ni	Fe	Cu	Mo	V	Deformation situation
	17.78	9.6	5.07	65.95	0.245	0.139	/	
	17.98	9.69	5.17	65.66	0.384	0.142	/	
Nan Hua Nan Ling company	17.35	10.0	5.45	66.42	0.196	0.095	0.082	
	17.39	9.77	5.3	65.42	0.247	/	/	
	17.61	9.78	5.34	65.7	0.222	0.161	/	
	17.22	9.95	5.42	66.42	0.347	0.11	0.141	

By comparing the Metal Materials Handbook and GB/T 24588–2009 "Stainless Spring Steel Wire," it can be observed that the fractured 4th spring on the upper column of the Fe12 C phase has a steel grade close to 1Mn18Cr10MoVB chromium-manganese austenitic steel, not belonging to stainless steel. The other 18 spring steel grades are 12Cr18Mn9Ni5N stainless steel. The composition requirements of various steel grades are provided in Table 2 in the metal handbook or national standard.

Table 2. Chemical Composition Standards for Various Steel Grades

Grades	Cr	Mn	Ni	C	Si	Mo	V
1Mn18Cr10MoVB	9.5–11.5	17.0–19.0		0.12–0.17	0.3–0.7	0.4–0.6	0.7–0.9
12Cr18Mn9Ni5N	17.0–19.0	7.5–10.0	4.0–6.0	≤0.15	≤1		
06Cr19Ni10	18.0–20.0	≤2	8.0–10.5	≤0.08	≤1		

3 Macroscopic Fracture Analysis

In order to determine the degree of fracture on the spring's broken surface, a macroscopic analysis of the fracture was conducted, and the macroscopic morphology of the fracture is shown in Fig. 1.

As shown in the figure, the fracture is located at the corner position of the spring's end, precisely at the transitional region where stress is most concentrated. The fracture surface exhibits an oblique cleavage at a 45-degree angle to the cross-section, which is consistent with the stress characteristics experienced by the spring. Therefore, it can be judged that the part of the spring where the force is concentrated excessively is the location of the fracture, showing the appearance of fatigue fracture.

4 Electron Microscopy Analysis

The microstructure of the fracture was examined using a scanning electron microscope, and the micrograph of the fracture is displayed in Fig. 2.

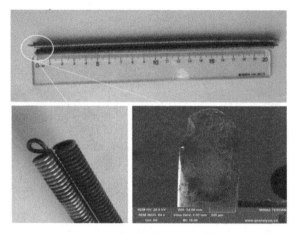

Fig. 1. Macroscopic Morphology of the Fractured Spring

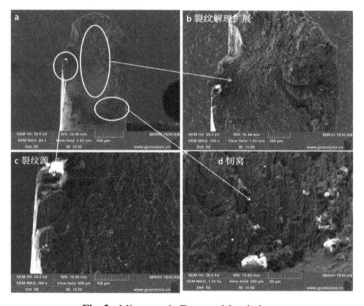

Fig. 2. Microscopic Fracture Morphology

As shown in the figure, the cause of the spring fracture is fatigue fracture, based on three points: First, the crack originates from the middle of the far left (as shown in Fig. 2c), with a large number of fatigue striations parallel to the outer surface of the spring, and microcracks propagate inward perpendicular to the fatigue striations. Second, the crack growth lines develop with the crack origin as the center (as shown in Fig. 2b). After the fatigue crack initiates, it subsequently propagates to the right in a cleavage-like manner. Third, in the lower right corner of the fracture surface, a dimple feature was found (as shown in Fig. 2d), while no similar dimple feature was observed in other regions. This dimple feature can only be generated in the final instantaneous fracture zone, where the remaining cross-section is under plane stress conditions with relatively less plastic deformation constraint, thus exhibiting a microvoid coalescence dimple feature. Subsequently, when magnifying the crack region, all show transgranular fracture characteristics, as shown in Fig. 3.

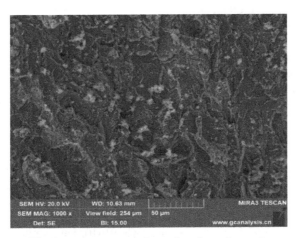

Fig. 3. Fracture morphology of perforated crystals

5   Energy Spectrum Elemental Analysis

Energy spectrum elemental analysis results of the outer surface near the crack source are shown in Fig. 4.

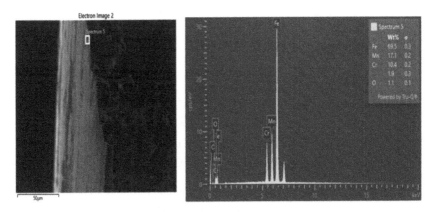

Fig. 4. Elemental analysis of the outer surface near the crack source

In order to figure out the composition of the elements on the outer surface near the crack source, the fracture parts were analyzed by energy spectrum elements, and no abnormal inclusions were found. As can be seen from Fig. 4 left, no obvious external damage was found on the outer surface of the spring; Fig. 4 right energy spectrum elemental analysis results are basically consistent with the spectral material analysis results, and no abnormal elemental composition was found.

6 Analysis and Discussion

Manganese is a weak deoxidizer in steel. An appropriate amount of manganese can effectively increase the strength of the steel, eliminate the harmful effects of sulfur and oxygen on the steel's hot brittleness, improve the hot workability of the steel, and reduce its susceptibility to cold brittleness, while not significantly decreasing the plasticity and impact toughness of the steel. The manganese content in ordinary carbon steel is typically around 0.3% to 0.8%. However, if the Mn content is too high (reaching 1.0% to 1.5% or higher), it will make the steel brittle and hard, reducing the fatigue resistance of the steel.

For comparison, the mechanical property indexes of the above three grades of steel were queried in the metal materials handbook, as shown in Table 3. The data in Table 3 are obtained from the steel samples after solid solution treatment and before undergoing cold working. It can be observed that the tensile strength of 1Mn18Cr10MoVB is significantly higher than that of 12Cr18Mn9Ni5N and 06Cr19Ni10, but its plasticity (elongation after fracture, reduction of area) is notably lower than that of 12Cr18Mn9Ni5N and 06Cr19Ni10.

Table 3. Mechanical Performance Indicators of Three Steel Grades

Grades	Specified Plastic Elongation Strength / MPa	Tensile Strength / MPa	Elongation at Break / %	Percentage of Sectional Contraction / %	Hardness / HRB
1Mn18Cr10MoVB	245	785	30	35	90–95
12Cr18Mn9Ni5N	275	520	40	45	100
06Cr19Ni10	205	520	40	60	90

Furthermore, relevant performance data of the three steel grades was investigated. However, no relevant information could be found for 12Cr18Mn9Ni5N. A comparison of the fatigue performance is presented in Table 4. It can be observed that the fatigue limit of 12Cr18Mn9Ni5N is significantly lower than that of 06Cr19Ni10.

In summary, the material of the 4th column head spring on the iron 12 C phase, which fractured abnormally before, is a type of chromium-manganese austenitic steel. The grade is similar to 1Mn18Cr10MoVB chromium-manganese austenitic steel. The elevated level of manganese (Mn) in the steel can lead to decreased plasticity, resulting in increased hardness and brittleness, thus reducing its fatigue performance.

Table 4. Fatigue Performance Comparison of Two Steel Grades

Grades	Fatigue Limit
1Mn18Cr10MoVB	After a 16-h exposure to 1150 °C followed by 800 °C, the high-temperature fatigue limit at 620 °C (N = 9.7 x 10^7) is 195 MPa
06Cr19Ni10	After solution treatment at 1080 °C–1130 °C, the fatigue limit represented by the specified plastic elongation strength is ≥ 900 MPa, and the fatigue limit represented by the tensile strength is ≥ 1100 MPa

From the appearance of the fracture surface, the spring fracture mode belongs to fatigue failure. No evidence of external force damage or abnormal elements was found. Fatigue cracks first initiate on the outer surface of the spring, leaving numerous fatigue striations at the initiation site. Subsequently, the cracks propagate inward and eventually lead to a micro-pore accumulation type of fracture at the instantaneous fracture zone.

Fig. 5. Spring Model

7 Stress Analysis

7.1 Three-Dimensional Model of the Spring

The spring of the substation is used as the reference object. Due to the operational nature of the substation, the spring undergoes continuous compression during its operation. Therefore, fixed disks are connected at both ends of the spring to facilitate subsequent operations. The three-dimensional model of the spring is shown in Fig. 5.

From the figure, it can be observed that a fixed fixture is installed at the bottom of the disk to connect the spring, and pressure is applied to the upper end while keeping the circular arc fixed. This connection is intended to allow the spring to move along the direction of force, stabilizing its force and preventing bending.

7.2 Finite Element Analysis

Prior to static stress analysis, the spring is first divided into a mesh, and a high-quality mesh division can improve the accuracy of the analysis. The mesh division is shown in Fig. 6.

Fig. 6. Mesh Division of Spring

Subsequently, static stress analysis was performed by applying four different pressures at the upper end of the disk, namely F1 = 135N, F2 = 225N, F3 = 352N, F4 = 474N. The maximum stress and displacement of the spring were measured. As shown in Fig. 7 (a b c d e represent the static stress and displacement plots for different loads F1 to F4).

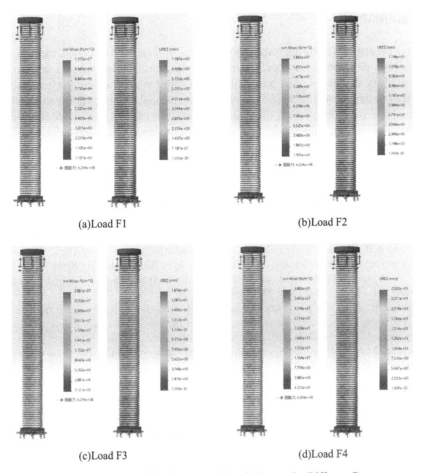

(a)Load F1   (b)Load F2

(c)Load F3   (d)Load F4

Fig. 7. Static Stress and Displacement of the Spring under Different Pressures

7.3 Conclusion

Based on the results from the four sets of finite element stress analyses, it can be observed that as the pressure increases, the stress and displacement experienced by the spring also increase. However, the deformation ratio of the spring becomes progressively smaller. When the deformation ratio reaches zero, the applied force is much higher than the forces shown in the above figures. Therefore, based on the conclusions mentioned earlier, it can be analyzed that the spring's failure is not due to external forces but rather a result of fatigue fracture, further confirming the accuracy of the fracture cause analysis mentioned earlier.

8 Summary

Through the analysis of spring fracture's optical spectral material detection, macroscopic morphology, electron microscope scanning analysis, and fracture energy spectrum element analysis, this paper proposes the primary factor for the fracture - fatigue fracture caused by an excessive Mn element content. Subsequently, the use of Solidworks' simulation section for finite element stress analysis validates the conclusions. Moreover, the simulation section includes modeling and analysis, providing quick and accurate results, which greatly facilitates the subsequent research on spring design and optimization.

References

1. Wang, J., Huang, R., Hu, H., et al.: Fracture Failure Analysis of the Energy Storage Spring of the Circuit Breaker in the 110kV Substation. J. Phys. Conf Ser. **2076**(1), 012094 (2021)
2. Rajesh, N.H., Sreekumar, M.: Design and simulation of a novel hybrid leaf spring with embedded cylindrical structures. Int. J. Heavy Veh. Syst. **23**(2), 131–154 (2016)
3. Yu, A., Yang, C.: Formulation and evaluation of an analytical study for cylindrical helical springs. Acta Mech. Solida Sin. **23**(1), 85–94 (2010)
4. Mulla, T.M.: Fatigue life estimation of helical coil compression spring used in front suspension of a three wheeler vehicle. In: Proceedings of the Modern Era Research in Mechanical Engineering-2016 (MERME-16), Urun Islampur, India, vol. 29 (2016)
5. Zaccone, M.A.: Failure analysis of helical suspension springs under compressor start/stop conditions. Pract. Fail. Anal. **1**, 51–62 (2001)
6. Gillner, K., Henrich, M., Münstermann, S.: Numerical study of inclusion parameters and their influence on fatigue lifetime. Int. J. Fatigue **111**, 70–80 (2018)
7. Günther, J., Leuders, S., Koppa, P., et al.: On the effect of internal channels and surface roughness on the high-cycle fatigue performance of Ti-6Al-4V processed by SLM. Mater. Des. **143**, 1–11 (2018)
8. Park, S.H., Lee, C.S.: Relationship between mechanical properties and high-cycle fatigue strength of medium-carbon steels. Mater. Sci. Eng. A **690**, 185–194 (2017)

Detection of Corrosion Areas in Power Equipment Based on Improved YOLOv5s Algorithm with CBAM Attention Mechanism

Wen Sun[1,2], Jian Zhang[3]([✉]), Wei Liao[1,2], Yanbing Guo[4], and Tengfei LI[1,3]

[1] Puneng Power Technology Engineering Branch, Shanghai Hengnengtai Enterprise Management Company, Limited, Shanghai 200437, China
[2] State Grid Shanghai Electric Power Research Institute, Shanghai 200437, China
[3] Shanghai Dianji University, Shanghai 201306, China
2496688297@qq.com
[4] College of Ocean Science and Engineering, Institute of Marine Materials Science and Engineering, Shanghai Maritime University, Shanghai 201306, China

Abstract. During long-term operation, power equipment can generate various forms of rust. Manual inspection consumes a significant amount of manpower and resources, and the results are not always satisfactory. The effectiveness and precision of traditional image processing methods in rust detection also have some shortcomings. In order to improve the timeliness and reliability of mobile devices such as drones in detecting rusty areas on power equipment, we propose an improved method based on the YOLOv5s object detection algorithm, which is optimized by integrating an attention mechanism. We use two different attention modules - CBAM and SE, to improve the YOLOv5s algorithm, enabling it to automatically identify rust areas. Experimental results show that Whether it is the original YOLOv5s, YOLOv5s + SE with the SE attention mechanism, or YOLOv5s + CBAM with the CBAM attention mechanism, all of them can effectively detect rust areas. However, by expanding the dataset and optimizing parameters, YOLOv5s + CBAM exhibits significant improvements in precision, recall, and mAP.

Keywords: Power Equipment · Rust Detection · YOLOv5s · CBAM Attention Mechanism

1 Introduction

The power industry uses metal extensively in a variety of electrical equipment. Since power equipment is usually installed outdoors and exposed to weather conditions, metal parts are susceptible to corrosion. Corrosion of electrical equipment can lead to transmission system failures and pose a threat to the stable operation of the power system

Technology Projects of State Grid Shanghai Municipal Electric Power Company (SGTYHT/21-JS-223)

J. Li et al. (Eds.): 6GN 2023, LNICST 553, pp. 277–284, 2024.
https://doi.org/10.1007/978-3-031-53401-0_25

[1–3]. Currently, the main solution is manual inspection, which not only consumes a lot of labor and material resources, but also has unsatisfactory results. To solve this problem and improve the efficiency of corrosion detection in electrical equipment, machine vision technology has become a promising approach. For example, Liao et al. [4] used the H component in the HSI color space to convert grayscale images into three groups. Each group utilizes a different detection method to automatically process the rust images. Zhao et al. [5] used the improved YOLOv7 algorithm for rust detection on shock absorbers. They used HSV model for color space transformation, introduced BiFPN for multi-scale feature fusion, and integrated GSConv module into the YOLOv7 network. Compared with the original network, the improved algorithm increased the detection accuracy by an average of 1.6% and the average speed by 17%.Yuan et al. [6] proposed a novel network called RPN-FCN, which combines an enhanced region-propositioning network (RPN) with a fully convolutional network (FCN) for equipment corrosion detection. The algorithm outperforms SIFT and Faster R-CNN in terms of accuracy, recall and detection speed.

Deep learning based image processing algorithms currently dominate the computer vision field [7, 8]. These methods outperform traditional object detection techniques in terms of recognition speed and accuracy [9]. Object detection techniques can be categorized into single-stage and two-stage algorithms. Typical one-stage algorithms include YOLO, SSD, etc. and two-stage algorithms include Faster R-CNN [10], Mask R-CNN, etc. Although two-stage algorithms are superior in accuracy, single-stage algorithms have faster processing speed.

2   Introduction and Improvement of YOLOV5s Algorithm

2.1   Introduction to the YOLOV5s Algorithm

This research uses the YOLOv5s algorithm to detect rust areas on electrical equipment. As one of the most widely used object detection algorithms currently, YOLOv5s combines deep learning technology with a Feature Pyramid Network and Spatial Pyramid Pooling structures. The Feature Pyramid Network allows the model to extract image features at different scales, which is particularly useful for handling rust detection tasks on electrical equipment given their irregular sizes and shapes. The SPP structure can generate output features of a fixed size, regardless of the input image size, increasing the model's scale robustness. YOLOv5s is made up of three parts: the Backbone Network, Neck Network, and Head Network. The Backbone Network is the core of the model. After completing the feature extraction from the input image, it combines different scale feature maps through cross-layer connections and channel compression to output a feature map with semantic information. The Neck Network strengthens the post-features, conducting pooling operations at different scales through the spatial pyramid structure to ensure the features have deformable invariance and multi-scale characteristics, preventing the loss of some information. The Head Network classifies and regresses the merged features for further classification and localization. Figure 1 shows the algorithmic structure of YOLOv5s.

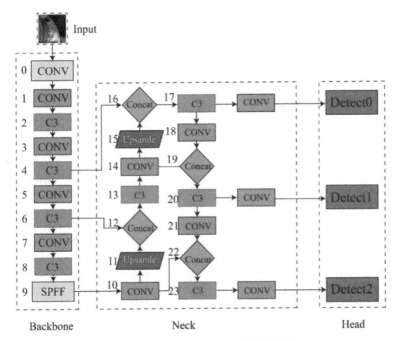

Fig. 1. Algorithm structure of YOLOV5s

2.2 YOLOV5s Algorithm Improvement

Introducing Attention Mechanisms. Many attention models are used in deep learning models. These models originate from research on human vision, which trains a mask layer with information inside the attention domain. By assigning different weight values, the feature extractor can ignore irrelevant information, focusing more on important information. Introducing an attention mechanism can help the model filter out important information from the input image, reducing the interference from irrelevant information. An attention mechanism module is added to the Backbone Network of YOLOv5s, enabling the model to pay more attention to important information during the feature extraction phase.

Introducing SE Attention Mechanism. This paper also enhances the network's capacity to extract features by adding an attention mechanism compression and excitation module between the main network's convolution layers. After computing the channel weight values, the SE module can determine the importance of different channels. This enables us to pay more attention to important features, thereby laying a solid foundation for the development of attention mechanisms [11]. The SE attention module is a channel attention module, commonly used in visual models. As illustrated in Fig. 2, it compresses each on-channel spatial feature into a global feature. Then it predicts the importance of each channel, and after getting the importance calculation it is excited to the corresponding channel. Lastly, the calculated channel weight values and corresponding two-dimensional matrices are multiplied to generate final output.

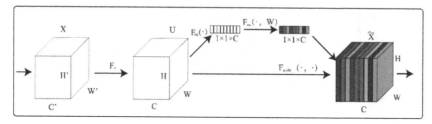

Fig. 2. Squeeze-and-Excitation Module

Introducing CBAM Attention Mechanism. The paper added the CBAM (Convolutional Block Attention Module) attention mechanism in the backbone network of YOLOv5s. The CBAM module can allocate weights to input feature maps in both spatial and channel dimensions. This weight allocation mechanism allows the model to focus more on important spatial locations and channel features, thereby further improving the detection of corrosion areas in power equipment.

The structure of CBAM is shown in the Fig. 3, which uses both spatial attention and channel attention to achieve an ordered attention structure from channel to space. Spatial attention allows the neural network to focus more on pixel regions that play a decisive role in the image, while ignoring unimportant regions. Channel attention is used to handle the allocation relationship of feature map channels and provides attention allocation in both dimensions, improving the performance of the attention mechanism.

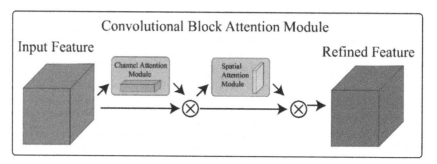

Fig. 3. Convolutional Block Attention module

Figure 4 shows the Channel Attention module. In the channel attention module, the input feature map is passed through max pooling and average pooling separately, and then input to the shared MLP. The output features of the shared MLP are element-wise added and passed through a sigmoid activation function to obtain the feature map of the channel attention module. Figure 5 shows the Spatial Attention module. In the spatial attention module, the feature map outputted by the channel attention module is used as input. First, a max pooling and average pooling based on channels are performed, and then the two layers are concatenated. Then, convolution is performed to reduce the number of channels to 1, and sigmoid activation is applied to obtain the feature map of the spatial attention module.

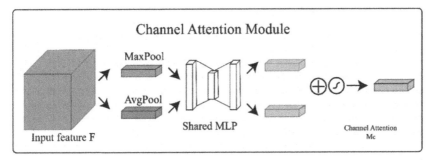

Fig. 4. Channel Attention module

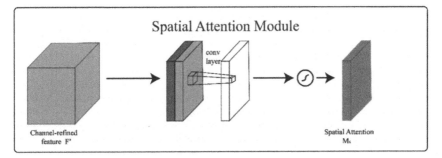

Fig. 5. Spatial Attention module

3 Experimental Results and Analysis

3.1 Data Sets and Environments

In this study, a total of 492 corrosion images of various power equipment, including cables, transformers, and distribution boxes, were collected through methods such as unmanned aerial vehicle photography, web scraping, and field collection. The size of the images was normalized to 416 × 416. In order to facilitate the extraction of corrosion features from the images by convolutional neural networks, all 492 images were manually annotated using the open-source annotation tool LabelImg. A standard dataset was established based on the VOC format.

3.2 Training Environment

The training environment for this experiment was a Windows 10 64-bit operating system, with an Intel i5-13490F processor, 2.50 GHz frequency, 32 GB RAM, Nvidia GeForce GTX 3060 Ti graphics card with 8 GB VRAM. The development environment used was Python 3.9.16 and torch-1.12.0 + cu113, with PyCharm as the IDE.

3.3 Experimental Analysis Results

As shown in Table 1, YOLOv5s with the CBAM attention mechanism achieves higher accuracy and recall rates compared to the standard YOLOv5s and YOLOv5s with the SE attention mechanism. Furthermore, the weight files of YOLOv5s with CBAM are smaller, which meets the storage requirements of industrial equipment. Overall, the YOLOv5s algorithm with the CBAM attention mechanism is the best detection algorithm.

Table 1. Analysis of the results of the three models

Model	Precision	Recall	Map@ 0.5	Weight size/MB	Time/ms
Yolov5s	0.784	0.831	0.747	13.7	9.4
Yolov5s + SE	0.77	0.793	0.695	13.8	11.1
Yolov5s + CBAM	0.803	0.847	0.759	13.2	10.0

The detection results of different algorithms are shown in Fig. 6. Comparing Fig. 6 (b), Fig. 6 (c), and Fig. 6 (d), it can be observed that there is some improvement in IoU, and all three models can detect corrosion areas well.

4 Dicussion

Significant progress has been made in rust detection using deep learning in the field of computer vision, as mentioned in the introduction. Several methods for rust detection using deep learning have been discussed. Utilizing the YOLO network for rust detection on mobile devices has proven to be effective.Compared to traditional two-stage algorithms like Faster R-CNN, YOLOv5s maintains high detection accuracy while being able to process images faster. This enables more efficient handling of large-scale datasets. The improved YOLOv5s algorithm is particularly suitable for rust detection in the power industry. Compared to the YOLOv7 model, the improved YOLOv5s algorithm has advantages in terms of model size and detection time, making it more suitable for drone-based applications in the power industry.

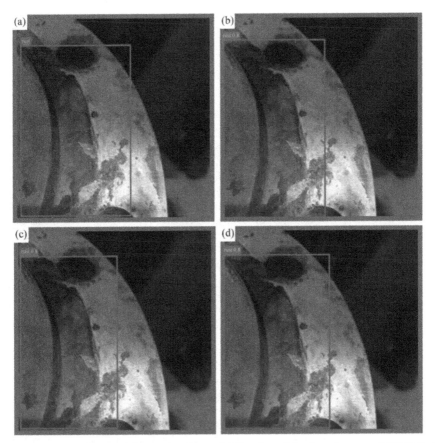

Fig. 6. Detection effect of different algorithms: (a) Framing of the detection area, (b) YOLOv5s detection results, (c) YOLOv5s + SE detection results, (d) YOLOv5s + CBAM detection results.

5 Conclusion

In this paper, a YOLOv5s-based algorithm with improved CBAM attention mechanism is proposed to address the problem of corrosion in power equipment. The following main conclusions are drawn:

1. 1.Under the condition of a small training dataset, whether it is the original YOLOv5s, YOLOv5s + SE with added SE attention mechanism, or YOLOv5s + CBAM with added CBAM attention mechanism, they can all detect corrosion areas well. However, by increasing the dataset size and optimizing the parameters, YOLOv5s + CBAM achieves significant improvements in accuracy, recall rate, mAP, and other metrics.
2. Overall, the improved YOLOv5s + CBAM model performs better in the task of corrosion area detection in power equipment compared to the original YOLOv5s and YOLOv5s + SE models. It not only effectively improves the accuracy and recall rate of the model but also can be deployed on mobile devices and achieves a detection

efficiency of 100fps on dedicated image processors, meeting the real-time monitoring requirements for corrosion areas in power equipment.

References

1. Chen, H., Qiao, X., Tian, F., Sun, Y.: Corrosion and protection of metallic components in power grid equipment. In: Proceedings of 2020 3rd International Conference on Electron Device and Mechanical Engineering (ICEDME), pp. 88–92 (2020)
2. Zhang, Y., Chen, W., Yan, H., Wang, X., Zhang, H., Wu, S.: The effect of atmospheric chloride ions on the corrosion fatigue of metal wire clips in power grids. Atmosphere **14**(2), 237 (2023)
3. Bondada, V., Pratihar, D.K., Kumar, C.S.: Detection and quantitative assessment of corrosion on pipelines through image analysis. Procedia Comput. Sci. **133**, 804–811 (2018)
4. Liao, K.W., Lee, Y.T.: Detection of rust defects on steel bridge coatings via digital image recognition. Autom. Constr. **71**, 294–306 (2016)
5. Zhao, Z., Guo, G., Zhang, L., Li, Y.: A new anti-vibration hammer rust detection algorithm based on improved YOLOv7. Energy Rep. **9**, 345–351 (2023)
6. Jiangru Yuana, B.X., Zhang, W.: RPN-FCN based rust detection on power equipment. Procedia Comput. Sci. **147**, 349–353 (2019)
7. Feizenszwalb, P.F., Girshick, R.B., McAllester, D., et al.: Object detection with discriminatively trained part-based models. IEEE Trans. Pattern Anal. Mach. Intell. **32**(9), 1627–1645 (2010)
8. LeCun, Y., Bottou, L., Bengio, Y., et al.: Gradient-based learning applied to document recognition. Proc. IEEE **86**(11), 2278–2324 (1998)
9. Agarwal, S., Terrail, J.C.D., Jurie, F.: Recent advances in object detection in the age of deep convolutional neural networks. Computer Vision and Pattern Recognition (2018). arXiv:1809.03193
10. Ren, S., He, K., Ross, G., et al.: Faster R-CNN: Towards real-time object detection with region proposal networks. In: Proceedings of the 28th International Conference on Neural Information Processing Systems, pp. 91–99 (2015)
11. Hu, J., Shen, L., Sun, G.: Squeeze-and-excitation networks. In: Proceedings of the IEEE Conference on Computer Vision and Pattern Recognition, pp. 7132–7141. IEEE, Los Alamitos (2018)

TASE-Net: A Short-Term Load Forecasting Model Based on Temperature Accumulation Sequence Effect

Lisen Zhao[1]([✉])[ID], Lihua Lu[1][ID], Xiang Yu[1][ID], Jing Qi[2][ID], and Jiangtao Li[1]

[1] School of Electronics and Information, Shanghai Dianji University,
Shanghai 201306, China
lisenmori@gmail.com, {lulihua,yux}@sdju.edu.cn
[2] College of Mathmatics and Information Science, Zhengzhou University of Industry,
Henan, China

Abstract. Electricity consumption forecasting plays an important role in ensuring efficient dispatch and reliability of the grid. The results are influenced by several factors at the same time. Inspired by the effect of temperature accumulation on load which the load forecast is effected by the temperature of the previous days in a specific temperature range, in this paper, we propose a model structure based on temperature accumulation sequence effects. It incorporates the temperature accumulation effects in a network: Temperature Accumulation Sequence Effects Network(TASE-net) which in a way generates a set of temperature accumulation sequences, uses a combined K-Shape-PSF method for feature extraction, and abstracts the sequence identity by Temporal Convolutional Network (TCN). To verify our proposed method, it is compared with other state-of-the-art methods for extracting similar sequences by using the datasets from three regions. The experimental results show that TASE-net reduces the error by 16% to the comparative method and achieve better MAPE.

Keywords: Electricity Consumption Forecasting · Temperature Accumulation · Temporal Convolutional Network · Pattern Sequence Similarity · K-Shape

1 Introduction

Electricity consumption forecasting refers to the estimation of electricity load in a certain region for a certain period. A highly accurate power load forecasting model can improve the security and stability of the local power system and bring great economic benefits to the peak and valley regulation of the power grid [1]. Electricity consumption is affected by many factors, among which temperature is an important factor. The load curve of continuous high temperature days is different from that of single high temperature days, which is considered to reflect the cumulative effect of temperature [2]. In addition to the current day's

J. Li et al. (Eds.): 6GN 2023, LNICST 553, pp. 285–298, 2024.
https://doi.org/10.1007/978-3-031-53401-0_26

temperature, the impact of previous days' temperature should also be considered. It is also important to deal with the seasonal periodicity and patterns of the data Therefore, feature extraction of input data is the key point to determine the prediction performance.

There are statistical methods and artificial intelligence methods for load forecasting [3]. The former are mainly ARIMA, Kalman Filter(KF) etc. [4]. The latter has various machine learning methods, such as support vector machines(SVM), clustering and neural network-based methods. Neural Network(NN) can model complex nonlinear curves and find the relationship between inputs and outputs. Martinez-Alvarez et al. [5] proposed a Pattern Sequence-Based Forecasting(PSF) algorithm to obtain similar pattern sequences by finding windows that are consistent with the sequence labels of the n days before the time step to be predicted. Based on the PSF method, Koprinska et al. [6] put the predicted values obtained from PSF into NN for secondary prediction to obtain more accurate predictions. Chi Zhang et al. [7] also proposed a decomposition selection model(DSF) based on Artificial Neural Networks(ANN), Empirical Mode Decomposition(EMD) and a certain feature selection method. It has better performance compared with other single methods. Li et al. [8] proposed a Densely Connected Network(DCN) to address load forecasting. The results indicate that forecasts using trend decomposition have improved behavior. Bin Li et al. [9] proposed a semi-parameter model based regression forecast considering weather and load information, in which the cumulative temperature effect is proposed for load application. John Paparrizos et al. [10] proposed a time series clustering method, K-Shape, which uses shape-based time series distance measurement and centroid calculation. Yu Zhou et al. [11] combined the K-Shape with CNN-LSTM for building energy consumption prediction. There are many other strategies that use K-Shape for load prediction [12–16]. Shaojie Bai et al. [17] proposed a generic Temporal Convolutional Network(TCN) used in sequence modeling, which has better performance compared with Recurrent Neural Networks(RNN) and Long Short-Term Memory(LSTM). There have been some incorporation of TCN into electricity consumption forecasting models [18–24]. Liu et al. [25] proposed a model Similarity Filtering(SFCL) based on the combined structure of CNN-LSTM.

Although many literature have several contributions on the load forecasting, they have various points can be improved in both the performance and accuracy. In our proposed model: Temperature Accumulation Sequence Effects Network(TASE-net) can optimize them further. The contributions of TASE-net mainly include:

(1) A feature extraction method under the influence of accumulated temperature effect: temperature accumulation sequence (TAS) is proposed, which can better detect the factors influencing the load by temperature change. It can improve the prediction accuracy and convergence speed. The Temperature Accumulated Sequence Influence Network (TASE-net) is proposed, which has better interpretability in feature exploitation.

(2) Filtering for load sequences at different time periods is also worth considering. The pattern similarity and seasonal periodicity of the data will be evaluated in the combined method K-Shape-PSF. It brings suitable data filtering performance, and identification of data characteristics.

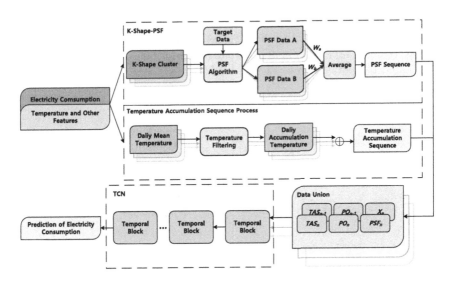

Fig. 1. Framework of TASE-net model.

2   Our Model: TASE-Net Model

The TASE-net model consists of three components: feature extraction, sequence integration and neural network training. The first two parts are completed by K-Shape-PSF method and TAS method, while TCN is used for neural network components. The whole framework of TASE-net is shown in Fig. 1. The data integration component will union the multiple feature sequences obtained above into a form that the network can recognize. The NN component uses the TCN, which is composed of multiple temporal blocks.

The input data are load: temperature and other features in which each of length 24 or 48 sampled values per day. These are used to generate two feature sequences by two methods respectively: K-Shape-PSF and temperature accumulation sequence process. In the former, the historical data are clustered into K categories by K-Shape. Then, the PSF algorithm is performed on the clustering labels of the target historical data. The data obtained by PSF filtering are separated into two groups: A and B. Group A has the same seasonal periodicity as the target, while B does not. The PSF series $PSF_n(n = 1, 2, 3...., N)$ are acquired by multiplying the A and B data with W_a and W_b weights and averaging them respectively. The latter is the integration of multi-day temperature data into a set of sequences $TAS_n (n = 0, 1, 2, 3...., N)$. Finally, the two feature sequences are mixed with the original historical temperature data PO_n and load data $X_n (n = 0, 1, 2, 3...., N)$ to get the training data.

The addition of TAS can improve the overall capacity and complexity of the model, which makes the model have better performance in multi-step short-term prediction. Meanwhile, the added one-dimensional TAS do not affect the overall duration of training. Observed from the results, the TASE-net model has a significant effect on reducing the forecast error. Also the prediction accuracy is relatively high. Especially it has more high prediction accuracy for continuous accumulated temperature and sudden load.

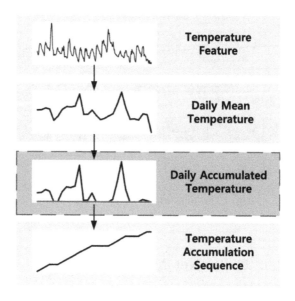

Fig. 2. Flow chart of TAS sequence generation.

3 Model Description

3.1 TAS Component

The influence of temperature on load takes different situation, as shown below: If the place is hot for a long time, the load in that area will be at a high level. In this case, even if the temperature drops, the load reduction is not significant. On the contrary, when the cool weather lasts for a period of time, even if the temperature suddenly rises, the degree of load increase is not significant. The phenomenon that the load lags behind the temperature change is called the temperature accumulation effect [9, 26].

Thus, the temperature accumulation effect occurs under the condition of perceiving a distinct temperature range. The intensity of the effect is influenced by the temperature change of the N days before the forecast date. In addition, the pattern exists in the difference between the temperature of the previous N days and the average temperature of the month. The accumulation temperature sequence(TAS) is proposed to be generated by simulating the influence of accumulation temperature on the forecast.

The multi-day temperature series have a lot of redundant information, which makes it difficult to extract positive features directly after adding to the NN. Moreover, it will lead to increase the training time. TAS pulls feature information from these temperature data, and thus generates only one column of feature values of length 24 or 48 (depending on the number of daily samples of load). The generation process of TAS is depicted in Fig. 2.

In Fig. 2, the average temperature T_{av} of the predict time within the same month are obtained. And the daily average temperature $T_n(n = 1, 2, 3, ..., N)$ of the previously N days before the forecast date are also taken.

Calculate the daily accumulation temperature Tac_n, which is obtained by subtracting the two averages. The formula is given in (1).

$$Tac_n = \begin{cases} 0, & (T_n - T_{av}) \leq 0 \\ \\ T_n - T_{av}, & (T_n - T_{av}) > 0 \end{cases} \tag{1}$$

In order to exclude the days with weak accumulated temperature influence, the accumulated temperature values calculated as negative are replaced by 0.

Tac_n is cumulated sequentially to obtain the TAS_n series. The formula is given in (2).

$$TAS_n = \sum_{n=1}^{n} Tac_n \tag{2}$$

The TAS_n series can guarantee its monotonically increasing character, which makes it always include the cumulative temperature influence. It has a length of N (usually 7 or 30, depending on the degree of influence of the historical temperature step).

To add TAS_n to the NN, the last cumulative temperature value of TAS_n is extended each time to the same dimension as the other sequences, generating an N-dimensional TAS (N size as above). The processing relies on a batch function that incorporates multiple days of historical training data into the network at a time. The differences in the multiple cumulative temperature values added to the NN each time reflect the cumulative effect.

3.2   Components of TCN

Temporal Convolutional Network (TCN) is a neural network model dedicated to time series [17]. It proposes causal convolution and dilation convolution based

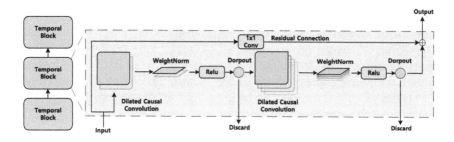

Fig. 3. Internal structure diagram of Temproal Block.

on the characteristics of time series. Causal convolution is a model that strictly restricts the time relationship. The value of the weight for a layer in the hidden layer at a certain time T is determined only by the time before the previous layer. The length of traditional convolution for time modeling is limited by the size of the convolution kernel. In order to track the relevance of distant moments and predictions, more hidden layers need to be stacked. However, by expanding the convolution, it allows gap between convolutions. As the number of stacked layers increases, the convolution gap becomes larger. The window of convolution grows exponentially with the number of layers, and a larger receptive field can be obtained with fewer layers.

TCN also packs each hidden layer like the above convolutional layer in a temporal block. But the stacking of blocks leads to overfitting phenomenon due to the deepening of the network structure. Therefore, residual connection and dropout layers are added in each time block. The structure within the temporal blocks is shown in Fig. 3.

The aim is to achieve a reduction in training error when the network is deepened or the number of training sessions is increased. But many times it is unsatisfactory. Shallow networks are able to get better results than deep networks. This problem is solved by introducing residual connection. It passes the parameters of the upper layer network directly to the next layer. The relationship between the network input and output mappings is translated into a sum of the input and the parameter mapping obtained in the block. A block function is shown by Eq. (3).

$$\phi(x_t^i) = M^i(x_t^i) + x_t^i \tag{3}$$

where x_t^i represents the tth training sample from the input domain X in the ith layer of the convolutional network. Accordingly, there also exists y_t belonging to the t th training sample in the target space Y. $M^i(\cdot)$ denotes the network parameters of this layer. $\phi(\cdot)$ stands for this output mapping. After several iterations, the export of the network $\Phi(x)$ can converge to y_t.

The latter relationship is more easily obtained by network optimization and the deeper network is able to learn the information.

The TAS-TCN combination can make full use of the sequential information of TAS. Its causal logic can disassemble the TAS implicit features in one direction. The negative information can be filtered out by the combined form of forward extended receptive fields and residual connection, so the gradient at the beginning of training decreases rapidly and stays around the minimum value for stable approximation.

3.3   K-shape-PSF Component

The K-Shape-PSF method is a combination of the K-Shape method [10] and the PSF algorithm [5]. The PSF algorithm has good performance in the preprocessing of time series prediction. It can track the similarity and potential differences of serial patterns in historical data for N consecutive days. However, it relies on the performance metrics of time series clustering. The K-Shape computes the offsets of time series within a class when computing the sequence centroids and iteratively clusters the previously computed clustering centroids by aligning them with the reference sequence. It also uses normalized cross-correlation to describe the correlation between sequences. K-Shape has a strong ability in identifying patterns and correlations of time series. Therefore the K-shape-PSF combination can be better preprocessed.

After combining these two methods, this paper also makes changes to the forecasting part of the PSF method. The PSF averages the historical loads consistent with the window to obtain the final forecast. However, the periodicity and seasonality of the data should also be a consideration. Differences in the structure of electricity consumption due to working days versus non-working days and different seasons can have an impact on the forecast. Therefore, in the PSF forecasting stage, the PSF window data are compared with the target data in terms of date and season, which are divided into two categories and then weighted and averaged to process. The Eq. (4) is as follows:

$$PSF = W_a * L_A + W_b * L_B \tag{4}$$

The PSF is a 24 or 48 dimensional vector. Where $L_A(L = X_1, X_2, X_3, ..., X_n)$ and $L_B(L = X_1, X_2, X_3, ..., X_m)$ are the sets of historical data extracted from the PSF algorithm respectively. X_N is a sequence of length 24 or 48. Data with the same weekend or non-weekend and the same season will be added to L_A, while L_B contains different weekends and seasons. A higher value of W_a is usually assigned to ensure that similar features are retained as much as possible, while a few remaining features of L_B are retained. In the selection of weights: W_a and W_b, the optimal is obtained by multiple experiments.

4   Experiment

4.1   Dataset

Electricity consumption datasets from three different regions for different years were used, namely the Sydney of Australia, the Midwest of USA and the GEFcom2012. These three datasets are sampled differently, with the Australian dataset being sampled every half hour and covering the years 2006 to 2009. The U.S. Midwest dataset and the GEFCom2012 dataset are sampled hourly, covering the years 2015 to 2022 and 2004 to 2008, respectively. The Australian and U.S. datasets are complete, while the GEFCom2012 dataset incomplete. The GEFCom2012 dataset was derived from open source data from the competition, and temperature data for Australia and the central United States were sampled from local weather websites. The visualization of these three datasets and the segmentation of the datasets are shown in Fig. 4 below.

The training of this model first needs to obtain the PSF sequence, which requires the K-shape-PSF method for data. The sliding historical window data in the method is not allowed to enter the next step. Thus, half of the historical data in the training dataset is used as needed to generate PSF sequences, while the other half is trained. The same goes for validation data.

The predominance of hot environments leads to a high impact of cumulative temperature effects. Therefore, February 2010 in the Australian dataset, August 2021 in the US Midwest dataset, and June 2008 in the GEFCom2012 dataset are selected for explore their performance.

In addition, for each dataset, each season in the data is validated to verify the generalization ability and reliability of the model.

4.2   The Datasets and Source Code

The details of the relevant datasets, model parameters and source code can refer to: https://github.com/Lisen-Zhao/TASE for reference.

4.3   Experimental Results

The error measure used in this paper: Mean Absolute Percentage Error(MAPE), is a relative error measure that uses absolute values to eliminate the effect between positive and negative errors.

The experimental results of our model will contrast to other two models: the SFCL model and K-Shape-PSF-TCN. The former focuses on extracting similar pattern sequences in preprocessing. They added the Similar Filter(SF) processed data into CNN-LSTM network for training. The latter is the baseline of TASE-net, which is the case without TAS.

From three datasets, we describe the experimental results of SFCL, K-Shape-PSF-TCN and TASE-net. The prediction curves of above three models for one consecutive month are shown in Fig. 5.

In the enlarged blocks of each subfigure in Fig. 5, all results of TASE-net have more higher accuracy of load prediction in continuous rise than the other models, which fits better with the actual curve. The close relationship between cumulative temperature and continuously rising load means that cumulative temperature has an advantage in prediction, and it can better predict the peak of short-term multi-step load. Its average error MAPE values are given by Table 1.

From Table 1, TASE-net shows the best performance in the Australian dataset, which exhibits continuous high and dry climate characteristics. The

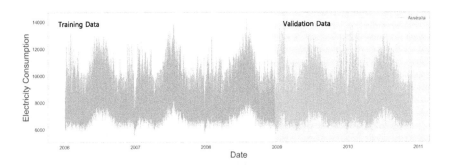

(a) Australia Dataset.

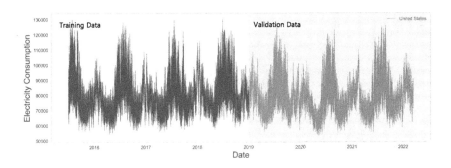

(b) American Dataset.

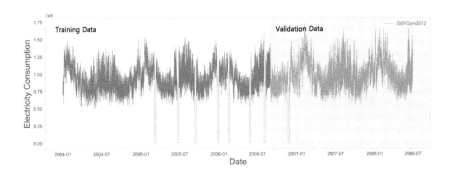

(c) GEFCom2012 Dataset.

Fig. 4. The visualization of three datasets and data set segmentation.

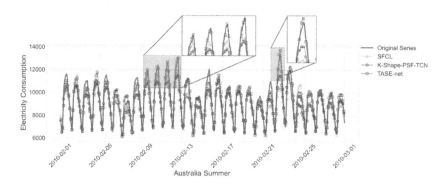

(a) Australian dataset with three models.

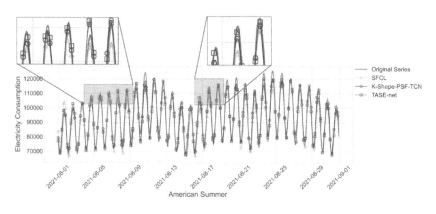

(b) American dataset with three models.

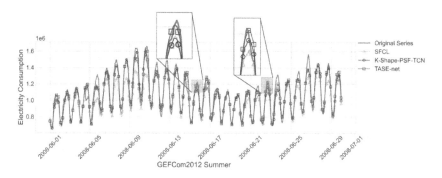

(c) GEFCom2012 dataset with three models.

Fig. 5. Comparison of prediction and observation of three methods in three datasets in summer.

Table 1. Mean MAPE values for three models across three datasets.

Summer Prediction			
Dataset	TASE-net	K-Shape-PSF-TCN	SFCL
Australia	**3.97**	5.24	4.30
American	2.82	**2.73**	3.11
GEFCom2012	**4.09**	4.36	6.02

model capture the influence of the temperature signature on the N days prior to the forecast. Based on several experiments, TASE-net produced the top performing TAS in the Australian dataset looking back 7 days. The benchmark K-Shape-PSF-TCN has the largest prediction error due to the large effect of temperature on the load, causing it to produce abrupt changes. In the U.S. dataset, TASE-net shows the same as the benchmark essentially and outperforms the SFCL model. This may be due to the presence of excessive or flat hot weather in the forecast months, resulting in an overall smoothing of the TAS series and less impact of the temperature characteristics on the forecast. In the GEFCom2012 dataset, TASE-net also outperforms SFCL and K-Shape-PSF-TCN.

TASE-net proves its stability in three datasets. It enhances the benchmark model K-Shape-PSF-TCN for hot weather smoothing and strengthens the underprediction of severe temperature changes. A comparison of the daily MAPE values of the different models in the Australian datasets for the summer forecast is shown in Fig. 6.

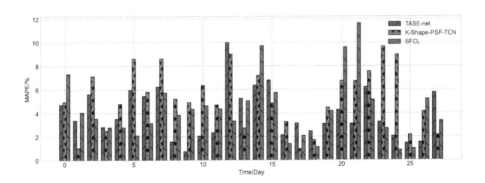

Fig. 6. Daily MAPE histogram of Australian dataset predicted by three models.

Figure 6. shows that TASE-net performs relatively well in terms of prediction error, with a lower peak prediction error than the SFCL model. Moreover, the TASE-net model significantly better than the other models in the situation of smaller prediction errors. The larger errors on individual days may be due to the effects of special continuous cold weather and holidays, which make TAS not read the temperature-load relationship well, when the correlation between temperature and load is low. When the temperature exceeds a certain range, the impact of temperature accumulation becomes weak, so the model shifts to the K-Shape-PSF-TCN. In Table 2, the prediction results for spring, autumn and winter datasets with different models are provided with error values.

Table 2. The MAPE values of three model in other seasons.

Season	Dataset	K-Shape-PSF-TCN	SFCL
Spring	Australia	2.87	**2.70**
	American	4.07	**2.90**
	GEFCom2012	**4.21**	4.23
Autumn	Australia	2.27	**1.71**
	American	**2.99**	3.39
	GEFCom2012	**3.95**	5.64
Winter	Australia	**2.91**	3.07
	American	**3.28**	4.87
	GEFCom2012	**6.72**	6.98

From Table 2, the overall prediction accuracy of the K-Shape-PSF-TCN model without TAS sequences is better than the SFCL. The prediction error is significantly lower in the fall and winter, and the overall error is flat in the spring. In the U.S. dataset, the seasonal nature of the spring data is not obvious leading to low adaptation of the TASE-net model predictions.

Overall, TASE-net has better performance compared to K-Shape-PSF-TCN and SFCL, with 9% and 16% error reduction respectively. The performance of the K-Shape-PSF-TCN and SFCL differ in terms of their performance for different datasets and seasons. After excluding individual biases, the former model reduced the overall error by 5.7% compared to the latter method.

5   Conclusions

TASE-net, by adding TAS features to the neural network for tracing the effect of summer temperature on load, generates a short time series with rich implicit features that can be recognized by the network to accelerate convergence The TCN component is predictive for the time series and has skip-connect properties, which makes TAS have good linkage with its strictly time-bound network characteristics. The features of TAS can be better presented in TCN. The K-Shape-PSF model of the underlying processing is also the key to accelerate the convergence of the network.

However, there are still many factors to be considered in load prediction, and the prediction performance of TASE-net for continuous low temperature in summer, interference by holidays, and generalization ability for different datasets are to be improved. We also need to add consideration to the effect of other environmental variables and additional features. Furthermore we also need to investigate whether such a model can be used in medium- and long-term forecasting.

References

1. Hadjout, D., Torres, J.F., Troncoso, A., Sebaa, A., Martínez-Álvarez, F.: Electricity consumption forecasting based on ensemble deep learning with application to the Algerian market. Energy **243**, 123060 (2022)
2. Genyong, C., Jingtian, S.: Study on the methodology of short-term load forecasting considering the accumulation effect of temperature. In: 2009 International Conference on Sustainable Power Generation and Supply, pp. 1–4. IEEE (2009)
3. Yang, D., Wang, W., Hong, T.: A historical weather forecast dataset from the European centre for medium-range weather forecasts (ECMWF) for energy forecasting. Sol. Energy **232**, 263–274 (2022)
4. Santhosh, M., Venkaiah, C., Vinod Kumar, D.M.: Current advances and approaches in wind speed and wind power forecasting for improved renewable energy integration: a review. Eng. Rep. **2**(6), e12178 (2020)
5. Alvarez, F.M., Troncoso, A., Riquelme, J.C., Ruiz, J.S.A.: Energy time series forecasting based on pattern sequence similarity. IEEE Trans. Knowl. Data Eng. **23**(8), 1230–1243 (2010)
6. Koprinska, I., Rana, M., Troncoso, A., Martínez-Álvarez, F.: Combining pattern sequence similarity with neural networks for forecasting electricity demand time series. In: The 2013 International Joint Conference on Neural Networks (IJCNN), pp. 1–8. IEEE (2013)
7. Zhang, C., Wei, H., Zhao, J., Liu, T., Zhu, T., Zhang, K.: Short-term wind speed forecasting using empirical mode decomposition and feature selection. Renew. Energy **96**, 727–737 (2016)
8. Li, Z., Li, Y., Liu, Y., Wang, P., Lu, R., Gooi, H.B.: Deep learning based densely connected network for load forecasting. IEEE Trans. Power Syst. **36**(4), 2829–2840 (2020)
9. Li, B., Mingzhen, L., Zhang, Y., Huang, J.: A weekend load forecasting model based on semi-parametric regression analysis considering weather and load interaction. Energies **12**(20), 3820 (2019)
10. Paparrizos, J., Gravano, L.: k-shape: efficient and accurate clustering of time series. In: Proceedings of the 2015 ACM SIGMOD International Conference on Management of Data, pp. 1855–1870 (2015)
11. Zhou, Y., Ren, B., Xue, X., Chen, L.: Building energy consumption forecasting based on k-shape clustering and CNN-LSTM. In: 2022 4th International Conference on Power and Energy Technology (ICPET), pp. 1147–1152. IEEE (2022)
12. Zhang, Y., et al.: Improving aggregated load forecasting using evidence accumulation k-shape clustering. In: 2020 IEEE Power & Energy Society General Meeting (PESGM), pp. 1–5. IEEE (2020)
13. Yang, L., Zhang, Z.: A deep attention convolutional recurrent network assisted by k-shape clustering and enhanced memory for short term wind speed predictions. IEEE Trans. Sustain. Energy **13**(2), 856–867 (2021)
14. Wen, L., Zhou, K., Yang, S.: A shape-based clustering method for pattern recognition of residential electricity consumption. J. Clean. Prod. **212**, 475–488 (2019)

15. Wang, B., Zhang, D., Yang, W., Leng, Z.: An intelligent forecasting model for building energy consumption using k-shape clustering and random forest. In: 2021 2nd International Conference on Artificial Intelligence and Information Systems, pp. 1–4 (2021)
16. Yang, J., et al.: k-shape clustering algorithm for building energy usage patterns analysis and forecasting model accuracy improvement. Energy Build. **146**, 27–37 (2017)
17. Bai, S., Kolter, J.Z., Koltun, V.: An empirical evaluation of generic convolutional and recurrent networks for sequence modeling. arXiv preprint arXiv:1803.01271 (2018)
18. Xiaoyan, H., Bingjie, L., Jing, S., Hua, L., Guojing, L.: A novel forecasting method for short-term load based on TCN-GRU model. In: 2021 IEEE International Conference on Energy Internet (ICEI), pp. 79–83. IEEE (2021)
19. Wang, H., Zhang, Z.: TATCN: time series prediction model based on time attention mechanism and TCN. In: 2022 IEEE 2nd International Conference on Computer Communication and Artificial Intelligence (CCAI), pp. 26–31. IEEE (2022)
20. Gopali, S., Abri, F., Siami-Namini, S., Namin, A.S.: A comparison of TCN and LSTM models in detecting anomalies in time series data. In: 2021 IEEE International Conference on Big Data (Big Data), pp. 2415–2420. IEEE (2021)
21. Zhang, Z., Chen, H., Huang, Y., Lee, W.J.: Quantile huber function guided TCN for short-term consumer-side probabilistic load forecasting. In: 2020 IEEE/IAS Industrial and Commercial Power System Asia (I&CPS Asia), pp. 322–329. IEEE (2020)
22. Wang, Y., et al.: Short-term load forecasting for industrial customers based on TCN-LightGBM. IEEE Trans. Power Syst. **36**(3), 1984–1997 (2020)
23. Zhao, Y., Jia, L.: A new hybrid forecasting architecture of wind power based on a newly developed temporal convolutional networks. In: 2020 IEEE 9th Data Driven Control and Learning Systems Conference (DDCLS), pp. 839–844. IEEE (2020)
24. Song, J., Peng, X., Yang, Z., Wei, P., Wang, B., Wang, Z.: A novel wind power prediction approach for extreme wind conditions based on TCN-LSTM and transfer learning. In: 2022 IEEE/IAS Industrial and Commercial Power System Asia (I&CPS Asia), pp. 1410–1415. IEEE (2022)
25. Liu, J., Lu, L., Yu, X., Wang, X.: SFCL: electricity consumption forecasting of CNN-LSTM based on similar filter. In: 2022 China Automation Congress (CAC), pp. 4171–4176. IEEE (2022)
26. Wang, M., Zixuan, Yu., Chen, Y., Yang, X., Zhou, J.: Short-term load forecasting considering improved cumulative effect of hourly temperature. Electric Power Syst. Res. **205**, 107746 (2022)

Predicting Time Series Energy Consumption Based on Transformer and LSTM

Haitao Wang, Jiandun Li$^{(\boxtimes)}$ (iD), and Liu Chang

School of Electronic Information, Shanghai Dianji University, Shanghai 202304, China
lijd@sdju.edu.cn

Abstract. Energy is crucial to economic and social development. Accurate energy consumption forecasting is essential to effective energy management, reasonable energy layout planning, and ensuring the sustainable and healthy development of the energy industry. Nevertheless, precisely and efficiently forecasting energy consumption remains to be a challenge. Previous studies have proposed solutions mainly from traditional machine learning and mathematical statistics, which can effectively forecast short-term energy consumption in small-scale data. However, it is still challenging to explore the characteristics of high-dimensional and large-scale energy data and predict medium- to long-term energy consumption and its fluctuation trends. In this paper, a new time series energy consumption prediction model is proposed, which combines the attention mechanism of the Transformer model with the natural language processing ability of LSTM, based on a dataset of Spain's energy production and climate change from 2015 to 2018. Compared with the state-of-the-art models such as RNN, GRU, and LSTM, our model achieves better performance in the 7-day energy consumption prediction task (RMSE = 0.7, MAE = 0.5).

Keywords: Energy Consumption Forecast · Deep Learning · Transformer · LSTM

1 Introduction

Energy is a fundamental material resource for human production and life, and its supply and consumption are closely related to the economic, social, and environmental well-being of countries and people. According to statistics, China's total energy consumption reached 5.41 billion tons of standard coal in 2022, a 2.9% increase from the previous year. With the continued growth of energy consumption, effective energy management has become increasingly crucial. Energy consumption prediction serves as an essential basis for reasonable and efficient energy planning, aiding governments and enterprises in developing realistic energy supply plans and mitigating the problems caused by supply-demand imbalances such as energy shortages, price fluctuations, etc. Besides, energy consumption prediction can help optimize energy use efficiency, and reduce energy wastage and emissions, thus contributing towards the attainment of sustainable development objectives.

© ICST Institute for Computer Sciences, Social Informatics and Telecommunications Engineering 2024
Published by Springer Nature Switzerland AG 2024. All Rights Reserved
J. Li et al. (Eds.): 6GN 2023, LNICST 553, pp. 299–314, 2024.
https://doi.org/10.1007/978-3-031-53401-0_27

Energy consumption prediction is challenging due to several factors. Firstly, high data noise affects prediction accuracy as it influences the quality of historical data used. Secondly, energy consumption is influenced by multiple interacting factors, including climate and energy production, making it difficult to quantify and analyze. Thirdly, uncertainty arises from the complexity of relationships among numerous factors and variables involved in energy consumption prediction, leading to difficulty in determining their individual effects and introducing uncertainty in predictions. Finally, existing solutions have limited accuracy and most of them fall short to accurately predict medium- and long-term energy consumption and their fluctuation trends.

Researchers have proposed various solutions for energy consumption prediction, including statistical model methods, machine learning methods, and deep learning methods. Unlike statistical models and traditional machine learning techniques, deep learning models such as Fully Connected Neural Networks and Recurrent Neural Networks have demonstrated promise in addressing the challenges posed by nonlinearity and high-dimensional data. These models excel in handling complex data and improving prediction accuracy. However, it is worth noting that deep learning models may demand larger datasets for optimal performance. Additionally, they can encounter issues like gradient vanishing and the extraction of time-dependent features, which can adversely affect the accuracy of long-term predictions. To address these challenges, this paper proposes a new time series model that integrates Transformer and LSTM models. The approach leverages Transformer's self-attention mechanism to capture long-term dependencies and LSTM's strengths in sequence modeling. Experimental results on the Spanish energy and climate dataset from 2015 to 2018 demonstrate the effectiveness of the proposed model after feature selection.

The contribution of this paper is threefold: (1) a feature selection method based on data change rate to identify factors affecting energy consumption; (2) integrating Transformer and LSTM models effectively; (3) introducing a time series energy consumption prediction model that predicts energy usage for the next 7 days with high accuracy.

The rest of this paper is organized as follows: In the "Related Work" section, we will present an overview of the current research status and progress in energy consumption forecasting. In the "Dataset and Data Preprocessing" section, we will introduce the dataset's origin, features, and conduct the preprocessing operations to prepare the data for training the deep learning model. The "Model Construction" section presents a detailed description of the proposed model and outlines its computational process. In the "Experimental and Discussion" section, we perform experiments to verify the performance of various models in medium- to long-term energy consumption prediction. We analyze and discuss the results obtained from these experiments. Finally, the "Conclusion" section summarizes the main content of this paper, discusses the shortcomings of the proposed approach, and outlines future directions for improvement.

2   Related Work

Regarding the challenge of low accuracy and efficiency in energy consumption prediction, researchers have proposed various models. These models can be broadly classified into mathematical statistical models, machine learning models, and deep learning models.

2.1 Statistical Models

Using historical energy consumption data, predictive models can be established through statistical methods such as time series Autoregressive Integrated Moving Average (ARIMA) models, Linear Regression models, etc. For example, Ewa et al. [1] predicted power load using ARIMA models and assessed the model's predictive ability under different noise conditions. Liu et al. [2] used the Gray Model (GM) to predict future residential electricity consumption based on 8 years of residential electricity consumption data from Jilin Province. Although the results have some errors, they have to some extent solved the problem of difficult prediction of energy consumption. However, statistical models require high data requirements and cannot consider nonlinear relationships, so the prediction effect is unstable and can only be used in specific scenarios.

2.2 Machine Learning Models

Traditional machine learning methods, such as Random Forest, Support Vector Machine (SVM), and Back Propagation Neural Networks (BPNN), have been widely used in energy consumption forecasting. These models are effective in dealing with high-dimensional and nonlinear problems. For instance, Peng et al. [3] used a prediction approach based on wavelet transform and random forest to analyze power load data from the Australian energy market operator, reducing data noise, and enhancing short-term prediction accuracy. Similarly, Wan et al. [4] employed SVM to forecast short-term power load within a specific field, addressing the problem of insufficient data fitting to some extent. Cao et al. [5] introduced an ASO-BPNN short-term prediction model based on atomic search optimization and selected vital characteristics using related formulas, achieving promising effects in the short-term prediction of photovoltaic power generation. Nonetheless, traditional machine learning methods require good function or distribution characteristics and only perform well for short-term prediction.

2.3 Deep Learning Model

Although short-term predictions are useful in several practical applications, they are insufficient for long-term energy planning. Long and medium-term forecasting of energy consumption and its volatility trends are necessary. To extract temporal dependent features from time series data and enhance prediction accuracy, a model with memory capability is required that can predict future energy consumption and its volatility trends over some time. Traditional machine learning models do not meet these requirements, but with the advent of the big data era, deep learning has rapidly developed. Lai et al. [6] augmented historical data to make use of a Fully Connected Neural Network (DNN) that accurately predicted the daily peak load of several countries and reduced prediction errors. The accuracy of DNN is limited in long- and medium-term forecasting due to its inability to extract temporal changes in time series data. Recurrent Neural Networks (RNNs) can be used for time series data prediction because of their recursive structure, but they suffer from the problems of gradient explosion and disappearance. These problems led to the development of Long Short-Term Memory (LSTM), which improves upon RNNs by effectively extracting temporal features and alleviating the gradient fading problem while also providing a memory function. Gated Recurrent Unit (GRU), which is based on LSTM improvements, is widely used in time series prediction fields. For instance, Ibrahim et al. [7] employed LSTM for the short-term prediction of household electricity consumption, while Lu et al. [8] utilized GRU to predict multi-energy-coupled short-term loads. Both studies achieved good results in short-term predictions, demonstrating feasibility for long-term predictions. Nam et al. [9] employed LSTM and BiL-STM for weekly electrical load forecasting based on weather and calendar data, achieving enhanced accuracy in medium to long-term predictions. However, LSTM suffers from drawbacks such as non-parallel computation and difficulty in extracting long-term dependent features, which limits its long-term predictive capability. Encoder-decoder, a model utilizing two LSTM networks (encoder and decoder) for sequence-to-sequence learning tasks, exhibits good feature extraction ability in time series data processing and can also be applied for medium to long-term prediction. For instance, Dorado et al. [10] proposed a deep learning method that combines Encoder-decoder with residual connections to predict short-term electricity loads, addressing the gradient disappearance issue and implementing pre-processing methods to enhance accuracy. Nonetheless, their approach still lacks precision and performance when it comes to medium to long-term predictions.

Recently, Transformers [11] have been widely applied in Natural Language Processing (NLP) tasks [12]. Its self-attention mechanism and encoder-decoder framework effectively extract features from data. Huang et al. [13] demonstrated the ability of the Transformer to process long-term sequential information by predicting wind power generation with this model. However, due to its high computational complexity and sensitivity to position encoding, the Transformer's generalization ability is relatively low, and its predictive accuracy for non-autoregressive problems is limited. To address these issues, this paper proposes an improved Transformer-LSTM model that combines the self-attention mechanism and framework of the Transformer with the natural language processing and memory capabilities of LSTM. This model achieves good performance in medium and long-term energy consumption prediction.

3 Datasets and Preprocessing

The dataset applied in this academic article comprises 4 years (2015–2018) of electricity consumption, generation, pricing, and weather data in Spain. The consumption and generation data were acquired from the public portal of the transmission system operator (TSO) called ENTSOE. Weather data for the five largest cities in Spain were obtained from an open weather API. Initially, the dataset contained 46 feature columns recording hourly electricity consumption, generation, pricing, and weather conditions. Preprocessing involved removing feature columns with missing data and subjective features. Weather conditions were transformed into one-hot vector encoding, and feature columns with unchanged values in over 80% of the data were removed. Subsequently, data were aggregated at a daily level, where energy data was summed, and weather data were averaged. The resultant dataset has 25 feature columns, and Table 1 describes some feature details.

Table 1. Dataset description

	mean ± std	min	max
gen_biomass	9,199.33 ± 2,017.68	4,432	13,877
gen_fossil_brown_coal/lignite	10,747.90 ± 8,222.25	0	23,467
gen_fossil_gas	134,876.42 ± 44,825.71	43,540	364,926
gen_fossil_hard_coal	102,093.14 ± 45,406.03	15,169	191,003
gen_fossil_oil	7,155.79 ± 1,128.02	3,501	10,117
gen_nuclear	150,260.88 ± 20,088.63	66,847	170,851
gen_solar	34,366.33 ± 16,925.25	3,537	65,545
gen_waste	6,463.34 ± 1,182.81	2,515	8,318
total_load_actual	688,019.46 ± 66,962.05	342,907	847,564
temp	289.66 ± 6.45	276.14	302.77
pressure	1,070.20 ± 1,366.15	981.81	46,387.73
humidity	68.28 ± 9.52	41.23	94.05
wind_speed	2.46 ± 0.95	0.91	9.63
wind_deg	166.64 ± 36.09	48.99	295.83
is_rain	0.09 ± 0.12	0	1

We select some features from all feature attributes to calculate their correlation with the predicted total energy consumption ("total_load_actual"), and the correlation heatmap is shown in Fig. 1. From the figure we can observe that the data of "total_load_actual" generated from sources other than solar energy exhibit a positive correlation with the data generated from other energy sources. At the same time, we found that "total_load_actual" is negatively correlated with climate feature data.

In Fig. 2, we showcase selected energy generation and weather feature data. The figure illustrates fossil fuel power generation, local atmospheric pressure, and temperature in the region for the period from 2015 to 2019. However, it is essential to note that several other features also influence the outcomes of energy consumption prediction, despite not being included in the figure.

To speed up the model training and improve accuracy, the data is first normalized using the normalization [14] method, so that all feature data belongs to the (0, 1) normal distribution. Then, a sliding time window is established for data sampling. This study uses 7 days of historical data to predict the future 7 days of energy consumption and its fluctuation trend. Therefore, the sequence length of the sliding time window is set to 7. Finally, the dataset is divided with the first three years of data as the training set and the remaining one year as the test set. With this, data preprocessing is completed.

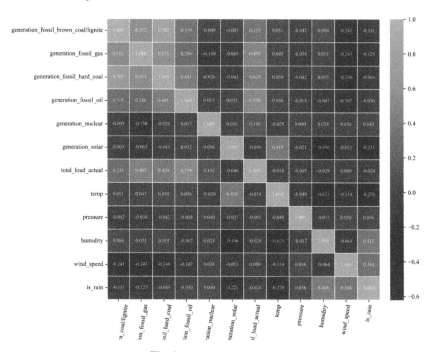

Fig. 1. Weather data heatmap

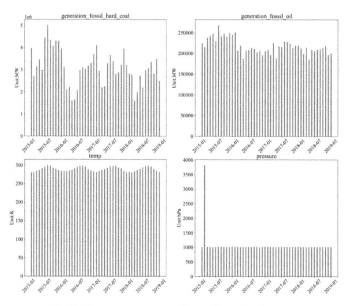

Fig. 2. Some energy generation and weather features

4 Methodology

4.1 Modeling

Drawing on the multi-head self-attention mechanism [15] and Residual Connections [16], as well as the layer normalization structures in the Transformer, this paper aims to better extract the multi-dimensional features of data and handle long-term dependencies between data. These techniques are effective in extracting time-dependent features of multi-dimensional data. However, the Transformer has relatively poor handling capabilities for autoregressive problems, and its accuracy in multi-step prediction is not satisfactory. In contrast, the LSTM has excellent memory capabilities and performs well in multi-step prediction of time series. Therefore, this paper proposes to process the encoder's input based on the Transformer using an LSTM or BiLSTM and then use a decoder composed of linear layers to achieve regression prediction. We propose a Transformer_LSTM model, and the model structure is depicted in Fig. 3.

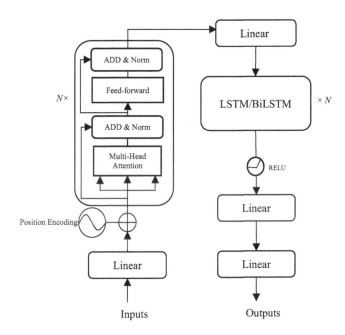

Fig. 3. The structure of the Transformer_LSTM Model

The Transformer is a neural network architecture that uses the self-attention mechanism to tackle sequence-to-sequence tasks, including machine translation, speech recognition, and summarization generation. In comparison to the traditional recurrent neural network (RNN)-based models, the Transformer outperforms in terms of performance and parallel computing capabilities. The central components of the Transformer include the multi-head attention mechanism and the feed-forward network. Through the multi-head attention mechanism, the input sequence undergoes multiple linear transformations, followed by the *softmax* operation to compute attention weights. Then, the input sequence is weighted and summed, generating context representations for each position. Meanwhile, the feed-forward network applies nonlinear transformations to every context representation, facilitating the production of new representations.

As shown in the model structure, when predicting energy consumption, the input sequence is first passed through a linear layer to expand each input sequence feature to a *d_model*-dimensional vector, *d_model* is a parameter of the model, representing the dimensionality extended by the PE operation. The input sequence is then position-encoded, and the resulting vector is added to the Positional Encoding (PE) operation. This operation accurately provides the position information of each time step to the model, allowing the model to understand the order relationship of the input sequence and achieve parallel computing. The calculation process of the PE operation is shown in Formula 4-1, where "*pos*" represents the absolute position of the feature in the data.

$$
\begin{cases}
PE(pos, 2i) = sin(\dfrac{pos}{10000^{\frac{2i}{d_{model}}}}) \\[4mm]
PE(pos, 2i+1) = cos(\dfrac{pos}{10000^{\frac{2i}{d_{model}}}})
\end{cases}
\tag{4-1}
$$

Next, each position in the input sequence is encoded through a multi-head attention mechanism with multiple attention heads. The self-attention mechanism can establish global dependencies, expand the receptive field of the model, and effectively extract long-term feature dependencies, which is a key part of the model. In the model, the input sequence undergoes multiple linear transformations, represented by Q, K, and V matrices, and then the attention weights are calculated using the *softmax* function. The input sequence is then weighted and summed to obtain the context representation of each position. The resulting context representation at each position is then used as the encoding representation for that position. The specific calculations are as follows:

Firstly, the query Q, key K, and value V matrices need to be calculated. W_Q, W_K, and W_V are weight matrices that are continually learned and updated. The embedded input sequence is multiplied with the weight matrices and undergoes linear transformations to obtain the matrices Q, K, and V, as shown in Formula (4-2).

$$
\begin{cases}
Q = Linear_q(X) = XW_Q \\
K = Linear_k(X) = XW_K \\
V = Linear_v(X) = XW_V
\end{cases}
\tag{4-2}
$$

Then the self-attention scores are calculated. For a word vector, the dot product is performed between its corresponding q and all k in the entire sequence, resulting in the corresponding score. The attention weights are then calculated using the *softmax* function, and finally, the input sequence is weighted and summed, as shown in Formula (4-3).

$$
Attention(Q, K, V) = softmax(\frac{QK^T}{\sqrt{d_{model}}})V
\tag{4-3}
$$

The multi-head attention mechanism divides the original data into different subspaces, calculates them separately, and then combines them for output. This improves the performance of the attention layer and better extracts the long-term feature dependencies of the data. Next, the encoding representation at each position undergoes layer normalization, and then each one is transformed through a feed-forward neural network to obtain a new representation, which is the final encoding representation at that position. Formula (4-4) shows the calculation process.

$$\begin{cases} Z = LayerNorm(Z + X) \\ Z_F = Relu(Z \cdot W_1 + b_1) \cdot W_2 + b_2 \end{cases} \tag{4-4}$$

After calculating through two layers of encoding, the encoding representation of the entire input sequence can be obtained. Then, a linear layer is applied to convert the d_model-dimensional vectors into X_t_size-dimensional vectors, where X_t_size represents the input vector dimension of the LSTM. The resulting vectors are then fed into a Long Short-Term Memory (LSTM) network.

The LSTM is a neural network model based on the traditional Recurrent Neural Network (RNN) that aims to solve the "long-term dependency" problem of traditional RNNs. The structure of the LSTM consists of a memory cell and input gates, forget gates, and output gates. The memory cell (C_t) is a state variable in the network used to store and transmit information. The input gate (i_t) and forget gate (f_t) control the input and retention of information, and the output gate (o_t) controls the output of information.

After LSTM computation, the output of the LSTM can be obtained from the output gate, and the output of the LSTM is then fed into a fully connected layer to obtain the prediction results with a dimension of $output\_size$. $Output\_size$ is a predetermined output vector dimension.

$$pred = Relu(O_t \cdot W_1 + b_1) \cdot W_2 + b_2 \tag{4-5}$$

4.2 Training and Testing

Before the training starts, the training dataset ($train\_loader$) is loaded, and the Adam optimizer and Mean Squared Error (MSE) loss function are used. The training parameters such as n_epochs and $batch\_size$ are also set. Then, the model is sent to the GPU, the optimizer is defined with a specific learning rate, and the loss function is defined. The training process starts, and the deep learning model is trained using the backpropagation (BP) algorithm [17]. The entire training process runs for the n_epochs round with several iterations per round equal to ($n\_epochs/batch\_size$). The training process is shown in Code Snippet 1.

Code Snippet 1. Model Training

```
Input: n_epochs, train_loader
01 for epoch in range(n_epochs):
02     for i, (inputs, label) in enumerate(train_loader):
03         inputs, label = inputs.to(device), label.to(device)
04         y_pred = Model (inputs)          # feedforward
05         loss = criterion (y_pred, label)    # calculate loss
06         optimizer.zero_grad()             # gradient to zero
07         loss.backward()                   # back propagation
08         optimizer.step()                  # update weights
Output: loss, Model weights
```

After the model is trained, the weights of the model are saved.

Then, model testing is performed by loading the model weights and evaluating the model using the test dataset. The obtained test data is compared with the actual values using the Root Mean Square Error (RMSE) and Mean Absolute Error (MAE) [18] evaluation metrics to assess the quality of the model. MAE can avoid the problem of error cancellation and thus accurately reflect the actual prediction error size.

RMSE is used to measure the deviation between observed values and true values, and smaller values indicate higher model accuracy. Similar to RMSE, MAE also evaluates the goodness of the model based on how small the difference is between predicted values and true values. The smaller the gap between predicted and true values, the better the model is.

4.3 Complexity Analysis

The time complexity of all the linear layers in the proposed Transformer_LSTM model is $O(n)$, where each encoder layer contains multi-head self-attention mechanisms and a feed-forward neural network. Considering the dot product operations between n keys and n queries, the computation complexity of self-attention for each layer is $O(n^2 \times d)$, where d is the dimension of the input vector. The time complexity of the LSTM layer mainly comes from the operations and parameter updates of the three gates, where the calculation of each gate has a complexity of $O(h^2)$, and h is the number of hidden units in the LSTM layer (i.e., *hidden_size*). Therefore, the entire time complexity of the Transformer_LSTM model is $O(n^2 \times d)$.

5 Experiment and Discussion

5.1 Experimental Platform

In this study, the deep learning framework PyTorch was used for modeling, and the experimental code was written in Python. The data processing and experiments were carried out on the Windows platform. The detailed configuration used in the experiments is shown in Table 2.

Table 2. Experimental configuration

Requirement Configuration	Details
Operating system	Windows11
Experimental Platform	PyCharm2022
Operating languages	Python3.9
Operating Framework	PyTorch1.12.1
GPU	Nvidia 2070s
Others	Sklearn, Pandas

All source codes in this paper are available at the website: (https://github.com/Luv U3OOO/EneryConsumptionForecast.git).

5.2 Experimental Design

The preprocessed dataset was used for comparative experiments of the proposed Transformer_LSTM model with RNN, GRU, LSTM, and encoder-decoder models. The models were fine-tuned and optimized, and their performance was evaluated by recording their predicted values, RMSE, and MAE. The experimental process is illustrated in Fig. 4.

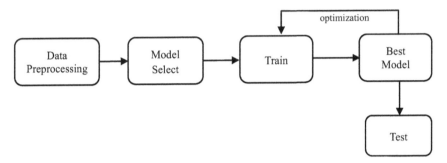

Fig. 4. Experimental flowchart

5.3 Experimental Operations and Results

For the experimental training parameters, this paper sets the training epochs for all models to be 200, dropout to be 0.2, batch size to be 20, and uses the Adam optimizer [19] and Mean Squared Error (MSE) loss function [20]. The learning rate is set to $1e-3$ and the weight decay is set to $1e-6$. This study conducted comparative experiments between the RNN, GRU, LSTM, BiLSTM, encoder-decoder models and the proposed Transformer_LSTM/BiLSTM models. The training loss of these models is shown in Fig. 5.

The model training parameters for Transformer-LSTM/BiLSTM include setting the multi-head attention mechanism to 8, setting the input dimension of the "PE" operation(d_model) to 32, setting the LSTM hidden layer size to 128, and setting the network layer number to 2. The detailed training loss is shown in Fig. 6.

After the models were trained and their weights were saved, this study used each model to predict the test set, compared the predicted results with the ground truth values, and evaluated the prediction performance of each model. This study used a 7-day history data to predict the energy consumption for the next 7 days, and the predicted results were integrated and denormalized to obtain accurate prediction values. Finally, the predicted values for the year 2018 were obtained and compared with the actual data. The comparisons between the ground truth and predicted results were visualized, as shown in Fig. 7.

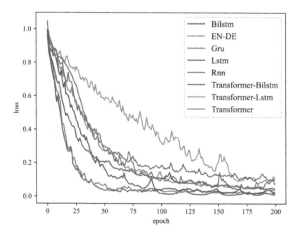

Fig. 5. All model losses

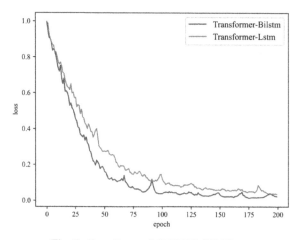

Fig. 6. Transformer-LSTM/BiLSTM Loss

Figure 7 presents a comprehensive comparison of the prediction results and actual ground truth data for commonly used deep learning models, including LSTM, Transformer, and the proposed Transformer-LSTM/BiLSTM model in this study. On the other hand, Fig. 8 and Fig. 9 focus specifically on the proposed Transformer-LSTM/BiLSTM model in this paper, providing a detailed comparison of its performance in predicting energy consumption and capturing its fluctuation trends.

Finally, we evaluated the prediction performance of various models on the test set, mainly using the RMSE and MAE metrics for evaluation. The evaluation results are shown in Table 3.

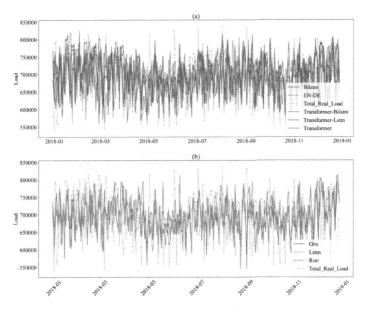

Fig. 7. Real and Predicted 7 Days Load Forecast

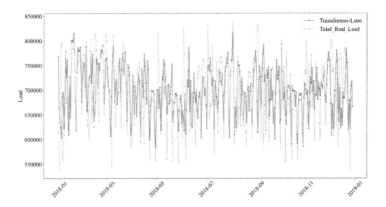

Fig. 8. Transformer-LSTM Real and Predicted 7 Days Load Forecast

5.4 Analysis of Experimental Results

From the perspective of training loss (Fig. 5), all models tend to converge after 200 training iterations. The BiLSTM, GRU, and Transformer-BiLSTM models exhibit similar loss curves, converging faster, and achieving the lowest loss. In contrast, the LSTM, Transformer-LSTM, and Transformer models display comparable loss curves. The RNN and encoder-decoder models demonstrate inferior performance compared to other models, possibly attributed to the limited feature extraction capability of RNNs for time-series data and the increased complexity of the encoder-decoder model, resulting in higher training loss and convergence challenges.

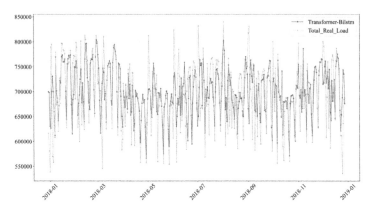

Fig. 9. Transformer-BiLSTM Prediction

Table 3. Evaluate Models

Models	RMSE	MAE
Transformer-BiLSTM	0.78	0.54
Transformer-LSTM	0.98	0.7
BiLSTM	1.15	0.88
LSTM	1.28	1.01
Transformer	1.33	1.03
LSTM-EN_DE	1.33	1.04
Gru	1.23	1.04
Rnn	1.36	1.12

From the comparison of the predicted results, it can be observed that both the proposed Transformer-LSTM/BiLSTM models and LSTM/BiLSTM models perform well in predicting future energy consumption fluctuations for the year 2018. These models exhibit lower prediction errors and accurately capture non-delayed fluctuation trends when compared to other models. It can be concluded that the Transformer-LSTM/BiLSTM models demonstrate good predictive ability in forecasting future energy consumption and its fluctuation trends.

Although the Transformer-LSTM/BiLSTM and BiLSTM models demonstrate good performance in terms of predictive accuracy, it is necessary to analyze the performance of each model using evaluation metrics such as RMSE and MAE. Figure 5 shows that the Transformer-BiLSTM and Transformer-LSTM models achieve improved RMSE and MAE metrics of (0.78, 0.54) and (0.98, 0.7), respectively, compared to LSTM (1.28, 1.01), BiLSTM (1.15, 0.88), and Transformer (1.33, 1.03). This indicates a noticeable improvement in prediction performance, with a difference of more than 0.3. Thus, the proposed Transformer-LSTM/BiLSTM models outperform the Transformer and LSTM models in terms of prediction.

Through analyzing the framework and calculation process of each model, I found several reasons why the Transformer-LSTM/BiLSTM performs better in medium-to-long-term energy prediction tasks:

(1) Self-attention mechanism: This model uses a self-attention mechanism to model sequence data, which can better capture long-term dependency relationships in the sequence and improve prediction accuracy.
(2) Multi-head attention mechanism: This model uses a multi-head attention mechanism to process input data, which can learn the relationships between different features and improve prediction accuracy.
(3) Using the Transformer encoder can map the input sequence to a high-dimensional vector space, thereby better representing the information in the sequence, improving prediction accuracy, and having better parallelism.
(4) LSTM decoder: Using LSTM and fully connected layers as the decoder can better capture information in the time series and improve prediction accuracy.
(5) Good robustness: The LSTM layer can control the flow and retention of information by adjusting the parameters of the gate mechanism, making the model highly robust and stable.

However, there is still some error in this experiment, which may be due to the selection of too many features that are difficult to fit completely, and the high complexity of the Transformer-LSTM/BiLSTM model, resulting in a longer training time.

Combined with the experimental results analysis, it can be concluded that the proposed Transformer-LSTM/BiLSTM model has significantly improved predictive performance compared to Transformer and LSTM models, indicating that this paper successfully combines the advantages of these two models and establishes a new model that improves the performance of medium-to-long term energy consumption prediction.

6 Conclusion

To address the challenge of accurately predicting medium- and long-term energy consumption, this paper proposes a neural network model that combines Transformer's self-attention mechanism and LSTM's memory capacity. The model effectively captures long-term dependencies and temporal patterns, leading to improved prediction accuracy for energy consumption over extended time horizons. However, the model does have some limitations, such as the high demand for computing resources and potential errors that are caused by redundant features. Our future work will focus on reducing model complexity, enhancing large-scale feature extraction capabilities, and validating the proposed approach on a wider range of electricity datasets. These efforts aim to optimize the model's performance and increase its applicability in real-world energy forecasting scenarios.

References

1. Ewa, C., Joanicjusz, N., Łukasz, N.: ARIMA models in electrical load forecasting and their robustness to noise. Energies **14**(23), 7952 (2021)
2. Liu, Y., Wang, Y., Yu, F.Y., et al.: Prediction and analysis of electricity demand in Jilin province 14th five-year plan based on GM (1, 1) method. J. Green Sci. Technol. **24**(18), 232–236 (2022)
3. Peng, L.L., Fan, G.F., Yu, M., et al.: Electric load forecasting based on wavelet transform and random forest. Adv. Theory Simul. **4**(12) (2021)
4. Wan, Q., Wang, Q.L., Wang, R.H., et al.: Short-term load forecasting of a regional power grid based on support vector machine. Power Syst. Clean Energy **32**(12), 14–20 (2016)
5. Cao, H.Z., Wang, T.L., Chen, P.D., et al.: Solar energy forecasting in short term based on the ASO-BPNN model. Front. Energy Res. (2022)
6. Lai, C.S., Mo, Z.Y., Wang, T., et al.: Load forecasting based on deep neural network and historical data augmentation. IET Gener. Transm. Distrib. **14**(24), 5927–5934 (2020)
7. Ibrahim, N.M., Megahed, A.I., Abbasy, N.H.: Short-term individual household load forecasting framework using LSTM deep learning approach. In: 2021 5th International Symposium on Multidisciplinary Studies and Innovative Technologies (ISMSIT), pp. 257–262. IEEE (2021)
8. Lu, C., Li, J., Zhang, G., et al.: A GRU-based short-term multi-energy loads forecast approach for integrated energy system. In: 2022 4th Asia Energy and Electrical Engineering Symposium (AEEES), pp. 209–213. IEEE (2022)
9. Jin, N.Y., Jo, H.H.: Prediction of weekly load using stacked bidirectional LSTM and stacked unidirectional LSTM. J. Korean Inst. Inf. Technol. **18**, 9–17 (2020)
10. Dorado Rueda, F., Durán Suárez, J., Del Real, T.A.: Short-term load forecasting using encoder-decoder WaveNet: application to the French grid. Energies **14**(9), 2524 (2021)
11. Vaswani, A., Shazeer, N., Parmar, N., et al.: Attention is all you need. In: Advances in Neural Information Processing Systems, vol. 30 (2017)
12. Chowdhary, K.R., Chowdhary, K.R.: Natural language processing. Fundam. Artif. Intell., 603–649 (2020)
13. Duong-Ngoc, H., Nguyen-Thanh, H., Nguyen-Minh, T.: Short term load forecast using deep learning. In: 2019 Innovations in Power and Advanced Computing Technologies (i-PACT), vol. 1, pp. 1–5. IEEE (2019)
14. Huang, L., Qin, J., Zhou, Y., et al.: Normalization techniques in training DNNs: methodology, analysis and application. IEEE Trans. Pattern Anal. Mach. Intell. (2023)
15. Voita, E., Talbot, D., Moiseev, F., et al.: Analyzing multi-head self-attention: specialized heads do the heavy lifting, the rest can be pruned. arXiv preprint arXiv:1905.09418 (2019)
16. He, K., Zhang, X., Ren, S., et al.: Identity mappings in deep residual networks. In: Leibe, B., Matas, J., Sebe, N., Welling, M. (eds.) Computer Vision–ECCV 2016: 14th European Conference, Amsterdam, The Netherlands, 11–14 October 2016, Proceedings, Part IV 14, vol. 9908, pp. 630–645. Springer, Cham (2016). https://doi.org/10.1007/978-3-319-46493-0_38
17. Cilimkovic, M.: Neural networks and back propagation algorithm. Institute of Technology Blanchardstown, Blanchardstown Road North Dublin, 15(1) (2015)
18. Hodson, T.O.: Root-mean-square error (RMSE) or mean absolute error (MAE): when to use them or not. Geosci. Mod. Dev. **15**(14), 5481–5487 (2022)
19. Zhang, Z.: Improved Adam optimizer for deep neural networks. In: 2018 IEEE/ACM 26th International Symposium on Quality of Service (IWQoS), pp. 1–2. IEEE (2018)
20. Popoola, S.I., Adetiba, E., Atayero, A.A., et al.: Optimal model for path loss predictions using feed-forward neural networks. Cogent Eng. **5**(1), 1444345 (2018)

Predicting Wind Turbine Power Output Based on XGBoost

Chang Liu, Jiandun Li$^{(\boxtimes)}$ (iD), and Haitao Wang

School of Electronic Information, Shanghai Dianji University, Shanghai 202304, China
lijd@sdju.edu.cn

Abstract. The prediction of wind power is crucial to ensuring the reliability and economic efficiency of wind power generation systems, as well as to maintaining balance and efficient operation of power systems. However, due to the non-stationary and chaotic nature of wind speeds, predicting wind power is a challenging task. Recently, various solutions have been proposed, e.g., SARIMA-based models and BP neural network-based models, which have successfully predicted periodicity and short-term wind power generation, but their performances are limited. In this paper, we select the top-eight most significant attributes from a public wind power dataset, i.e., wind direction, hub temperature, bearing shaft temperature, gearbox bearing temperature, gearbox oil temperature, rotor speed, reactive power and active power. We then train eight supervised machine learning models, i.e., Linear Regression, Support Vector Machine (SVM), eXtreme Gradient Boosting (XGBoost), K-Nearest Neighbors (KNN), Ridge and Lasso Regression, Decision Tree, and Gradient Boosting, to predict the wind power output of the next 70 days. Experimental results showed that the XGBoost model outperforms others (R-squared score = 0.96, accuracy = 95.39%, MAE = 39.43, and cross-validation score = 0.98). Compared to the state-of-the-art performance achieved by the Random Forest model, XGBoost has improved the prediction accuracy by 4.69% points.

Keywords: Machine learning · ensemble learning algorithms · time series forecasting · wind power generation

1 Introduction

In recent years, as the contradiction between the increasingly scarce traditional petrochemical energy and the strong demand for energy from social and economic development has intensified [1], the world has been vigorously developing clean energy to reduce energy dependence, optimize energy structure, ensure stable energy supply, and promote sustainable economic and social development. Among these clean energy sources, wind power has become the preferred choice for most countries to develop renewable energy due to its high industrial maturity, low generation cost of electricity, and significant contributions to the physical and social environment. According to the "Global Wind Energy Report 2022" released by the Global Wind Energy Council (GWEC), the global

J. Li et al. (Eds.): 6GN 2023, LNICST 553, pp. 315–330, 2024.
https://doi.org/10.1007/978-3-031-53401-0_28

wind power installed capacity increased by 93.6 GW in 2021, and wind power bids reached 88 GW, representing an increase of 12% and 153%, respectively, compared to the previous year.

In the process of promoting and developing wind power generation, accurate output prediction plays a vital role. Complex weather changes can lead to inaccurate wind power prediction, which poses multiple risks. For example, the wind power system may continue to operate or cause shutdown damage if it cannot predict real-time changes in wind speed, resulting in the waste of wind energy resources or missed opportunities for wind energy peaks. Moreover, inaccurate wind power prediction can further affect the stability of the power grid, as the volatility of wind power can cause changes in grid frequency and instability. These issues can affect the efficiency and reliability of wind power generation, reducing its economic and environmental benefits. Therefore, accurate wind power prediction is indispensable, and it requires combining a large amount of meteorological data and advanced statistical methods to effectively predict wind energy production. This is particularly important for wind power plants as it allows managers to take proactive measures. By adopting accurate wind power prediction technology, wind energy resources can be maximized, energy costs can be reduced, and wind power generation stability can be improved, contributing to sustainable development.

Wind power generation forecasting is challenging due to incomplete datasets, nonlinear and chaotic time series, and external factors like seasonal changes and equipment maintenance. Statistical methods like Autoregressive Integrated Moving Average Model (ARIMA), Vector Autoregressive Model (VAR), and ETS are effective for some problems but may not capture complex nonlinear relationships. Deep learning models like Long Short Term Memory (LSTM) and Gate Recurrent Unit (GRU) can capture more complex temporal relationships, but may accumulate errors over long-term time series. Combining multiple techniques such as cross-validation, feature engineering, and anomaly detection improves accuracy and robustness. Time series study methods include ARIMA, LSTM, and Convolutional Neural Networks (CNN).

Based on the shortcomings of existing research, this paper utilizes the Pearson correlation coefficient method with the support of a public dataset to perform feature selection, selecting 8 factors that affect ActivePower. Then, 9 machine learning models were trained, including Linear Regression, Support Vector Machine (SVM), Extreme Gradient Boosting algorithm (XGBoost), K-nearest neighbors algorithm (KNN), Ridge regression, Lasso regression, Decision Tree, Gradient Boosting, and Random Forest. Among them, XGBoost achieved the highest prediction accuracy.

The contribution of this paper is threefold: (1) Using data visualization techniques to analyze the correlation between features, perform data cleaning, including feature selection, missing value filling, and outlier removal. (2) Improving the accuracy and efficiency of wind power generation forecasting. (3) Comparing multiple models and verifying the superiority of the XGBoost model in terms of prediction accuracy and training time.

The rest of the paper is organized as follows: Section 2 surveys the relevant methods for predicting the power generation of turbines that have already been proposed. Section 3 describes the statistical analysis of public datasets. Section 4 elaborates on the steps of data preprocessing, feature selection, and model training in detail. Section 5 introduces the experimental situation and conducts in-depth analysis and discussion of the results. Section 6 concludes the paper.

2 Related Work

Studies on wind turbine production forecasting can be categorized into short-term and medium-to-long-term forecasting. For short-term forecasting, algorithms such as BP neural network, SVM, and Random Forest are commonly used by researchers. On the other hand, for medium-to-long-term forecasting, the ARIMA algorithm is frequently utilized to process time series data, while the LSTM algorithm is well-suited for predicting long-term sequence data.

In short-term forecasting, Bianco [2] used a Linear Regression model to predict electricity consumption in Italy. Although it solved simple linear problems, this model's prediction accuracy significantly decreased for more complex or nonlinear problems. Shi [3] used the BP neural network algorithm to predict short-term wind power. This model decomposed wind speed and power sequences at different scales using wavelets and used multiple BP neural networks to predict each frequency component. However, this model required a considerable amount of historical wind speed data for training, and its prediction accuracy was negatively affected by poor data quality, and its computational complexity was high. Although the BP neural network had a simple structure and robust problem-solving abilities, its learning algorithm used a gradient-based approach to adjust the network and weights, which gradually reduced the error. The objective function, however, was only the sum of the square differences between each input and output, leading to the network over-correcting data errors, resulting in an overfitting phenomenon. Ma [4] used the SVM algorithm to solve the difficulty of input selection encountered in transient stability analysis, achieving good results. However, the model was relatively single, making it challenging to compare with other algorithms and highlight the model's advantages. Li [5] also studied the SVM method based on short-term load forecasting. SVM was originally used to solve pattern recognition problems to discover decision rules with good generalization performance. The main idea of this method was to divide data into different categories by maximizing the distance between data. By using kernel functions, SVM could handle non-linear classification problems, and it is widely used in practical applications such as image classification, text classification, speech recognition, and other fields. Wang [6] used the random forest algorithm to predict power generation. Random forest is an ensemble learning algorithm based on decision trees that models data using multiple decision trees and combines their results into a single overall prediction result. It is simple to implement, computationally efficient, and exhibits strong performance in many practical tasks. The Random Forest algorithm [7] applies randomness and ensemble learning ideas to decision trees, making it better at handling large-scale, high-dimensional data with better performance and robustness. However, with only a 5-day dataset, the prediction on such a small dataset does not have an advantage, and only simple methods were used to decompose historical meteorological data, with a relatively single feature, resulting in poor prediction accuracy.

In mid-term time series forecasting, Zhao [8] utilized the ARIMA algorithm to predict the output power of photovoltaic systems and compared multiple models. However, the prediction accuracy was not satisfactory, leaving room for improvement. ARIMA builds mathematical models based on historical data of time series using methods such as autoregression and moving average to predict future trends. However, it is difficult to determine whether the analyzed sequence is linear or nonlinear, which limits the accuracy of the method. In particular, when the data displays seasonality, the error is even greater. Additionally, ARIMA requires significant prior knowledge to determine model parameters, further limiting its applicability in wind power generation forecasting. Wang [9] chose the LSTM [10] recurrent neural network as the model for wind power generation prediction. By using the random dropout method to randomly discard nodes in the hidden layer, the generalization ability of the neural network has been enhanced. The network transforms the output through a fully connected layer and continuously adds the latest meteorological and power generation data to the dataset to achieve rolling updates. However, as the prediction step length increases, the accuracy of the model gradually declines. The model requires complex optimization and training, which takes a long time and is easily affected by the gradient disappearance or explosion phenomenon, leading to slow or non-convergence of the model.

Short-term wind turbine power generation prediction using BP neural networks and ARIMA models is susceptible to significant prediction errors due to sudden events and seasonality. In contrast, SVM and LSTM models are less sensitive to feature selection and may use irrelevant features, thereby compromising prediction accuracy. García Hinde [11] proposed a hybrid model that combines ARIMA and SVM to predict the nonlinear sensitive part of the load. However, this approach is computationally expensive and has limitations in dealing with multidimensional data. To address these issues, XGBoost, which incorporates regularization terms to prevent overfitting and reduce variance, has been proposed. Ye [12] developed a short-term traffic flow prediction algorithm using CNN-XGBoost, and Sun [13] used XGBoost to identify complex carbonate rock lithology logging. Zhu [14] proposed a hybrid prediction model that uses pattern sequence matching and Extreme Gradient Boosting for holiday load forecasting. To address the challenges posed by incomplete data, seasonal changes, and long-term prediction for wind energy production capacity, we propose an algorithm model based on XGBoost [15] for wind turbine [16] power generation prediction. Additionally, missing or outlier data can affect the accuracy and reliability of data analysis results, limiting the robustness of the data.

3   Dataset

The dataset used in this paper comes from data sourced from a wind turbine's wind turbine generator (https://www.kaggle.com/datasets/theforcecoder/wind-power-foreca sting), with data features including: active power, ambient temperature, bearing shaft temperature, blade 1 pitch angle, blade 2 pitch angle, blade 3 pitch angle, control box temperature, gearbox bearing temperature, gearbox oil temperature, generator speed, generator winding 1 temperature, generator winding 2 temperature, hub temperature, mainframe box temperature, nacelle position, reactive power, rotor speed, turbine status, wind direction, and wind speed. Table 1 outlines the count, mean, and variance of several attributes.

Table 1. Data description

Attribute	Count	Mean ± Standard Deviation
Active Power	94,624	(619.04 ± 610.92)
Ambient Temperature	93,698	(28.78 ± 4.36)
Bearing Shaft Temperature	62,380	(43.11 ± 5.17)
Blade1PitchAngle	41,987	(9.757 ± 20.65)
Blade2PitchAngle	41,882	(10.04 ± 20.27)
Blade3PitchAngle	41,882	(10.04 ± 20.27)
Control Box Temperature	62,022	(0.00 ± 0.00)
Gearbox Bearing Temperature	62,402	(64.38 ± 10.02)
Gearbox Oil Temperature	62,430	(57.56 ± 6.32)
Generator RPM	62,287	(110,2.15 ± 527.97)
GeneratorWinding1Temperature	62,419	(72.46 ± 22.63)
GeneratorWinding2Temperature	62,441	(71.83 ± 22.65)
Hub Temperature	62,268	(36.98 ± 4.88)
Main Box Temperature	62,369	(39.64 ± 5.43)
Nacelle Position	72,108	(196.31 ± 88.28)
Reactive Power	94,622	(88.07 ± 116.50)
Rotor RPM	62,119	(9.91 ± 4.72)
Turbine Status	62,759	(228,4.02 ± 359,028.79)
Wind Direction	72,108	(196.31 ± 88.28)
Wind Speed	94,469	(5.88 ± 2.62)

4 Methodology

We apply XGBoost to predict the wind turbine electrical energy output, and the approach can be divided into three steps:

(1) The original dataset is visualized and processed, which includes addressing missing values, detecting and correcting outliers, and performing data transformations. Data cleaning and transformation are often required due to the large number of abnormal data points in wind farm records. For instance, Zhao's study [17] followed this approach.

(2) In the feature engineering phase, feature selection and extraction are performed to determine which features significantly impact the prediction. Feature transformation or dimensionality reduction may also be applied. For example, Deng [18] incorporated residual learning into the model to avoid gradient disappearance, and feature selection was employed to improve the model's generalization ability.

(3) Multiple algorithms are employed to establish a power generation prediction model. The model is trained, and the data is fitted and optimized to achieve good results. Wu's model [19] learned the potential features of sequence data and mapped them into feature vectors, and the GRU-NN was employed to establish a power load prediction model. Xie's model [20] integrated and recalibrated features using a fusion component, demonstrating the model's ability to capture multi-scale temporal patterns.

4.1 Data Preprocessing

The dataset involved in this study contains a large number of missing values and outliers. Specifically, Fig. 1 shows outliers in the *ActivePower* variable, and data points with *ActivePower* < 0 need to be removed. Furthermore, many features in the dataset have missing values, and as depicted in Fig. 2, the missing values of *ActivePower* coincide with *WindSpeed* values, which are recorded every 15 min. This indicates that the missing data may be caused by a lack of wind during that period. Since this dataset is used for time series prediction, deleting too much data can have a significant impact on the time series. Therefore, data imputation techniques such as mean, median, and interpolation can be used to fill missing values. In this study, the *fillna* method was employed with left and right padding to avoid disrupting the data distribution. Using other methods could result in data skewness and lead to inaccurate results.

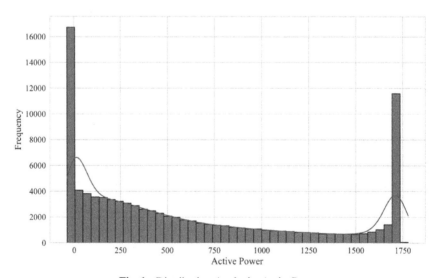

Fig. 1. Distribution Analysis: ActivePower

The *WTG* attribute in the dataset has only one value *G01*, which is not related to the output attribute *ActivePower*. Therefore, this attribute is not selected for analysis. Similarly, the *Control Box Temperature* attribute only contains missing or zero values and is also not relevant to the *ActivePower* attribute. Thus, it is removed from the dataset.

4.2 Feature Selection and Analysis

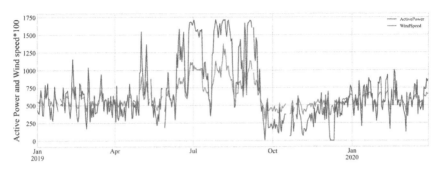

Fig. 2. Daily Average of Active Power and Wind Speed * 100 Graph

In order to facilitate better comparison between the daily averages of *ActivePower* and *WindSpeed*, and to highlight the monthly variations more effectively, we apply a scaling factor of 100 to the wind speed. This adjustment aims to achieve smoother curves and enhanced visual contrast (Fig. 3).

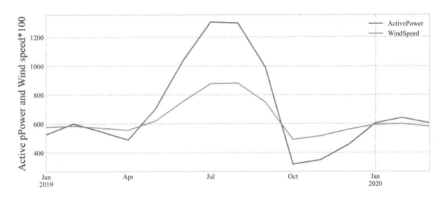

Fig. 3. Monthly Average of Active Power and Wind Speed * 100 Graph

The objective of this study is to employ both single-feature and multi-feature selections to train the dataset. During the single-feature selection process, data visualization revealed a linear correlation between *WindSpeed* and *ActivePower*. As a result, we selected *WindSpeed* as the input feature for single-feature prediction. Subsequently, we gradually introduced multiple features for prediction purposes (Fig. 4).

The degree of correlation between different turbine features can be assessed using the Pearson correlation coefficient. This method allows for sorting the remaining features based on their correlation with the predicted value of *ActivePower*. Only those features with a correlation coefficient above 0.3 are selected as input features for the model. Finally, the processed data is standardized to speed up the model training speed.

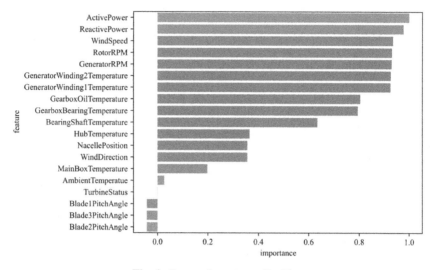

Fig. 4. Feature Importance Ranking

4.3 XGBoost Model

XGBoost uses weak classifiers to create an additive model of CART regression trees. An objective function is minimized to determine tree structure and leaf node values. XGBoost simplifies optimization by expressing the objective function in terms of output values, then solving for optimal values using derivatives. The objective function is decomposed into leaf node contributions, and the optimization problem is transformed into an output value optimization. Optimal values predict new samples and correspond to objective function value. Formula (1) shows XGBoost's objective function.

$$
\begin{aligned}
Obj^{(t)} &\simeq \sum_{i=1}^{n} \left[g_i f_t(x_i) + \tfrac{1}{2} h_i f_t^2(x_i) \right] + \Omega(f_t) \\
&= \sum_{i=1}^{n} \left[g_i w_q(x_i) + \tfrac{1}{2} h_i w_q^2(x_i) \right] + \gamma T + \lambda \tfrac{1}{2} \sum_{j=1}^{T} w_j^2 \\
&= \sum_{j=1}^{T} \left[\left(\sum_{i \in I_j} g_i \right) w_j + \tfrac{1}{2} \left(\sum_{i \in I_j} h_i + \lambda \right) w_j^2 + \gamma T \right]
\end{aligned}
\tag{1}
$$

The XGBoost algorithm follows a series of steps, which include initializing the samples, defining the loss function, building a new decision tree, and training and predicting the model, as depicted in Fig. 5.

4.4 Training

Given its high predictive performance, XGBoost is the primary focus of this study. XGBoost is an ensemble learning algorithm that uses gradient boosting to iteratively optimize the model while employing pruning and regularization techniques to prevent overfitting. This algorithm is highly efficient, scalable, supports multiple languages, and

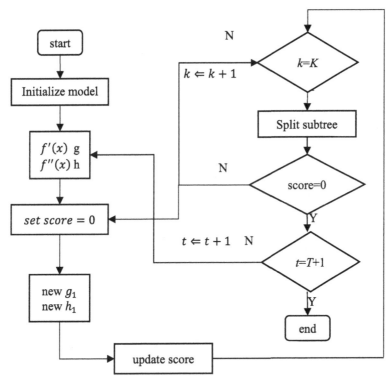

Fig. 5. Algorithm flow chart

provides feature importance evaluation and strong interpretability. In this study, XGBoost is utilized to preprocess incomplete and anomalous data and predict the electrical energy output of wind turbines. XGBoost and Random Forest are two classic models commonly employed in machine learning tasks, with the hyperparameters of all models set to default values during training. As LSTM is also employed for model prediction, the TensorFlow framework is utilized to set up the environment required for LSTM. Finally, the hyperparameters of XGBoost are continuously adjusted to achieve optimal performance. During the model training process, we take into account comprehensive data preprocessing (as discussed in the preceding section). Nevertheless, the initial performance of the XGBoost model falls short of the desired standards. Fortunately, through the diligent adoption of advanced methodologies, including cross-validation, we conducted targeted hyperparameter tuning. As a result, we successfully identify an optimal parameter combination, leading to a significant enhancement in the model's performance, thereby meeting our expectations.

4.5 Validation

To assess the accuracy of the algorithm, this paper uses three common evaluation metrics: Root Mean Square Error (RMSE), Mean Absolute Error (MAE), and Coefficient of

Determination (R2 score), these indicators are widely used in regression analysis and provide a comprehensive measure of the performance of the algorithm.

4.6 Time Complexity

XGBoost is a tree-based ensemble learning method with time complexity influenced by tree number/depth and feature count. Despite higher complexity than traditional ML algorithms, XGBoost handles large datasets and features efficiently with parallel computing. Complexity is $O(n \log(n))$ for n training samples, and $O(n \log(n))$ for each decision tree. Optimizations like distributed computing, sparse matrix storage, and caching lower practical complexity below theoretical complexity.

5   Experiment and Discussion

5.1 Experimental Design

The proposed method for predicting the electrical energy output of wind turbines in this paper is outlined in Fig. 6 in five steps:

(1) Obtain the turbine dataset and select the basic and wind data as the primary dataset;
(2) Preprocess the data by filling missing values, handling abnormal values, and scaling features;
(3) Divide the dataset into a training and testing set, and perform feature extraction;
(4) Train the dataset using the XGBoost model and other models;
(5) Test subsequent datasets using the XGBoost model and other models, predict future power generation, and identify equipment that requires maintenance.

5.2 Experimental Operation

This experiment employed the XGBoost algorithm to predict the electrical energy output of wind turbines. The XGBoost algorithm model was implemented on the TensorFlow machine platform using Jupyter notebook. Nine other algorithms, including Random Forest, and LSTM, were also used to predict turbine power generation, and their performances were compared with that of the XGBoost model. The training set was set to 80% of the dataset, and the test set was set to 20%. Table 2 shows the model parameters used in the training set.

5.3 Experimental Result

We present the comparison of XGBoost, Random Forest, LSTM, and other algorithms for single-feature time series analysis in Table 3. This initial experiment aims to compare their errors. The primary objective of this experiment is to assess the differences and errors in single-dimensional features and, through a second experiment, observe if further differences arise when considering the impact of multi-dimensional features.

In this experiment, represented in Table 4, we include additional algorithms for comparison and take into account the element of time in our analysis. Notably, we introduce

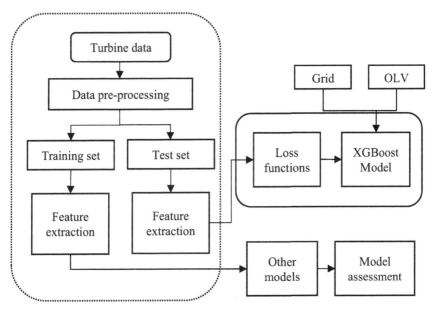

Fig. 6. Experimental procedure

Table 2. Parameters of three algorithms

Model	Parameter
RF	*n_estimators*:100
	max_depth:None
	*random_state:*None
LSTM	3 LSTM layers
	1 dense layer
	Learning_rate: 0.3
XGBoost	*max_depth*:6
	n_estimators:720

Table 3. Single-Feature Time-Series Model Error Comparison.

Module	MAE	RMSE	R2score
XG Boost	40.66	52.21	0.926
RF	41.61	53.25	0.923
LSTM	187.76	162.71	−0.19

Table 4. Multi-Feature Time-Series Model Error Comparison

Module	MAE	RMSE	R2score	time
Linear SVR	74.56	215.62	0.814	0.04 s
Linear Regression	77.97	195.90	0.847	0.00 s
Ridge	77.96	195.88	0.847	0.00 s
Lasso	76.54	191.90	0.853	0.01 s
K Neighbors Regressor	52.35	116.78	0.946	0.03 s
Decision Tree Regressor	71.56	152.93	0.890	0.28 s
Gradient Boosting	61.42	130.75	0.932	6.76 s
XGBoost	46.91	107.46	0.954	2.16 s
Random Forest Regressor	59.66	152.56	0.907	15.75 s
LSTM	104.78	162.18	0.911	682.18 s

the LSTM algorithm, a type of long short-term memory recurrent neural network, which provides a distinct contrast and enables further detailed analysis.

We display the ActivePower Forecasting results using the XGBoost algorithm in Fig. 7. Through this figure, we gain insights into XGBoost's performance in predicting ActivePower. Subsequent sections will further discuss the model's performance, comparing it with other algorithms.

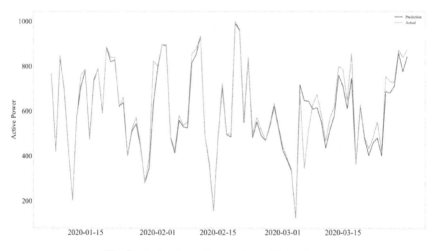

Fig. 7. ActivePower Forecasting with XGBoost

In Fig. 8 below, we present a comparison of ActivePower Forecasting with Random Forest. The performance disparities between the two algorithms in terms of ActivePower prediction become clearly observable through this figure.

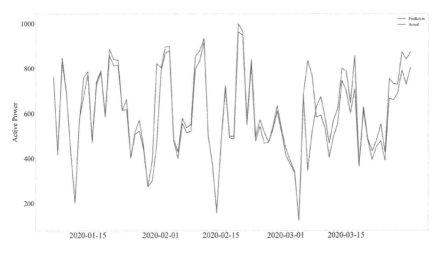

Fig. 8. ActivePower Forecasting with Random Forest

Figures 7, to 9 depict the fitting curves of predicted and actual values for each model, placing particular focus on the comparison of discrepancies between the two algorithms. The primary aim of this study revolves around identifying variations in the amplitude of fluctuations between the algorithms. Detailed exposition of the comparative findings awaits in the subsequent analysis section.

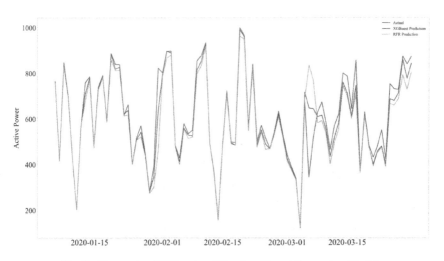

Fig. 9. Comparing XGBoost and Random Forest Regression Models

5.4 Experimental Comparison and Discussion

This paper compares XGBoost with 10 supervised models, which are the linear regression, support vector machine algorithm (SVM), K-nearest neighbor (KNN), ridge regression, Lasso regression, decision tree, gradient boosting, random forest, and long short-term memory (LSTM).

- LSVR: vector autoregressive model with support vector regression objective function.
- LR: Used to establish a linear relationship model between one or more independent variables and dependent variables.
- Ridge: A vector autoregressive model with L2 regularization term, commonly used for multivariate time series prediction problems.
- Lasso: Used to calculate the least squares method model, which can be used for variable selection and regression analysis.
- KNN: A non-parametric supervised learning algorithm used for regression tasks. The algorithm is suitable for nonlinear relationships and large amounts of training data.
- DTR: A supervised learning algorithm based on decision trees used for regression problems. It can handle nonlinear relationships and missing values, and is also effective for multiple outputs and mixed data types.
- GBR: It establishes a powerful predictive model by gradually improving the prediction accuracy of the decision tree.
- RFR: It is based on multiple decision trees to establish a model and predicts the target value of new data points by voting.
- LSTM: Long Short-Term Memory networks, as a special type of RNN, its temporal memory can solve sequence data to a certain extent.

This paper's objective is to predict the power output of wind turbines. In order to assess the prediction accuracy of the XGBoost model, the paper conducts experiments comparing it with ten other models, including Random Forest and LSTM. The prediction accuracy of each model is presented below:

The initial experiment aimed to predict using a single feature. The data was fed into XGBoost, Random Forest, and LSTM models for prediction, followed by an evaluation of these models using performance metrics. The resulting experimental findings have been summarized in Table 3.

According to Table 3, the R2 scores of XGBoost, Random Forest, and LSTM are 0.926, 0.923, and −0.19, respectively. Both XGBoost and Random Forest show a high degree of fit, while LSTM's R2 score is very poor, at −0.19. The MAE of XGBoost and Random Forest are 40.66 and 41.61, respectively. Compared to Random Forest, XGBoost reduces the MAE by 0.99. The RMSE of XGBoost and Random Forest are 52.21 and 53.25, respectively. XGBoost reduces the RMSE by 0.96 compared to Random Forest. Based on the results of experiment one, it can be concluded that the prediction accuracy of XGBoost and Random Forest is similar, and there is not much difference in the results. Experiment one uses a single-dimensional feature, so experiment two will select an appropriate number of features to further improve the prediction accuracy. In order to better compare these two models and select the better one, experiment two will increase the number of features. Since Random Forest's training time is longer, the evaluation indicator of training time will be added, and more models will be selected for comparison.

In the second experiment, a dimensionality reduction technique was applied, and 10 out of 20 input attributes were selected. The XGBoost model achieved an R2 score of 0.954, outperforming the Random Forest model's R2 score of 0.907. The *KNeighborsRegressor*, GradientBoostingRegressor, and LSTM models achieved R2 scores of 0.946, 0.932, and 0.911, respectively. Moreover, Linear Regression, Ridge, and Lasso had relatively shorter training times compared to XGBoost. The training time of XGBoost was 2.16 s, while LSTM took even longer. Additionally, the XGBoost model achieved lower MAE and RMSE values of 46.91 and 107.46, respectively, compared to other models. The comparison between the first and second experiments indicates that selecting an appropriate number of features can improve the algorithm's prediction accuracy. XGBoost outperforms Random Forest in terms of evaluation indicators, such as MAE, MSE, and training time, and has better robustness.

Figures 7 and 8 display the fitting curves of the two models with the true values. As the dataset has a data collection interval of 15 min, the graphs appear relatively chaotic. Therefore, to better visualize the error between predicted and actual values, the time interval is set to 1 day using the mean method to display the graph, as shown in Fig. 9. It is evident from the figure that XGBoost outperforms Random Forest in terms of the degree of fit.

The investigation demonstrates XGBoost's clear superiority over Random Forest and LSTM for long-term prediction. And possibles driving forces behind this phenomenon are:

(1) Handling Structured Data: XGBoost excels in processing tabular data with numerous features and intricate interdependencies. This capability significantly enhances prediction accuracy.
(2) Efficiency: Compared to LSTM, XGBoost shows much faster training and prediction speeds on large-scale datasets. This efficiency allows it to handle extensive historical data and make predictions quickly.
(3) Robustness: XGBoost's ensemble learning strategy with multiple weak classifiers makes it robust against noise and outliers. In long-term prediction tasks, where data quality is crucial, this feature ensures heightened reliability.

6 Conclusions

To overcome the challenge of low prediction accuracy in wind turbine power generation, this paper proposes an approach based on XGBoost, and employs the Pearson correlation coefficient method for feature selection using public datasets. The simulation experiment demonstrates that this approach achieves the highest performance and outperforms others. XGBoost outperforms other models with superior prediction accuracy and robustness by utilizing sample data to conduct feature selection. Multi-model comparison demonstrates its effectiveness in wind turbine production for important references. However, limitations exist due to the lack of more accurate deep and hybrid models. Future research will focus on optimizing and integrating multiple deep learning models to improve generalization capability.

References

1. Yude, B.: Scheduling optimization and emission analysis of clean energy under the background of dual carbon. Electric. Measur. Instrum., 1–10 (2022)
2. Bianco, V., Manca, O., Nardini, S.: Electricity consumption forecasting in Italy using linear regression models. Energy **34**(9), 1413–1421 (2009)
3. Shi, H., Yang, J., Ding, M., Wang, J.: A short-term wind power prediction method based on wavelet decomposition and BP neural network. Dianli Xitong Zidonghua/Autom. Electr. Power Syst. **35**(16), 44–48 (2011)
4. Ma, Q., Yang, Y.-H., Liu, W.-Y., Qi, Z., Guo, J.-Z.: Power system transient stability assessment with combined SVM method mixing multiple input features. Zhongguo Dianji Gongcheng Xuebao/Proc. Chin. Soc. Electric. Eng. **25**(6), 17–23 (2005)
5. Li, Y.-c., Fang, T.-j., Yu, E.-k.: Study of support vector machines for short-term load forecasting. Proc. CSEE, 55–59 (2003)
6. Yangguang, W., Min, X., Xiaoliang, D., Cheng, B.: Random forest model of wind power forecasting based on wavelet transform. Electr. Eng. **542**(8), 48–52+52 (2021)
7. Breiman, L.: Random forests. Mach. Learn. **45**(1), 5–32 (2001)
8. Binbin, Z., et al.: Photovoltaic power prediction in distribution network based on ARIMA model time series. Renew. Energy Resour. **37**(6), 820–823 (2019)
9. Wei, W., et al.: Wind power forecast based on LSTM cyclic neural network. Renew. Energy Resour. **38**(9), 1187–1191 (2020)
10. Hochreiter, S., Schmidhuber, J.: Long short-term memory. Neural Comput. **9**(8), 1735–1780 (1997)
11. García Hinde, Ó., Gómez Verdejo, V., Martínez-Ramón, M.: Forecast-informed power load profiling: a novel approach. Eng. Appl. Artif. Intell. **96**, 103948 (2020)
12. Jing, Y., Li-juan, L., Zhen-xu, T.: Short-term traffic flow forecasting based on CNN-XGBoost. Comput. Eng. Des. **41**(4), 1080–1086 (2020)
13. Sun, C., Lu, Q., Zhu, S., Zheng, W., Cao, Y., Wang, J.: Ultra-short-term power load forecasting based on two-layer XGBoost algorithm considering the influence of multiple features. Gaodianya Jishu/High Voltage Eng. **47**(8), 2885–2895 (2021)
14. Zhu, K., Geng, J., Wang, K.: A hybrid prediction model based on pattern sequence-based matching method and extreme gradient boosting for holiday load forecasting. Electr. Power Syst. Res. **190**, 106841 (2021)
15. Chen, T., Guestrin, C.: XGBoost: a scalable tree boosting system. In: Proceedings of the 22nd ACM SIGKDD International Conference on Knowledge Discovery and Data Mining (2016)
16. Chen, Z., Guerrero, J.M., Blaabjerg, F.: A review of the state of the art of power electronics for wind turbines. IEEE Trans. Power Electron. **24**(8), 1859–1875 (2009)
17. Zhao, Y., Ye, L., Zhu, Q.: Characteristics and processing method of abnormal data clusters caused by wind curtailments in wind farms. Dianli Xitong Zidonghua/Autom. Electr. Power Syst. **38**(21), 39–46 (2014)
18. Deng, C.-L., et al.: Prediction of the efficacy of radiotherapy and chemotherapy for cervical squamous cell carcinoma based on random forests. Ruan Jian Xue Bao/J. Softw. **32**(12), 3960–3976 (2021)
19. Xiaogang, W., Jie, Y., Chang, G., Yajie, T., Chouwei, N., Qingfeng, J.: Ultra-short-term power integration forecasting method for wind-solar-hydro based on improved GRU-CNN. Electr. Power, 1–9 (2023)
20. Xie, G.-C., Duan, L., Jiang, W.-P., Xiao, S., Xu, Y.-F.: Pedestrian volume prediction for campus public area based on multi-scale temporal dependency. Ruan Jian Xue Bao/J. Softw. **32**(3), 831–844 (2021)

Loop Closure Detection Based on Local and Global Descriptors with Sinkhorn Algorithm

Wei Xiao$^{(\boxtimes)}$ and Dong Zhu

School of Electronic and Information, Shanghai Dianji University, Shanghai, China
xiaow@sdju.edu.cn

Abstract. This paper presents a novel loop closure detection pipeline for SLAM systems, addressing the limitations of current deep learning methods in maintaining 3D point cloud structure and extracting high-quality semantic features. We utilize U-Net and FPN for feature extraction, with a descriptor generator that learns from local descriptors. The Sinkhorn algorithm is incorporated for 6DOF transformation matching between point clouds, effectively managing occlusions and aligning source and target clouds. Our method, evaluated on the KITTI dataset, outperforms traditional and other deep learning methods in computational efficiency and real-time performance.

Keywords: U-Net · Feature Pyramid Networks (FPN) · Sinkhorn algorithm · 6DOF transformation · Unbalanced optimal transport

1 Introduction

Simultaneous Localization and Mapping (SLAM) serves as a cornerstone in the fields of robotics and computer vision, facilitating autonomous navigation and path planning in unexplored environments. The SLAM process typically encompasses three stages: (1) consecutive scan alignment, which aligns successive scans utilizing odometry or inertial measurement unit data, (2) loop detection to recognize previously visited areas, and (3) loop closure, which adjusts the current scan to coincide with previously visited positions, thereby rectifying the map. In stage (1), due to the sequential nature of scans, errors can accumulate over time. Stages (2) and (3) mitigate accumulated errors by introducing new constraints to the pose graph when a loop is detected.

In the context of visual SLAM, the identification of loop closures is a critical task that involves discerning when a robot revisits a scene it has previously navigated. The precise detection of loop closures is instrumental in correcting cumulative errors, enhancing the global consistency of the map, and refining the robot's pose estimation. Loop closure detection is typically accomplished by comparing visual features from the current camera frame with those stored in a map or a database of previously observed frames.

© ICST Institute for Computer Sciences, Social Informatics and Telecommunications Engineering 2024
Published by Springer Nature Switzerland AG 2024. All Rights Reserved
J. Li et al. (Eds.): 6GN 2023, LNICST 553, pp. 331–346, 2024.
https://doi.org/10.1007/978-3-031-53401-0_29

Nevertheless, loop closure detection in visual SLAM presents several challenges. These include appearance variations due to alterations in lighting conditions, occlusions, or dynamic objects, which may lead to significant discrepancies between the visual features extracted from the same location at different times. Additionally, the computational complexity of feature matching can be substantial, particularly when dealing with large-scale environments and long-term operation. Another challenge is the potential for false loop closures, where unrelated places are incorrectly identified as the same location, which can result in catastrophic failures in the SLAM algorithm.

Bag of Words (BoW) use image feature descriptors to create a bag of words model, converting images into sparse vector representations. By calculating the similarity between images, potential loop closures can be identified. Common BoW methods include FAB-MAP [8], DBoW2 [18], etc. Global feature-based methods use global features from the entire image to represent a scene, such as GIST [20] and VLAD [12] descriptors. They usually strike a balance between computational efficiency and detection accuracy. In the realm of deep learning, intricate neural structures like Convolutional Neural Networks (CNNs) and Recurrent Neural Networks (RNNs) are employed for the purpose of feature extraction from visual data. Through specialized training of these neural networks, they become adept at formulating more advanced scene descriptors, thereby enhancing the fidelity and robustness in the identification of loop closures. A case in point is the NetVLAD algorithm [2].

Despite the numerous deep learning methods proposed for loop closure detection, which are generally faster than traditional handcrafted methods, they often cannot retain the 3D structural features of point clouds and fail to learn local fine-grained features while obtaining a sufficiently large or even global receptive field. Due to the sparse nature of point clouds, when a point cloud is occluded, the source point cloud may not have a match in the target point cloud, and in this case, the accuracy and real-time performance of other methods will significantly decrease. Although traditional handcrafted methods can achieve high accuracy when facing backward loop closures, their real-time performance is poor and practically meaningless, and they cannot detect backward loop closures in dynamic environments in real-time. Another difficulty in loop closure detection is point cloud registration, as changes in lighting conditions, seasons, and dynamic objects in the environment may all lead to failures in aligning source and target point clouds.

Point cloud registration aligns source point cloud and target point cloud from different viewpoints to obtain the complete 3D information of a scene. Common registration methods include ICP (Iterative Closest Point) [4,7] and RANSAC (Random Sample Consensus) [9]. ICP is an iterative point cloud registration method that aligns point clouds by finding the nearest neighbor point pairs between two sets of point clouds and calculating the transformation matrix based on these point pairs. The performance of ICP depends on the accuracy of the initial transformation matrix. RANSAC is a robust point cloud registration algorithm that can handle the influence of noise and outliers. RANSAC randomly samples a set of point pairs and calculates their transformation matrix. This process is repeated multiple times, ultimately selecting the transformation matrix with the most inliers as the final result.

In recent years, researchers have proposed several deep learning-based point cloud registration methods [1,28]. These methods typically use neural networks to learn meaningful features from point cloud data, thereby improving the accuracy and robustness of registration. [1] presents a novel point cloud registration method called PointNetLK that based on the PointNet architecture. Which is a deep learning-based method for processing unordered 3D point clouds that combined PointNet with the Lucas-Kanade (LK) algorithm, a widely used optical flow estimation technique, to create a robust and efficient point cloud registration method. By integrating these two approaches, PointNetLK learns local and global features from point cloud using deep learning, which ultimately improves registration accuracy.

In this article, a novel loop closure detection pipeline is proposed, which achieves real-time detection of loop closures and registration of point clouds in dynamic environments. Our method uses both U-Net [23] and FPN [16] networks to construct a shared feature extractor to extract point-wise local descriptors of point clouds. The U-Net architecture is fully convolutional, meaning that it does not contain any fully connected layers. This makes it more efficient and allows it to process images of varying sizes. Skip connections between the encoder and decoder paths help the network to recover fine-grained spatial information and produce more accurate segmentation results. U-Net can be trained end-to-end using relatively few training samples. Since its introduction, U-Net has been widely adopted and adapted for various segmentation tasks, and several variations of the original architecture have been proposed, such as V-Net [17] and Attention U-Net [19]. The Feature Pyramid Network (FPN) is constructed atop a Convolutional Neural Network (CNN), serving as its foundational backbone. The architecture is bifurcated into two primary routes: an ascending pathway and a descending pathway. The ascending route corresponds to the standard feedforward operations of the backbone CNN, generating a collection of feature maps across varying scales. Conversely, the descending route is tasked with the upscaling of spatially less-detailed but semantically richer feature maps, which are then integrated with their ascending pathway counterparts. This amalgamation is facilitated through lateral connections, thereby synthesizing a multi-scale feature pyramid that merges high-resolution features with weaker semantic information and low-resolution features with stronger semantic content.

2 Related Work

Our methodology is a synthesis of several key studies in the domain of 3D point cloud processing and Simultaneous Localization and Mapping (SLAM). We have assimilated and refined these seminal works to address the intricate challenges in this field.

Feature Extraction: The PV-RCNN model, as introduced by Shi et al. [25], has played a significant role in shaping our feature extraction mechanism. The model's capability to encode multi-scale features and generate superior 3D proposals through voxel-based operations and PointNet-based set abstraction operations is remarkable. However, we have expanded this approach by merging the architectures of U-Net and ResNet into our feature extractor. This integration amplifies the feature information of the point cloud, generates high-resolution feature maps, and enhances the efficiency of information propagation while minimizing computational overhead.

Global Descriptor: The strategy of generating global descriptors in our approach is inspired by the Generalized-Mean (GeM) pooling technique employed in MinkLoc3D [14] and PLReg3D [22]. These studies have demonstrated the effectiveness of the GeM pooling layer in creating a global descriptor vector that encapsulates the comprehensive information of the input point cloud. We have embraced this technique but have further refined the process by meticulously extracting receptive field features from the input point cloud data using a sophisticated convolutional neural network. This enhancement enables us to encapsulate a more comprehensive spatial and semantic representation of the point cloud data, thereby providing robust support for subsequent tasks such as place recognition.

Relative Pose Estimation: The work of Pham et al. [21] has been a cornerstone in shaping our relative pose estimation module. They advocated the use of the Sinkhorn algorithm for an efficient and differentiable approximation of optimal transport theory. Their approach offers a flexible matching strategy that transcends simple one-to-one correspondence, which is crucial for precise loop closure detection. We have incorporated this strategy but have also introduced the concept of unbalanced optimal transport (UOT) to accommodate the inherent variability and unevenness of real-world point cloud data. This innovation significantly alleviates the challenges posed by inaccurate point matching and bolsters resilience against the inherent randomness in keypoint sampling, thereby enhancing the accuracy and reliability of our loop closure detection.

In summary, our approach amalgamates and enhances the strengths of these prior works, while also introducing innovative improvements to tackle the challenges in 3D point cloud manipulation and SLAM.

3   Method

This chapter presents a thorough explanation of our approach, which utilizes local and global descriptors and incorporates the Sinkhorn algorithm. The flow of our proposed method can be seen in the form of a flowchart in Fig. 1. Our method consists of three parts: shared feature extractor, global descriptor generator, and relative pose estimator.

3.1   Feature Extraction

The input to our shared feature extractor is a 3D point cloud $P \in R^{J*4}$, J points with 4 values: x, y, z and reflection intensity that collected by rotating

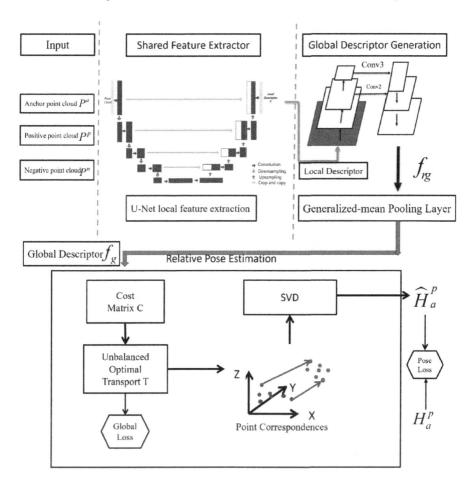

Fig. 1. Our advanced methodology is composed of multiple elements: a communal feature extraction unit, a universal descriptor generator, and a module for estimating relative pose. The communal feature extraction unit is responsible for isolating local attributes from the incoming data. Concurrently, the universal descriptor generator formulates global descriptors. The relative pose estimation module is tasked with determining the spatial transformation between two sets of point clouds. Throughout the training phase, we employ a trio of loss functions: the triplet loss, the global loss, and the pose loss.

LiDAR. The first step in this process is to voxelize the input point cloud, which means converting the continuous 3D point cloud data $P = \{(x_i, y_i, z_i)\}$ into 3D sparse tensor $P = \{(\hat{x}_i, \hat{y}_i, \hat{z}_i, r)\}$ (r is intensity). This transformation allows for easier processing and analysis of the 3D data. The skip connections of U-Net achieve efficient information propagation while maintaining high performance and reducing computational cost. The output of the shared feature extractor is point cloud P's $\mathbf{N}$ keypoints's feature set $fr = \{fr_1, fr_2, \ldots, fr_n\}$, $fr_i \in R^D$ is the D dimensional local descriptor of the i-th point.

PV-RCNN [25] combines voxel-based operations and PointNet-based set abstraction operations. Voxel-based operations effectively encode multi-scale features and generate high-quality 3D proposals. PointNet's set abstraction allows for flexible receptive fields and preserves precise position information. Integrating these two types of feature learning frameworks enables the model to learn more discriminative features and obtain finer-grained feature information. Our design concept is inspired by PV-RCNN and incorporates the architecture of U-Net and ResNet [10] network to construct the feature extractor. In the implementation process, the acquired point cloud data is first converted into polar coordinates and quantified, and then input into the feature extractor. Our network architecture adopts a U-shaped structure, composed of four convolution blocks from top to bottom $Conv0$, $\ldots$, $Conv4$ (i.e., the contraction path) and four deconvolution blocks from bottom to top $TConv0$, $\ldots$, $TConv4$ (i.e., the expansion path). Following each convolutional stage, a Rectified Linear Unit (ReLU) is employed as the designated activation mechanism. Within the compressive trajectory, the feature attributes of the point cloud are augmented, culminating in the generation of high-fidelity feature maps. Conversely, the expansive trajectory is responsible for harvesting global attributes and spatial configurations of the point cloud. This is achieved through the execution of deconvolutional procedures and the subsequent linkage to the high-definition feature maps originating from the compressive trajectory.

We unify the bottom-up and top-down information flow through lateral connections using $1 \times 1 \times 1$ convolutional kernels, adjusting the channel numbers before merging. This approach creates feature maps with higher spatial resolution and larger receptive fields, enabling us to effectively capture and understand the complex structures and semantic information of the input point cloud data. With this architectural configuration, our model is proficient not just in the robust extraction and manipulation of point cloud data features, but also in the retention of the spatial structural attributes inherent to the point cloud. This dual capability culminates in enhanced accuracy in the processing and analytical tasks related to point clouds.

The pointwise local descriptors fr generated by the shared feature extractor are utilized as input to the FPN network to obtain the feature frg of a larger receptive field. Firstly, the local descriptor fr is introduced as an input to the bottom-up pathway of the Feature Pyramid Network (FPN). This pathway is essentially the feedforward computation of a Convolutional Neural Network (CNN). The fr descriptor initially enters the first convolutional layer, where it is subjected to a series of convolutions, non-linear activations, and pooling operations, culminating in the generation of the first-level feature map, C_1. C_1 is then passed onto the second convolutional layer where a similar process is undertaken, thereby yielding the second-level feature map, C_2. This process repeats until the final layer, resulting in a series of feature maps $\{C_1, C_2, ..., C_n\}$, which together create a feature pyramid where each level corresponds to a network stage. Subsequently, the outputs from the bottom-up pathway are integrated via a top-down pathway and lateral connections, leading to the generation of the feature frg with a larger receptive field. In detail, a 1×1 convolution operation is first conducted on the highest-level feature map, C_n, to generate the coarsest feature map, P_n. P_n is then upsampled and added element-wise to the feature map $C_{(n-1)}$ following a 1×1 convolution, resulting in $P_{(n-1)}$. This process is carried out down to the bottom layer, producing a series of feature maps $\{P_1, P_2, ..., P_n\}$, which together constitute our feature frg with a larger receptive field. In the context of FPN, all feature map levels share the same feature dimension, or the number of channels. In our experiment, this dimension is set to 256, indicating that all additional convolutional layers output 256-channel feature maps. These additional layers do not utilize non-linear activation functions, as we have discovered that the impact of such functions on the results is minimal. Through this process, the local descriptor fr is transformed into the feature frg, which has a larger receptive field. These features encapsulate information from fr across multiple scales, providing a more comprehensive reflection of the image's global structure and contextual information. Thus, FPN provides us with an effective method to convert the local descriptor fr into the feature frg with a larger receptive field, enhancing our ability to address tasks that necessitate consideration of a broad scope of contextual information.

3.2 Global Descriptor

In our work, we propose an innovative strategy for generating global descriptors, which utilizes receptive field features, frg, to produce the final global descriptor, fg. Our approach is inspired by the GeM (Generalized-Mean pooling) technique found in MinkLoc3D [14] and PLReg3D [22], and further optimizes and improves upon it. Initially, we take the receptive field features frg as input. These features are meticulously extracted from the input point cloud data using an advanced convolutional neural network, encapsulating rich spatial and semantic information of the point cloud data. Subsequently, these receptive field features are processed through a GeM pooling layer. The core idea of the GeM pooling layer is to perform specific mathematical operations on the input features (take the p-th power, sum, and then take the p-th root), thereby generating a global descriptor vector. This global descriptor vector is capable of capturing the global information of the input point cloud, thereby providing robust support for subsequent tasks, such as place recognition. The formula for our GeM pooling layer is:

$$f_g^{(k)} = \left(\frac{1}{N} \sum_{j=1}^{N} [f_{rg}^{(j,k)}]^p \right)^{\frac{1}{p}}$$

where $f_{rg}^{(j,k)}$ is the k-th feature of the j-th feature point in receptive field features frg, p is a learnable pooling parameter.

3.3   Relative Pose Estimation

In this section, we elaborate on our unique relative pose estimation module, an integral part of our loop closure detection system within the framework of point cloud-based SLAM. We are given two point clouds, source point cloud P and target point cloud Q, and our objective is to determine the 6-DoF transformation that brings these point clouds into alignment. This is accomplished by comparing the global descriptors $f_g(P)$ and $f_g(Q)$, derived from the features identified in Sect. 3.1 and subsequently transformed into global descriptors in Sect. 3.2.

When dealing with point cloud data, a point in P might not have an exact match in Q due to factors such as occlusions, changes in viewpoint, or dynamic elements within the scene. This necessitates a more adaptable matching strategy that goes beyond simple one-to-one correspondence, which is particularly important for precise loop closure detection. To tackle this challenge, we employ the Sinkhorn algorithm, known for its efficient and differentiable approximation of optimal transport (OT) theory. However, the traditional OT problem requires a mass preservation constraint, which is often not feasible in the context of point cloud-based loop closure detection.

In order to bypass the limitations of traditional optimal transport (OT) theory, we employ the concept of unbalanced optimal transport (UOT) [21]. This innovative approach allows for the manipulation of mass, enabling both its creation and annihilation. This flexibility significantly mitigates the challenges posed by inaccurate point matching and bolsters resilience against the inherent randomness in keypoint sampling. As a result, it enhances the accuracy and reliability of our loop closure detection by accommodating the inherent variability and unevenness of real-world point cloud data.

We utilize a recent extension of the Sinkhorn algorithm to approximate the UOT. The cost matrix, denoted as C, is defined as the cosine distance between the keypoints' features. If we denote the feature vectors of the i-th keypoint in P and the j-th keypoint in Q as $f_g p^{(i)}$ and $f_g q^{(j)}$ respectively, then the element C_{ij} of the cost matrix C is given by:

$$C_{ij} = 1 - \frac{f_g p^{(i)} \cdot f_g q^{(j)}}{\|f_g p^{(i)}\| \cdot \|f_g q^{(j)}\|}$$

where C_{ij} is the element of the cost matrix C for the i-th keypoint in P and the j-th keypoint in Q. $f_g p^{(i)}$ and $f_g q^{(j)}$ are the global descriptor of the i-th keypoint in P and the j-th keypoint in Q, respectively. Upon estimating the unbalanced optimal transport, which represents the soft correspondences between keypoints' global descriptor $f_g p^{(i)}$ and $f_g q^{(j)}$, denoted as T, is computed by solving an optimization problem that is a variant of the classical optimal transport problem with entropy regularization. The problem can be formulated as follows:

$$\min_{T} \sum_{i,j} C_{ij} T_{ij} - \varepsilon H(T) + \lambda \text{KL}(T|P)$$

where C_{ij} represents the cost matrix, defined as the cosine distance between the features of keypoints. T_{ij} is the transport plan, indicating the amount of mass transported from the i-th point in P to the j-th point in Q. $H(T)$ is the entropy of T, promoting a more spread out transport plan. $KL(T|P)$ is the *Kullback-Leibler* divergence between transport plan T and the product measure P. The product measure P is defined as the product of the distributions of points in the two point clouds, denoted as μ and ν. ϵ and λ are regularization parameters. The Sinkhorn algorithm (Algorithm 1) is typically employed to solve this optimization problem.

Once the T is estimated, the projected coordinates p'_i for every keypoint p_i in P within Q are calculated using a weighted sum of the keypoints in Q, guided by the values in T:

$$p'_i = \sum_j T_{ij} \cdot q_j$$

where p'_i represents the projected coordinates of the i-th keypoint in P within Q. q_j represents the j-th keypoint in Q.

In our approach, the selection of regularization parameters ϵ and λ is crucial. These parameters control the balance of the optimization problem, where ϵ modulates the dispersion of the transport plan, and λ regulates the divergence between the transport plan and the product measure. In practice, we select the optimal values for ϵ and λ through k-fold cross-validation on the training set. We have observed that for our problem, the optimal values for ϵ and λ typically lie within a small range, indicating the robustness of our method to the choice of these parameters.

Finally, we estimate the 6-DoF transformation that aligns the original point cloud P with its projection in Q. This is done using a differentiable version of the weighted SVD method, allowing us to train our relative pose estimator in an end-to-end manner by comparing the predicted transformation with the ground truth. This robust and efficient approach significantly enhances the accuracy of loop closure detection in our SLAM system.

Algorithm 1. Sinkhorn algorithm for Unbalanced Optimal Transport

---

Require: $C \in \mathbb{R}^{n \times n}$, $\varepsilon > 0$, $\lambda > 0$, $P \in \mathbb{R}^{n \times n}$
Ensure: $T \in \mathbb{R}^{n \times n}$
1: Initialize $k = 0$, $u_0 = v_0 = 0$, $\eta = \varepsilon/U$
2: **while** $k \le \tau U/\varepsilon + 1$ **do**
3:     Compute $a_k = B(u_k, v_k)1_n$
4:     Compute $b_k = B(u_k, v_k)^T * 1_n$
5:     **if** k is even **then**
6:         Compute $u_{k+1} = (u_k + \eta \log(a) - \log(a_k))(\eta\tau/(\eta + \tau))$, $v_{k+1} = v_k$
7:     **else**
8:         Compute $v_{k+1} = (v_k + \eta \log(b) - \log(b_k))(\eta\tau/(\eta + \tau))$, $u_{k+1} = u_k$
9:     **end if**
10:     Increment $k = k + 1$
11: **end while**
12: Compute $T = B(u_k, v_k)$
13: **return** T

---

where C is $n \times n$ matrix, representing the cost matrix between point clouds P and Q. ϵ and λ is a positive number used to control the balance of the optimization problem. P is an $n \times n$ matrix, representing the product measure of the distributions of point clouds P and Q. T is an $n \times n$ matrix, representing the transport plan from point cloud P to Q. k is Iteration count. u_0 and v_0 are Vectors initialized to 0, used in the iterative process of the Sinkhorn algorithm. η is a parameter, equal to ϵ/U, U is an upper bound. τ is used to determine the stopping condition of the algorithm. a_k and b_k are the row and column averages of the optimal transport matrix $B(u_k, v_k)$, given u_k and v_k. a and b are represent the distributions of points in the point clouds P and Q.

3.4   Loss Function

In our approach, the loss function is composed of three main parts: the triplet loss, the global descriptor loss, and the relative pose loss. Firstly, the triplet loss is designed to ensure that in the feature space, the distance between the anchor point cloud and the positive point cloud is less than the distance between the anchor point cloud and the negative point cloud. This design is similar to the loss function used in LCDNet [5]. The triplet loss is defined as:

$$L_{\text{triplet}} = \max(0, m + D(f_a, f_p) - D(f_a, f_n)) \tag{1}$$

where f_a, f_p, f_n represent the features of the anchor, positive, and negative point clouds respectively, D is the distance function, and m is the margin parameter.

Secondly, the global descriptor loss is designed to ensure that the global descriptors of the anchor and positive point clouds are similar, while the global descriptors of the anchor and negative point clouds are dissimilar. This is inspired by the global descriptor loss used in PLReg3D [22]. The global descriptor loss is defined as:

$$L_{\text{global}} = \max(0, m + D(f_{g_a}, f_{g_p}) - D(f_{g_a}, f_{g_n})) \tag{2}$$

where f_{g_a}, f_{g_p}, f_{g_n} represent the global descriptors of the anchor, positive, and negative point clouds respectively.

Lastly, the relative pose loss is designed to ensure that the estimated relative pose correctly aligns the anchor and positive point clouds. This design is similar to the relative pose loss used in LCDNet [5]. The relative pose loss is defined as:

$$L_{\text{pose}} = \|T_{ap} - T_{ap}^*\|_2 \tag{3}$$

where T_{ap} is the estimated relative pose and T_{ap}^* is the ground truth relative pose. The total loss is a weighted sum of these three parts:

$$L_{\text{total}} = \alpha L_{\text{triplet}} + \beta L_{\text{global}} + \gamma L_{\text{pose}} \tag{4}$$

where α, β and γ are the weights for the triplet loss, the global descriptor loss, and relative pose loss respectively. These weights are hyperparameters that can be tuned during training.

This particular loss function motivates the model to acquire both local and global descriptors that are highly discriminative for point clouds. Additionally, it aids in the precise computation of the relative spatial orientation between different point clouds. These aspects are pivotal for the effective execution of loop closure detection as well as point cloud alignment.

4   Experimental and Discussions

4.1   Loop Closure Detection

The effectiveness and robustness of our proposed method were thoroughly evaluated through a series of comprehensive experiments conducted on the SemanticKITTI [3] dataset, which is due to its rich and diverse content. For our experiments, we divided the dataset into training and testing sets. Sequences 00 and 08 were used for testing, as they contain the highest number of loops and reverse loops, respectively. While sequences 05, 06, 07, and 09 were reserved for training.

Drawing inspiration from LCDNet [5], for each individual scan, referred to as i within the sequence, we evaluate the affinity between its global descriptor, denoted as $f_g(i)$, and the descriptors of all antecedent scans. The scan j, manifesting the highest affinity, is earmarked as a likely loop candidate. Should the affinity between the pair of descriptors surpass a predefined threshold, symbolized as th, the tuple (i, j) is designated as a loop. Under such conditions, we proceed to authenticate the ground truth spatial closeness between the scans in question: if this closeness falls below five meters, it is classified as a true positive; otherwise, it is considered a false positive. On the flip side, if the affinity is beneath the threshold yet a scan within a five-meter radius around the current scan i is discernible, it is labeled as a false negative.

Our primary metric for this evaluation is Average Precision (AP). The results obtained from our method are juxtaposed with those from other state-of-the-art methods for a comprehensive comparison. The detailed comparison, including the AP scores of all methods, is presented in Table 1.

In our experiments, we evaluated our method on two distinct datasets. On the 00 dataset, which represents standard loop closure scenarios, our method achieved an impressive AP of 98%. This result signifies a substantial improvement over other methods we benchmarked against.

On the other hand, we also tested our method on the 08 dataset, which is specifically designed for reverse loop closure scenarios. Despite the increased complexity and challenge presented by reverse loop closures, our method still managed to achieve a commendable AP of 80%. This demonstrates the robustness and versatility of our approach, as it can effectively handle both forward and reverse loop closure detection. Beyond the Average Precision (AP), we further analyzed the performance of our method by plotting the precision-recall curves for both datasets. The two curves are illustrated in Fig. 2(a) and (b). These curves, which display high precision across a broad spectrum of recall levels, underscore the robustness of our method in loop closure detection tasks.

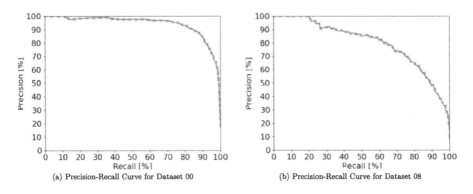

(a) Precision-Recall Curve for Dataset 00   (b) Precision-Recall Curve for Dataset 08

Fig. 2. Precision-Recall Curves

Table 1. Comparison of Average Precision with State-of-the-Art Methods on the KITTI 00 and KITTI 08 Datasets

Method	KITTI 00	KITTI 08
M2DP [11]	0.93	0.05
Scan Context [13]	0.96	0.65
ISC [26]	0.83	0.31
LiDAR-Iris [27]	0.96	0.64
OverlapNet [6]	0.95	0.32
SG_PR [15]	0.49	0.13
Ours	0.98	0.80

From the results, it is evident that our method delivers superior performance in the task of loop closure detection. This superior performance can be primarily attributed to the innovative design of our feature extractor and relative pose estimation module. The global descriptor effectively captures the holistic information of the point cloud, enabling accurate and robust loop closure detection even in challenging scenarios. On the other hand, the 6DOF transformation provides precise alignment between point clouds, further enhancing the reliability of our loop closure detection.

Moreover, our method demonstrates robustness against various adverse conditions, such as occlusions and dynamic changes in the scene, outperforming other state-of-the-art methods in terms of average precision. This robustness is a testament to the effectiveness of our method in handling real-world variability and complexity in point cloud data. Furthermore, our method achieves these results while maintaining computational efficiency. This is largely due to the efficient design of our network architecture and the optimization strategies employed during training.

4.2 Point Cloud Registration

We conduct an evaluation of our proposed methodology specifically within the realm of point cloud alignment. The metrics employed for this assessment include the success rate (defined as the proportion of pairs that are accurately aligned), along with the average translation error *(TE)* and rotation error *(RE)* calculated over both successful and all positive pairs. A pair is considered to be accurately aligned if the resulting errors in both rotation and translation fall below five degrees and two meters, respectively. We contrast the performance of our approach with that of existing cutting-edge methods, and the detailed results are presented in Table 2.

Table 2. Comparative analysis of relative pose errors, both rotational and translational, among positive pairs within the KITTI 00 and 08 datasets.

Approach	Sequence 00			Sequence 08		
	Success	TE[m](succ./all)	RE[deg](succ./all)	Success	TE[m](succ./all)	RE[deg](succ./all)
Scan Context* [13]	97.66%	-/-	1.34/1.92	98.21%	-/-	1.71/3.11
ISC* [26]	32.07%	-/-	1.39/2.13	81.28%	-/-	2.07/6.27
LiDAR-Iris* [27]	98.83%	-/-	0.65/1.69	99.29%	-/-	0.93/1.84
ICP(P2p) [31]	35.57%	0.97/2.08	1.36/8.98	0%	-/2.43	-/160.46
ICP(P2pl) [31]	35.54%	1.00/2.11	1.39/8.99	0%	-/2.44	-/160.45
RANSAC [24]	33.95%	0.98/2.75	1.37/12.01	15.61%	1.33/4.57	1.79/37.31
FGR [32]	34.54%	0.98/5972.31	1.2/12.79	17.16%	1.32/35109.13	1.76/28.98
TEASER++ [29]	34.06%	0.98/2.72	1.33/15.85	17.13%	1.34/3.83	1.93/29.19
OverlapNet* [6]	83.86%	-/-	1.28/3.89	0.10%	-/-	2.03/65.45
RPMNet [30]	47.31%	1.05/2.07	0.60/1.88	27.80%	1.28/2.42	1.77/13.13
DCP [28]	50.71%	0.98/1.83	1.14/6.61	0%	-/4.01	-/161.24
Ours	90.82%	0.99/1.19	1.19/1.52	80.72%	1.12/1.58	1.37/1.88

Inspired by LCDNet [5], we evaluated our method in comparison to state-of-the-art handcrafted point cloud registration techniques that LCDNet [5] has previously explored. For handcrafted methods such as ICP, RANSAC, Fast Global Registration (FGR), and TEASER++, we downsampled the point clouds using a voxel size of 0.3 m. For DNN-based approaches like RPMNet, Deep Closest Point (DCP), and our own method, we perform point cloud registration using 4096 sampled points. We conducted this comparison under the same dataset. Upon scrutinizing the outcomes, it becomes evident that our method demonstrates exceptional performance in the realm of point cloud registration. This can be primarily attributed to our unique feature extractor and relative pose estimation module.

5 Conclusion

In this manuscript, we introduced an innovative methodology for loop closure detection and point cloud registration, encompassing four primary components:

a shared feature extractor, a global descriptor, a relative pose estimation module, and a loss function. The shared feature extractor, founded on a U-Net architecture, transforms the input 3D point cloud data into a 3D sparse tensor, facilitating more straightforward processing and analysis. The global descriptor, drawing inspiration from the GeM (Generalized-Mean pooling) technique, is further refined to encapsulate the comprehensive information of the input point cloud. The relative pose estimation module applies the concept of unbalanced optimal transport (UOT), circumventing the constraints of traditional optimal transport (OT) theory, thereby boosting the precision and dependability of our loop closure detection. The loss function motivates the model to learn distinguishing local and global descriptors for point clouds and to accurately estimate the relative pose between point clouds, which are indispensable for successful loop closure detection and point cloud registration.

We have carried out thorough evaluations of our methodology on semanticKitti datasets, demonstrating that our method establishes a new standard in the field. Our approach successfully identifies loops even under challenging conditions and surpasses handcrafted methods in terms of precision and robustness. Furthermore, our proposed relative pose estimation module exhibits impressive results, outperforming existing approaches for point cloud registration and loop closure detection.

In summary, we have substantiated the performance of our proposed algorithm through empirical tests on real-world datasets acquired by our robotic platform, attaining favorable results. As future work, we intend to scrutinize both the efficacy and the generalizability of our approach in the domain of global localization, particularly when applied to more extensive map configurations.

Acknowledgment. This work was supported by the National Natural Science Foundation of China (grant 61802247).

References

1. Aoki, Y., Goforth, H., Srivatsan, R.A., Lucey, S.: PointNetLK: robust & efficient point cloud registration using PointNet. In: Proceedings of the IEEE/CVF Conference on Computer Vision and Pattern Recognition, pp. 7163–7172 (2019)
2. Arandjelovic, R., Gronat, P., Torii, A., Pajdla, T., Sivic, J.: NetVLAD: CNN architecture for weakly supervised place recognition. In: Proceedings of the IEEE Conference on Computer Vision and Pattern Recognition, pp. 5297–5307. IEEE (2016)
3. Behley, J., et al.: SemanticKITTI: a dataset for semantic scene understanding of LiDAR sequences. In: Proceedings of the IEEE/CVF International Conference on Computer Vision (ICCV) (2019)
4. Besl, P.J., McKay, N.D.: Method for registration of 3-D shapes. In: Sensor Fusion IV: Control Paradigms and Data Structures, vol. 1611, pp. 586–606. SPIE (1992)
5. Cattaneo, D., Vaghi, M., Valada, A.: LCDNET: deep loop closure detection and point cloud registration for lidar slam. IEEE Trans. Rob. **38**(4), 2074–2093 (2022)
6. Chen, X., Läbe, T., Milioto, A., Röhling, T., Behley, J., Stachniss, C.: OverlapNet: a Siamese network for computing LiDAR scan similarity with applications to loop closing and localization. Auton. Robot **46**, 1–21 (2022)

7. Chen, Y., Medioni, G.: Object modelling by registration of multiple range images. Image Vis. Comput. **10**(3), 145–155 (1992)
8. Cummins, M., Newman, P.: FAB-MAP: probabilistic localization and mapping in the space of appearance. Int. J. Rob. Res. **27**(6), 647–665 (2008)
9. Fischler, M.A., Bolles, R.C.: Random sample consensus: a paradigm for model fitting with applications to image analysis and automated cartography. Commun. ACM **24**(6), 381–395 (1981)
10. He, K., Zhang, X., Ren, S., Sun, J.: Deep residual learning for image recognition. In: Proceedings of the IEEE Conference on Computer Vision and Pattern Recognition, pp. 770–778 (2016)
11. He, L., Wang, X., Zhang, H.: M2DP: a novel 3D point cloud descriptor and its application in loop closure detection. In: 2016 IEEE/RSJ International Conference on Intelligent Robots and Systems (IROS). IEEE (2016)
12. Jégou, H., Perronnin, F., Douze, M., Sánchez, J., Pérez, P., Schmid, C.: Aggregating local image descriptors into compact codes. IEEE Trans. Pattern Anal. Mach. Intell. **34**(9), 1704–1716 (2011)
13. Kim, G., Choi, S., Kim, A.: Scan Context++: structural place recognition robust to rotation and lateral variations in urban environments. IEEE Trans. Rob. **38**(3), 1856–1874 (2021)
14. Komorowski, J.: MinkLoc3D: point cloud based large-scale place recognition. In: Proceedings of the IEEE/CVF Winter Conference on Applications of Computer Vision, pp. 1790–1799 (2021)
15. Kong, X., et al.: Semantic graph based place recognition for 3D point clouds. In: 2020 IEEE/RSJ International Conference on Intelligent Robots and Systems (IROS), pp. 8216–8223. IEEE, October 2020
16. Lin, T.Y., Dollár, P., Girshick, R., He, K., Hariharan, B., Belongie, S.: Feature pyramid networks for object detection. In: Proceedings of the IEEE Conference on Computer Vision and Pattern Recognition, pp. 2117–2125 (2017)
17. Milletari, F., Navab, N., Ahmadi, S.A.: V-Net: fully convolutional neural networks for volumetric medical image segmentation. In: 2016 Fourth International Conference on 3D Vision (3DV). IEEE (2016)
18. Mur-Artal, R., Tardós, J.D.: ORB-SLAM2: an open-source slam system for monocular, stereo, and RGB-D cameras. IEEE Trans. Rob. **33**(5), 1255–1262 (2017). https://doi.org/10.1109/TRO.2017.2705103
19. Oktay, O., et al.: Attention U-Net: learning where to look for the pancreas. arXiv preprint arXiv:1804.03999 (2018)
20. Oliva, A., Torralba, A.: Modeling the shape of the scene: a holistic representation of the spatial envelope. Int. J. Comput. Vision **42**(3), 145–175 (2001)
21. Pham, K., Le, K., Ho, N., Pham, T., Bui, H.: On unbalanced optimal transport: an analysis of Sinkhorn algorithm. In: International Conference on Machine Learning, pp. 7673–7682. PMLR, November 2020
22. Qiao, Z., Wang, H., Zhu, Y., Wang, H.: PLReg3D: learning 3D local and global descriptors jointly for global localization. In: 2021 27th International Conference on Mechatronics and Machine Vision in Practice (M2VIP), pp. 121–126. IEEE (2021)
23. Ronneberger, O., Fischer, P., Brox, T.: U-Net: convolutional networks for biomedical image segmentation. In: Navab, N., Hornegger, J., Wells, W.M., Frangi, A.F. (eds.) MICCAI 2015. LNCS, vol. 9351, pp. 234–241. Springer, Cham (2015). https://doi.org/10.1007/978-3-319-24574-4_28

24. Rusu, R.B., Blodow, N., Beetz, M.: Fast point feature histograms (FPFH) for 3D registration. In: 2009 IEEE International Conference on Robotics and Automation. IEEE (2009)

25. Shi, S., et al.: PV-RCNN: point-voxel feature set abstraction for 3D object detection. In: Proceedings of the IEEE/CVF Conference on Computer Vision and Pattern Recognition, pp. 10529–10538 (2020)

26. Wang, H., Wang, C., Xie, L.: Intensity scan context: coding intensity and geometry relations for loop closure detection. In: 2020 IEEE International Conference on Robotics and Automation (ICRA), pp. 2095–2101. IEEE, May 2020

27. Wang, Y., Sun, Z., Xu, C.Z., Sarma, S.E., Yang, J., Kong, H.: LiDAR Iris for loop-closure detection. In: 2020 IEEE/RSJ International Conference on Intelligent Robots and Systems (IROS), pp. 5769–5775. IEEE, October 2020

28. Wang, Y., Solomon, J.M.: Deep closest point: learning representations for point cloud registration. In: Proceedings of the IEEE/CVF International Conference on Computer Vision, pp. 3523–3532 (2019)

29. Yang, H., Shi, J., Carlone, L.: TEASER: fast and certifiable point cloud registration. IEEE Trans. Rob. **37**(2), 314–333 (2020)

30. Yew, Z.J., Lee, G.H.: RPM-Net: robust point matching using learned features. In: Proceedings of the IEEE/CVF Conference on Computer Vision and Pattern Recognition, pp. 11824–11833 (2020)

31. Zhang, Z.: Iterative point matching for registration of free-form curves and surfaces. Int. J. Comput. Vision **13**(2), 119–152 (1994)

32. Zhou, Q.-Y., Park, J., Koltun, V.: Fast global registration. In: Leibe, B., Matas, J., Sebe, N., Welling, M. (eds.) ECCV 2016. LNCS, vol. 9906, pp. 766–782. Springer, Cham (2016). https://doi.org/10.1007/978-3-319-46475-6_47

Stainless Steel Crack Detection Based on MATLAB

Wei Liao[1,2], Yongheng Wang[3(✉)], and Yanbing Guo[4(✉)]

[1] State Grid Shanghai Electric Power Research Institute, Shanghai 200437, China
liaowei@sh.sgcc.com.cn
[2] Puneng Power Technology Engineering Branch, Shanghai Hengnengtai Enterprise
Management CO., Ltd., Shanghai 200437, China
[3] School of Mechanical Engineering, Shanghai Dianji University, Shanghai, China
2536621496@qq.com
[4] College of Ocean Science and Engineering, Institute of Marine Materials Science and
Engineering, Shanghai Maritime University, Shanghai 201306, China
yanbingg1984@126.com

Abstract. Crack detection plays a vital role in ensuring the health and safety of stainless steel structures, and also provides sufficient information for subsequent stainless steel welding, so efficient detection of cracks is essential. The methods usually used to detect cracks include manual inspection, liquid penetration testing and ultrasonic testing. The most common detection method is through manual inspection, which makes the process of manual crack detection very time-consuming and laborious due to the presence of a large amount of interfering information in the crack picture. In order to solve the problems such as unclear crack identification that may be faced by manual inspection, this study uses image processing techniques to create an efficient crack detection system. The system includes steps such as image grayscaling, image enhancement, image denoising, image binarization crack recognition and GUI design. Through these steps, automatic measurement of crack features such as width, length, area and direction is realized.

Keywords: Crack Detection · Image Processing · GUI Design

1 Introduction

Stainless steel is an alloy that has the property of resisting air and other corrosive media, so it doesn't need to be treated with color plating etc. It has excellent corrosion resistance and a smooth surface. 316 stainless steel belongs to the austenitic stainless steel, with the addition of the element Mo, which significantly improves its corrosion resistance and high temperature strength, and can withstand high temperatures as high as 1,200–1,300 °C and is suitable for use in harsh environments [1]. It is widely used in chemicals,

Technology Projects of State Grid Shanghai Municipal Electric Power Company (SGTYHT/21-JS-223).

J. Li et al. (Eds.): 6GN 2023, LNICST 553, pp. 347–352, 2024.
https://doi.org/10.1007/978-3-031-53401-0_30

seawater equipment, fuel oxalic acid production equipment, and photographic and other precision industries. 316 stainless steel over a long period of time can be affected by a variety of external forces and environmental factors, resulting in cracks that can have serious effects on equipment, structure and safety. Identifying, detecting and repairing cracks prevents them from expanding, thereby extending the life of a material, component or device and reducing replacement and repair costs.

In this experiment, the macroscopic form surface at the corner of the bend was found to be flat and there were no significant cracks as observed by enlarging the graph of the 316 stainless steel sample. However, upon further zoom-in observations, some tiny chips and cracks can be observed. These cracks will lead to stress concentrations at the corners of the bend and may further expand under external forces, affecting its subsequent use. In this paper, we want to implement crack detection on 316 stainless steel samples by building a crack detection system.

2 Image Preprocessing

2.1 Image Grayscaling

A grayscale image is one with only photometric information and no color information. Image grayscale is the transformation of a color image into a grayscale image. The value of each pixel in a grayscale image is an integer from 0 to 255, which represents the pixel brightness level, 255 for white and 0 for black [2].

Most of the spectrum in nature can be represented by different ratios and intensities of red, blue and green light [3]. Our common color images are usually saved in RGB format. In MATLAB, an RGB image can be represented as a matrix $M \times N \times 3$. M and N represent the number of rows and columns of the image, respectively, and 3 represents the three channels of red (R), green (G) and blue (B). In the preprocessing of images, grayscaling techniques are generally used in order to improve the processing efficiency of images and to remove as much redundant information as possible. This approach can better solve the above problem, so that only the brightness information is retained in the image. There are many ways to transform an RGB image into a grayscale image in MATLAB, and the following methods are commonly used:

1. rgb2gray function: this is a MATLAB built-in function, using 0.2989 weighted average, 0.5870 and 0.1140 weighted summation of the red, green and blue three-channel value.
2. Average method: add up the values of the red, green and blue channels and divide by 3.
3. Maximum value method and minimum value method: select the maximum or minimum value of the red, green and blue channels as the grayscale value.

In this test, we choose to use the weighted average method for grayscaling. By this method, the grayscale image is obtained as shown in Fig. 1.

2.2 Histogram Equalization

Histogram equalization is a technique used to improve the contrast of an image. This method enhances the visual effect and readability of an image by redistributing the

Original drawing   Grayscale image

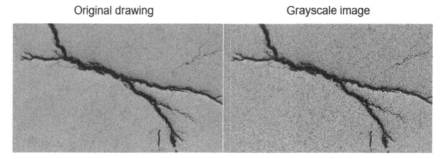

Fig. 1. Comparison of original image and the grayscale image

pixel intensity values of the image such that the pixel values in the image are uniformly distributed over the entire intensity range. However, applying histogram equalization may bring some side effects such as increasing the visibility of noise and changing the overall brightness of the image [4]. Therefore, when using histogram equalization, subsequent adjustments and processing may be required to achieve better results, and Fig. 2 illustrates the comparison between the histogram equalized image and the original image.

Original drawing   Histogram equalization image

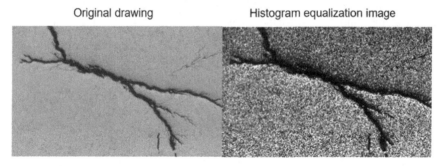

Fig. 2. Comparison of original image and Histogram equalization image

2.3   Image Binarization

Image binarization is the process of converting a grayscale image into a black and white binary image [5], the binarized image is compared with the original image as shown in Fig. 3. The basic principle is to classify the gray level in the image into two regions: one with the gray value set to black, and the other with the gray value set to white. In the process of binarization, we select a threshold, and then all the gray values below the threshold are set to 0, and all the gray values equal to or higher than the threshold are set to 1, and ultimately we get an image with only black and white (0 and 1) gray values. Through binarization, the original gray scale range of the image is changed from 0–255 to only two gray scales, which greatly improves the speed of the CPU to process the image.

Original drawing   Binarization image

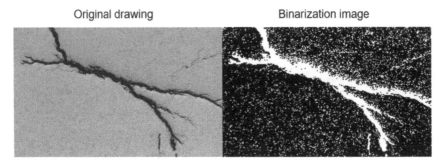

Fig. 3. Comparison of t original image and the binary image

2.4 Image Filtering

Image filtering is a method of processing an image that focuses on modifying the image to eliminate noise [6]. By applying various filters, the image can be made smooth while enhancing the details of the image. Common image filtering methods include mean filtering [7], median filtering [8] and Gaussian filtering [9].

In this experiment, we used the median filtering method, and the comparison between the filtered image and the original image is shown in Fig. 4. Median filtering is a nonlinear filtering technique that can effectively suppress noise while maintaining image details. It uses a sliding window in which the pixels are sorted and its median is used as the value of the pixel in the center of the window.

Original drawing   Filtered image

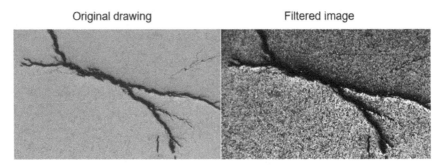

Fig. 4. Comparison of original image and the filtered image

2.5 Image Enhancement

When processing photographs of stainless materials, the contrast between the cracks and their surroundings should be increased to make it more intuitive to show the desired information because of the high degree of similarity between the cracks and their surroundings. Contrast indicates the degree of difference in brightness between different pixels in an image, and by enhancing the contrast, the details of the image can be made more prominent [10]. There are several contrast enhancement methods including histogram equalization, adaptive histogram equalization, and contrast stretching.

In this experiment, histogram equalization is adopted, and the image enhancement results are shown in Fig. 5, after contrast enhancement, the crack region is more obvious compared with the original image. By enhancing the contrast, we are able to better analyze and understand the crack situation in the stainless steel image.

Original drawing   Enhanced image

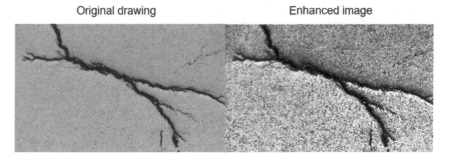

Fig. 5. Comparison of original image and the enhanced image

3   GUI Design

In this paper, an image processing graphical user interface based on MATLAB GUI is designed, which consists of graphical objects such as windows, menus, icons, cursors, buttons, dialog boxes and text. The basic functions of crack detection are realized through this interface.

This system involves many techniques, mainly image graying, histogram equalization, median filter denoising, contrast enhancement, binarization processing, binary image filtering, crack identification, crack judgment, and other techniques. Specific parameter information is demonstrated in the crack detection results shown in Fig. 6.

4   Summary

The research objective of this project is to take 316 stainless steel cracks as the research object, build a crack detection system in MATLAB and realize the crack detection of 316 stainless steel. After a series of processing steps, the system can identify the cracks of 316 stainless steel cracks, and get the length, width, area and shape of the cracks and other information, and successfully realize the detection of 316 stainless steel cracks, to achieve the expected results of the experiment. The whole process of detection focuses on the noise generated by image preprocessing. The imperfect performance of the camera equipment, interference in the image transmission process or defects in the image processing algorithm itself will bring complex noise and increase the processing complexity.

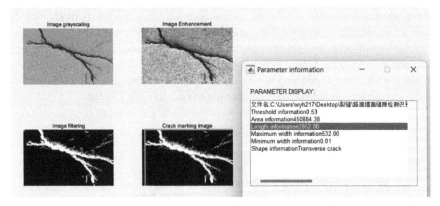

Fig. 6. Crack Detection Results

With the development and innovation of science and technology, convolutional neural network, as a kind of deep learning technology, has been very successful in the field of image detection. With the improvement of hardware performance and algorithm level, the deep learning model will become stronger and more effective, and the image detection accuracy and efficiency will be further improved.

References

1. Marshall, P.: Austenitic stainless steels: microstructure and mechanical properties (1984)
2. Kumar, T., Verma, K.: A theory based on conversion of RGB image to gray image. Int. J. Comput. Appl. **7**(2), 7–10 (2010)
3. Maxwell, J.C.: On the theory of compound colours, and the relations of the colours of the spectrum. Philos. Trans. R. Soc. Lond. **150**, 57–84 (1860)
4. Pizer, S.M., Amburn, E.P., Austin, J.D., et al.: Adaptive histogram equalization and its variations. Comput. Vis. Graph. Image Process. **39**(3), 355–368 (1987)
5. Sauvola, J., Pietikäinen, M.: Adaptive document image binarization. Pattern Recogn. **33**(2), 225–236 (2000)
6. Singh, G., Mittal, A.: Various image enhancement techniques-a critical review. Int. J. Innov. Sci. Res. **10**(2), 267–274 (2014)
7. Pan, J.J., Tang, Y.Y., Pan, B.C.: The algorithm of fast mean filtering. In: 2007 International Conference on Wavelet Analysis and Pattern Recognition, vol. 1, pp. 244–248. IEEE (2007)
8. Huang, T., Yang, G., Tang, G.: A fast two-dimensional median filtering algorithm. IEEE Trans. Acoust. Speech Signal Process. **27**(1), 13–18 (1979)
9. Deng, G., Cahill, L.W.: An adaptive Gaussian filter for noise reduction and edge detection. In: 1993 IEEE Conference Record Nuclear Science Symposium and Medical Imaging Conference, pp. 1615–1619. IEEE (1993)
10. Kaur, M., Kaur, J., Kaur, J.: Survey of contrast enhancement techniques based on histogram equalization. Int. J. Adv. Comput. Sci. Appl. **2**(7) (2011)

A New Combination Model for Offshore Wind Power Prediction Considering the Number of Climbing Features

Lei Yin[1], Weian Du[1], Peng Leng[1], Xiaoyan Miao[1], Xiaodong Yang[1], Zhiyuan Zhao[1], Jinrui Lv[1], Shuai Shi[2(✉)], and Hao Zhang[2]

[1] Clean Energy Branch of Huaneng (Zhejiang) Energy Development Co., LTD., Beijing 310000, Zhejiang, China
[2] Shanghai University of Electric Power, Shanghai 200090, China
shishuai@shiep.edu.cn

Abstract. The accurate identification of offshore wind power ramp events has great effects on wind power forecast. In order to improve the prediction accuracy of offshore wind power, this paper proposes an XGBoost-GRU combined forecasting model considering the number of climbing features. Firstly, the adaptive revolving door algorithm is used to identify the wind power climbing event, as well as data compression and feature extraction. Then, the XGBoost decision tree and gating loop unit are used to make preliminary power prediction. In case studies, the results are weighted and combined in detail. It is proved that the proposed model has a terrific performance on the offshore wind power prediction.

Keywords: Wind power ramp events · XGBoost · gated recurrent unit · wind power prediction

1 Introduction

With the introduction of the dual carbon strategy in China, the penetration of renewable energy like wind power in the power grid has increased. However, due to their inherent volatility and randomness, the difficulty of wind power forecasting has increased, and then affect the stable operation of the power system.

Currently, common wind power forecasting methods include physical methods, statistical methods, and artificial intelligence methods. Physical methods rely on the forecast results of numerical weather prediction (NWP) [1] and convert them into power according to the power curve of the turbine. They are suitable for medium to long-term forecasting but require a large amount of meteorological information and are limited by geographical conditions. Statistical methods [2] include linear autoregressive models and Kalman filtering, which predict the next wind speed based on the current and previous wind speeds. However, their prediction accuracy is not sufficient to meet the requirements of stable operation of the current power system. Artificial intelligence

J. Li et al. (Eds.): 6GN 2023, LNICST 553, pp. 353–362, 2024.
https://doi.org/10.1007/978-3-031-53401-0_31

methods include machine learning models and deep learning models. Machine learning models such as random forest, support vector machine, and XGBoost can directly explore the coupling relationship between input features and output features, with fast fitting speed. However, these models have a shallow structure and cannot capture deep features, resulting in lower prediction accuracy. Deep learning models include convolutional neural networks, long short-term memory neural networks [3], and gated recurrent units [4, 5].

In reference [6], two machine learning algorithms, gradient boosting machine (GBM) and support vector machine (SVM) [7], as well as the regression model of multivariate adaptive regression splines (MARS), were used to predict the long-term wind power generation. Reference [4] proposed a hybrid model based on multi-stage principal component extraction, kernel extreme learning machine (KELM), and gated recurrent units (GRU) network. Reference [8] employed five machine learning algorithms to predict long-term wind power based on daily wind speed data. Several case studies were conducted to reveal the performance of different algorithms, each exhibiting distinct characteristics.

To address the above issue, this paper proposes an XGBoost-GRU wind power combination prediction model [9] that considers ramp feature quantities. Firstly, the definition of ramp events is introduced, as well as the representation of ramp feature quantities and methods for detecting ramp events. The parameter adaptive algorithm is used to identify wind power ramp events and extract ramp feature quantities. Then, an XGBoost-GRU combination model is established. Based on analysis of the calculation results, the XGBoost-GRU combination model considering ramp feature quantities has the highest prediction accuracy, and thus it satisfies the requirements of power system scheduling, achieving the standard for wind power prediction.

2 Wind Power Ramp Events

2.1 Ramp Feature Quantities

Wind power ramp events [11] refer to the sharp increase or decrease in wind power generation within a short period of time, typically between a few minutes to an hour. They are often triggered by sudden changes in meteorological conditions, such as wind speed, wind direction, turbulence, and atmospheric stability. Wind power ramp events pose significant challenges to the stability and reliability of the power system. However, there is currently no unified standard for the definition of wind power ramp events. This paper lists the following four ramp feature quantities to characterize their existence:

1) Ramp event duration Δt: $\Delta t = t_2 - t_1$, t_1 represents the start time of the ramp, and t_2 represents the end time of the ramp;
2) Ramp amplitude Δp: $\Delta p = p_2 - p_1$, p_1 represents the power value at the time of the ramp event, and p_2 represents the power value at the end of the ramp event;
3) Ramp direction: Ramp events can be divided into upward ramps and downward ramps based on the increase or decrease in power during the ramp;
4) Ramp rate Δk: $\Delta k = \frac{\Delta p}{\Delta t}$.

2.2 Definition of Ramp

Though occur with a small probability, ramp events have an impact on the accuracy of power prediction and the stable operation of the power system, especially in high wind penetration scenarios. Based on the listed ramp characteristics in Sect. 2.1, the following two definitions are proposed as shown in Fig. 1:

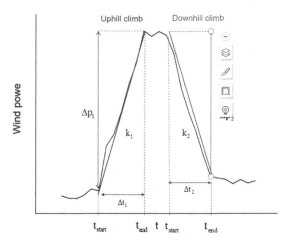

Fig. 1. Characteristic quantities of wind power ramp events

Definition 1: A ramp event is determined to occur when the absolute difference $|\Delta p|$ between the initial power value and the ending power value within the time interval (t_1, t_2) exceeds a given threshold.

$$p_2 - p_1 > p_{threshold} \tag{1}$$

Here, p_1 and p_2 represent the wind power at time t_1 and t_2, respectively, and $p_{threshold}$ is the given power threshold.

Definition 2: A ramp event is determined to occur when the ratio between the absolute difference $|\Delta p|$ between the initial power value and the ending power value within the time interval (t_1, t_2) and the duration Δt exceeds a given threshold.

$$\frac{|p_1 - p_2|}{\Delta t} > k_{threshold} \tag{2}$$

Here, $k_{threshold}$ is the given ramp rate threshold.

3 Ramp Event Detection

In this study, the parameter adaptive rotation gate algorithm is used to identify wind power ramp events and extract ramp characteristics. The power changes are shown in Fig. 2, where the point O represents the stored point of the previous compression segment, i.e., the starting point of that segment. Taking the segments at a distance of r from the midpoint O as the gates, when there is only one data point, the gates are in a closed state. As the number of data points increase, the gates rotate and open outward (without closing in the opposite direction), and the width of the gates can increase until the angle between the two gates and the y-axis exceeds 180°. The previous data point is stored, and this process continues.

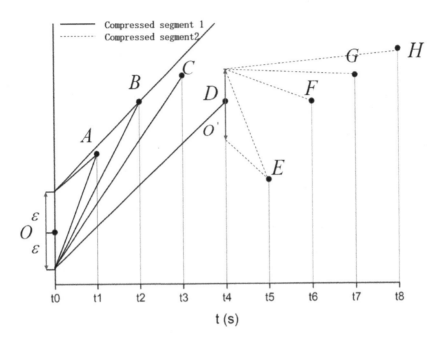

Fig. 2. Schematic diagram of the revolving door algorithm

In the figure, the last stored point is returned, and the data is processed according to Eq. (3) to (5) obtain the ramp characteristics, which provide the basis for subsequent analysis.

$$\begin{cases} k_1 = (P_a - P_o - \varepsilon)/(t_a - t_o) \\ k_2 = (P_a - P_o + \varepsilon)/(t_a - t_o) \end{cases} \tag{3}$$

Here, k_1 and k_2 are the slopes of the upper and lower gates, respectively, P_o and P_a are the power values corresponding to points O and A, t_o and t_a represent the corresponding time moments of the power, ε is the door width parameter.

$$\begin{cases} k_{1m} = (P_m - P_o - \varepsilon)/(t_m - t_o) \\ k_{2m} = (P_m - P_o + \varepsilon)/(t_m - t_o) \end{cases} \tag{4}$$

Here, k_{1m} and k_{2m} are the upper and lower door slopes between the point to be sought and the previous storage point, respectively, P_o and P_a are the power values corresponding

O to A the two points, t_o and t_a are the moment of the corresponding power moments.

$$\begin{cases} k_1 = \max(k_1, k_{1n}) \\ k_2 = \min(k_2, k_{2n}) \end{cases} \tag{5}$$

By comparing the size of k_1 and k_2, so that it meets the decision rule, when $k_1 \geq k_2$, continue to search; when $k_1 < k_2$, store and use it as a new point.

4 Based on the XGBoost-GRU Combinatorial Model

4.1 XGBoost Decision Tree Model

The core idea of XGBoost [12, 13] is taking decision trees as base learners and training the model with the Gradient Boosting algorithm. In XGBoost, each decision tree is composed of multiple decision tree nodes and corresponding decision rules, and each node's decision rule is based on a feature value. During training, the leaf nodes of each decision tree are assigned a predicted score, and the final prediction result is the weighted sum of all predicted scores.

In XGBoost, training of each decision tree is performed employing the predicted residual of previous tree (i.e., the difference between the true value and the predicted value) as the new label. Therefore, for each sample, the model first predicts its label value and then calculates the difference. Next, these residuals are used as new labels to train the next decision tree, so that the error between the model's prediction results and the true values is continuously reduced. This iterative training method can effectively improve the performance of the model and reduce the risk of overfitting.

4.2 Gated Recurrent Unit Model

The core of the Gated Recurrent Unit (GRU) [4] is a recurrent neural network unit that utilizes gating mechanisms to regulate information flow. The unit consists of two gates: the update gate and the reset gate.

The update gate is calculated based on the current input and the output from the previous time step. It adjusts the degree to which new input should be integrated into the current state. On the other hand, the reset gate controls the extent to which the past hidden state should be forgotten and the amount of new input to be considered.

The output of GRU unit is obtained by combining the new input with the unit's hidden state, which is computed using the update and reset gates. The output then will be passed to the next time step as the previous hidden state. GRU neuron structure, as shown in the following Fig. 3, the calculation formulas are:

$$z_t = \sigma\left(W_z \cdot \left[h_{t-1}, x_t\right]\right) \tag{6}$$

$$r_t = \sigma\left(W_r \cdot \left[h_{t-1}, x_t\right]\right) \tag{7}$$

$$\tilde{h}_t = \tanh\left(W \cdot \left[r_t * h_{t-1}, x_t\right]\right) \tag{8}$$

$$h_t = (1 - z_t) * h_{t-1} + z_t * \tilde{h}_t \tag{9}$$

Here, z_t is the update gate; r_t for the reset door; x_t is the current input; $\tilde{h}_t$ is the Summary of inputs and past hidden layer states; $\tilde{h}_t$ is the hidden layer output; W_z, W_r, W are trainable parameter matrices.

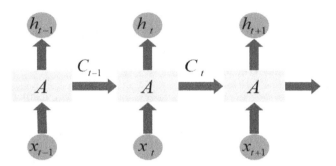

Fig. 3. GRU neuron

Compared to other RNN models, the GRU structure has several advantages, including simplicity and high computational efficiency. It has fewer parameters than the Long Short-Term Memory (LSTM) network, making it faster to train and less prone to overfitting.

4.3 XGBoost-GRU Combined Model

To improve prediction accuracy, this paper proposes a combined model of XGBoost decision trees and GRU. As shown in the diagram 4 (Fig. 4).

The original wind power data is first utilized for pre-training the XGBoost model. The resulting predictions are then used for feature extraction and transformed into new input features, which are collectively input into the GRU neural network. Finally, a weighted algorithm is applied to obtain the final prediction result. The calculation steps are as follows:

Obtain the deviation matrix E.

$$E = \begin{bmatrix} \sum_{t=1}^{N} e_{1t}^2 & \sum_{t=1}^{N} e_{1t} e_{2t} \\ \sum_{t=1}^{N} e_{2t} e_{1t} & \sum_{t=1}^{N} e_{2t}^2 \end{bmatrix} \tag{10}$$

Here, N is the total number of power point samples; e_{1t} and e_{2t} are the prediction error of the XGBoost model and GRU neural network prediction model.

The weight vector P is calculated by the Lagrangian multiplier method.

$$P = \begin{bmatrix} P_1 & P_2 \end{bmatrix}^T = \frac{E^{-1}R}{R^T E^{-1}R} \tag{11}$$

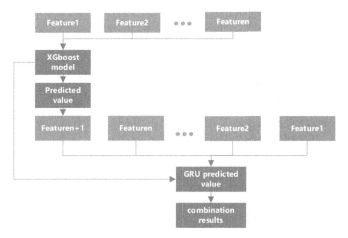

Fig. 4. XGBoost-GRU combined model structure diagram

$$P_1 + P_2 = 1 \tag{12}$$

$$R = \begin{bmatrix} 1 , 1 \end{bmatrix}^T \tag{13}$$

Here, P_1 and P_2 are the weights of the XGBoost model and the GRU neural network prediction model respectively.

Calculate the final power prediction:

$$f = P_1 f_1 + P_2 f_2 \tag{14}$$

Here, f_1 and f_2 are the predicted values of the XGBoost model and the GRU neural network prediction model.

5   Case Study Analysis

5.1   Experimental Data Description

The case study in this paper selected wind power data from a wind farm located on the east coast of China for one year to verify the proposed wind power prediction model. The wind farm has a total of 34 wind turbines, each with an installed capacity of 3MW. The wind power data features include wind speed, wind direction, temperature, and pressure. To enhance the prediction accuracy, the data features are expanded by introducing wind speed maximum, mean, and minimum features. To validate the effectiveness of the proposed model, the data is divided into a training set and a test set in an 8:2 ratio. To eliminate the influence of different feature dimensions on the prediction results, the data is normalized using the min-max method. The model is based on the TensorFlow 2.0 platform. The number of neurons in the initial layer of the hidden layer of the GRU model is 200, the number of neurons in the second layer is also 200, the number of training times is 100, and the learning rate is 0.005; the maximum depth of the XGBoost tree is 4, the learning rate is 0.1, the number of submodels is 225, the loss function is logistic.

5.2   Comparison of Prediction Results

In this experiment, the model was constructed using the XGBoost module from the machine learning toolbox and the GRU-based wind power prediction model built on the Keras module. The Adam optimizer was chosen, and the Rectified Linear Unit (ReLU) function was selected as the activation function for the GRU.

The training and testing datasets were divided in a 8:2 ratio, with a time resolution of 15 min. To validate the experimental performance of the proposed model, several control groups were established, including a single XGBoost decision tree model, a single GRU model, and the XGBoost-GRU combination model. Additionally, a control experiment considering the ramp-up feature was included. The experimental results are shown in the Table 1 and Fig. 5–6.

Table 1. Comparison of the performance of different predictive models

Algorithm	RRMSE	MAPE
XGBoost	14.125	12.193
GRU	12.382	10.324
XGBoost-GRU	9.847	8.214
GRU2	10.108	8.841
XGBoost-GRU2	9.142	7.692

Here, GRU2 represents a single model considering the climbing feature quantity, and XGBoost-GRU2 represents the combined model considering the climbing feature quantity. From the figure, it can be observed that the prediction performance of the single GRU model is superior to that of the single XGBoost decision tree model. Compared to the single models, the combined XGBoost-GRU model exhibits improved prediction performance. The RRMSE and MAPE predicted by the combined model were 9.847% and 8.214%, respectively, which were reduced by 4.278% and 3.979% compared with the single model XGBoost, and 2.535% and 2.11% respectively compared with the single prediction model GRU. Furthermore, when comparing the control group with the models that did not consider the ramp-up feature, the prediction performance of the control group is superior to the corresponding models, the comparative experiment considering the climbing feature was added to increase the power prediction information feature with a large degree of contribution, and the RRMSE and MAPE prediction errors proposed in this paper reached 9.142% and 7.692%, respectively, which were 0.966% and 1.149% lower than the prediction errors of the experimental model without considering the climbing feature quantity. This indicates that the ramp-up feature has an impact on the accuracy of power prediction, thus validating the effectiveness of the proposed prediction method.

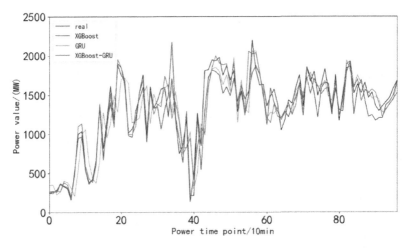

Fig. 5. Prediction performance comparison chart without considering climbing features

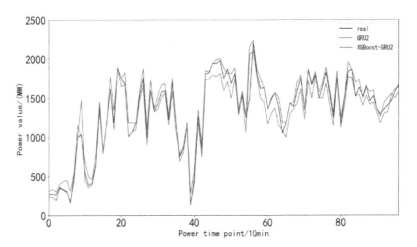

Fig. 6. Prediction performance comparison chart considering the amount of climbing features

6   Conclusion

Currently, the prediction accuracy of single model cannot satisfy the criteria of production. This paper proposes an XGBoost-GRU combined model considering ramp feature quantities. By optimizing the weights of each model, the prediction accuracy is improved while retaining the advantages of each model. The adaptive rotation gate algorithm is applied to extract ramp features from wind power sequences, expanding the dimension of the input feature matrix. Through case analysis, it is concluded that ramp feature quantities can affect the prediction accuracy of wind power. After experiments, it is found that the characteristic amount of wind power climbing will have an impact on the prediction accuracy of wind power power, and the subsequent wind power can start from this. This proves the effectiveness of the proposed XGBoost-GRU combined model in this paper.

Acknowledgment. This work was supported by Science and Technology Project of China Hua-neng Group Co., Ltd. (NO. HNKJ20-H66, Offshore Wind Power Site Selection and Associated Technologies for the Deep-sea areas).

References

1. Zhang, H., Yan, J., Liu, Y., et al.: Multi-source and temporal attention network for probabilistic wind power prediction. IEEE Trans. Sustain. Energy **12**, 2205–2218 (2021)
2. Ahmadi, A., Nabipour, M., Mohammadi-Ivatloo, B., et al.: Long-term wind power forecasting using tree-based learning algorithms. IEEE Access **8**, 151511–151522 (2020)
3. Sun, Z., Zhao, M.: Short-term wind power forecasting based on VMD decomposition, ConvLSTM networks and error analysis. IEEE Access **8**, 134422–134434 (2020)
4. Zou, F., Fu, W., Fang, P., et al.: A hybrid model based on multi-stage principal component extraction, GRU network and KELM for multi-step short-term wind speed forecasting. IEEE Access **8**, 222931–222943 (2020)
5. Dong, X., Sun, Y., Li, Y.: Spatio-temporal convolutional network based power forecasting of multiple wind farms (2021)
6. Ahmed, S.I., Ranganathan, P., Salehfar, H.: Forecasting of mid- and long-term wind power using machine learning and regression models. In: IEEE Kansas Power and Energy Conference. IEEE (2021)
7. Jiang, D.: Study on short-term load forecasting method based on the PSO and SVM model. Int. J. Control Autom. **8**(8), 181–188 (2015)
8. Demolli, H., Dokuz, A.S., Ecemis, A., et al.: Wind power forecasting based on daily wind speed data using machine learning algorithms. Energy Convers. Manage. (2019)
9. Li, K., Huang, D., Tao, Z., et al.: Short-term wind power prediction based on integration of feature set mining and two-stage XGBoost (2021)
10. Li, J., Song, T., Liu, B., et al.: Forecasting of wind capacity ramp events using typical event clustering identification. IEEE Access **8**, 176530–176539 (2020)
11. Yang, C., Yh, A., Xiong, X.C., et al.: Algorithm for identifying wind power ramp events via novel improved dynamic swinging door. Renew. Energy (2021)
12. Phan, Q.T., Wu, Y.K., Phan, Q.D.: A comparative analysis of XGBoost and temporal convolutional network models for wind power forecasting. In: International Symposium on Computer, Consumer and Control. IEEE (2020)
13. Ma, Z., Chang, H., Sun, Z., et al.: Very short-term renewable energy power prediction using XGBoost optimized by TPE algorithm. In: 2020 4th International Conference on HVDC (HVDC) (2020)

Fault Diagnosis Method of Gas Turbine Combustion Chamber Based on CNN-GRU Model Analysis

Xinyou Wang[1], Yulong Ying[1(✉)], Xiangyan Li[2], and Zaixing Cui[1]

[1] College of Energy and Mechanical Engineering, Shanghai University of Electric Power, Shanghai 200090, China
yingyulong060313@163.com
[2] College of Engineering Science and Technology, Shanghai Ocean University, Shanghai 201306, China

Abstract. The safety, reliability and economy of gas turbines all depend on the fault diagnosis of gas turbines. In order to solve the fault diagnosis accuracy problems of false alarm and false alarm in gas turbine combustion chamber, a fault diagnosis method of gas turbine combustion chamber based on gated recurrent unit (GRU) optimization convolutional neural network (CNN) model analysis is proposed. First, a sample set of combustion chamber failure data is generated by constructing a gas turbine thermodynamic model. The CNN model is then optimized using GRU to extract the spatial and temporal features of the data, using small convolution kernels and 2D convolution methods. Finally, the extracted features are fused and fed into a fully connected layer for fault type identification. Experimental results show that the proposed method is highly practical and feasible compared to the traditional combustor threshold-defined fault diagnosis methods and other artificial intelligence fault diagnosis methods, with an average diagnosis accuracy of 97.66%, which is higher than the identification accuracy.

Keywords: Gas turbine · Combustion chamber fault diagnosis · CNN · GRU · Visual analysis

1 Introduction

Gas turbine is a kind of rotating impeller thermal engine, which is widely used in power generation, aviation and other fields [1]. Compressor, combustor and turbine are the main components of gas turbine, among which combustor is considered as the "heart" of gas turbine [2]. A gas turbine can be operated regularly only by ensuring that the combustion chamber is in a steady state of combustion. At present, the increasing working time of gas turbines brings great risks to the health of components, and easily leads to performance degradation or damage of core components [3, 4]. Fault diagnostics can accurately and effectively detect problems in gas turbine operation and enhance operational data analysis capabilities with the goal of improving the accuracy of gas

J. Li et al. (Eds.): 6GN 2023, LNICST 553, pp. 363–374, 2024.
https://doi.org/10.1007/978-3-031-53401-0_32

turbine combustion fault diagnostics. The exhaust temperature at the end of the turbine is an effective indicator for fault diagnosis of the gas turbine combustor. Thermocouples with a uniform distribution are arranged in the exhaust section of the gas turbine to probe the exhaust temperature of the turbine section. The exhaust temperature profile is homogeneous when the combustion chamber is in a lively state and inhomogeneous when the chamber is in a faulty state.

Aiming at the problems and shortcomings of the above methods, this paper applies the Gated Recurrent Unit Optimized Convolutional Neural Network (CNN-GRU) model to the fault diagnosis of gas turbine combustor for the first time. The fault simulation experiments were performed using an established combustion chamber thermodynamic model combined with actual operational data from the power plant, and the CNN model optimized by the GRU was used for data training and comparison with other intelligent algorithms. For the model results, the Unified Manifold Approximation and Projection (UMAP) method is used for visual analysis and radar plot comparison. Finally, the simulation and comparison show that the method proposed in this paper can accurately analyze the health (failure) of each combustion unit in the combustion chamber.

2 Theory

2.1 CNN

Convolutional neural network (CNN) is a deep feedforward neural network model. The usual structure of it consists of five layers, namely input layer, convolution layer, pooling layer, fully connected layer and output layer (classification layer) [5]. Figure 1 shows a plot of the structure of the convolutional neural network.

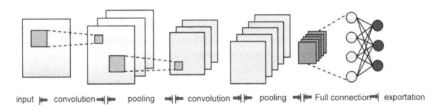

input |← convolution →| |← pooling →| |← convolution →| |← pooling →| |← Full connection →| exportation

Fig. 1. Structure diagram of CNN

The main role of the convolutional layer is to perform convolution calculation with the input feature map through the convolution kernel to extract features. The convolution process can be summarized as follows:

$$X_j^k = f\left(\sum_{i \in M_j} X_i^{k-1} \times W_{ij}^k + b_j^k\right) \tag{1}$$

where X_j^k and X_j^{k-1} are the output and input of the kth layer network, respectively. M_i is the characteristic set; W_{ij}^k is the weight matrix of the convolution kernel; b_j^k is the bias term; $f(\cdot)$ is the activation function. Activation functions can make the neural network nonlinear, improve its goodness-of-fit, and make it linearly separable by transforming the multidimensional features. In this paper, we use the commonly used nonlinear activation function ReLu, which is denoted as follows:

$$f(x) = \begin{cases} x(x > 0) \\ 0(x \le 0) \end{cases} \tag{2}$$

It can be seen from formula (2) that the nonlinear transformation equation provided by ReLu is especially simple, with rapid convergence speed and simple gradient calculation.

The pooling layer is mainly used for feature extraction, which is equivalent to sample sampling according to a certain rule under the premise of ensuring the rotation unchanged. In this experiment, the convolution results are maximally sampled. In the face of complex data samples, more feature information can be extracted, gradient descent is faster, model structure is simplified, and efficiency is improved. If the pooling window is $m \times n$, then the output of the kth convolution kernel corresponding to the S-layer is as follows.

$$X_{S,j,k} = \frac{\sum\limits_{i=mn-n+1}^{mn} X_k}{m} \tag{3}$$

where $X_{S,j,k}$ is the jth output of the $S - th$ layer corresponding to the kth convolution kernel.

Fully connected layers are commonly placed at the end of convolutional neural networks and are used to convert the output of 2D features after convolution and pooling into 1D vectors. The advantage of adding this layer is to reduce the influence of location on the classification results. Although the learning rate is reduced compared to the regularization layer, the stability of the entire network is improved. It serves as a feature extractor and classifier in the model.

2.2 GRU

Gate Recurrent Unit (GRU) is a kind of recurrent neural network (RNN), which combines cell state and hidden state, and can solve the problem of long-term memory and gradient in backpropagation in RNN. Figure 2 shows a plot of the structure of the GRU.

The GRU model contains two gates: a reset gate and an update gate. The reset gate determines how the updated information is combined with the previous memory, while the update gate defines how much of the previous memory is updated at the current time step. The mathematical description of a GRU network is as follows:

$$Z_t = \sigma\left(W_Z \cdot \left[h_{t-1}, x_t\right]\right) \tag{4}$$

$$R_t = \sigma\left(W_r \cdot \left[h_{t-1}, x_t\right]\right) \tag{5}$$

$$\tilde{h}_t = \tanh\left(W_{\tilde{h}} \cdot \left[R_t \cdot h_{t-1}, x_t\right]\right) \tag{6}$$

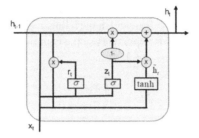

Fig. 2. Diagram of the GRU network structure [6].

$$h_t = (1 - z_t) \times h_{t-1} + z_t \times \tilde{h}_t \tag{7}$$

where σ is a *sigmoid* function, which can be used to transform the data into a value in the range 0–1 and thus act as a signal for the gate. Z_t is the output of the update gate; R_t is the output of the reset gate; $\tilde{h}_t$ is the output of the candidate hidden state; let h_t be the implied state at time t; W_Z, W_R, $W_{\tilde{h}}$, respectively, are the corresponding quantity weight matrices.

2.3 Gas Turbine

The working environment of heavy-duty gas turbine combustion chamber is extraordinarily strict, and continuous high temperature and pressure environment work is normal, especially the outlet exhaust temperature can reach 1300 °C to 1700 °C. Due to the elevated temperature inside the combustion chamber of a gas turbine, it is difficult to directly measure the internal temperature changes of the gas turbine combustor using temperature sensors on a regular basis. In general, the exhaust gas temperature at the outlet of the turbine is used as a parameter index to indirectly judge the anomalous condition of the gas turbine combustor. However, as the terminal parameter of a gas turbine, the exhaust temperature is subject to interference from multiple parameters, so that an accurate analysis of the exhaust temperature becomes crucial for current gas turbine combustor fault diagnostics.

In ideal conditions for gas turbine operation, the structure and combustion state of all computers are essentially the same, and the temperature measured by each thermocouple is approximately the same. However, when there is a fault in the combustion unit, its exit temperature is affected, showing local high or local low temperatures. This anomalous signal will be transmitted to the thermocouple and the corresponding thermocouple readout will be different from the other thermocouples. As a result, when the gas turbine is in normal operation, the exhaust gas temperature is uniformly and continuously distributed; This represents an operational fault in the combustion chamber when the exhaust temperature is continuously and inhomogeneously distributed. The greater the temperature difference between thermocouple readings at different temperatures, the greater the probability that the combustion chamber will fail. Specific diagnostic methods are as follows:

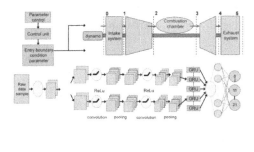

Fig. 3. Fault diagnosis method of gas turbine combustor based on CNN-GRU model analysis.

1) Thermal modeling was performed to construct a thermal model of the object to be diagnosed, based on relevant data from the Mitsubishi M701F4 gas-vapour combined cycle heating unit. The established model is used as a reference model for the simulation of various combustor faults.

2) The parameters of the inlet boundary conditions of the thermal model are set and adjusted according to the corresponding environment, operating conditions, and various faults of the combustion chamber. The large amount of prediction data obtained is used to train the classification accuracy of the CNN-GRU model, which can be used to improve the fault diagnosis accuracy of the CNN-GRU model for operational data.

3) The CNN-GRU model is used for fault diagnosis, the results of the model analysis are passed to the UMAP model, and the radar maps are combined for visual analysis to effectively distinguish the normal operation data and the fault data of the combustion chamber from the data features.

3 Simulation Experiment and Result Analysis

3.1 Model Structure and Parameter Design

The subject of this paper is the Mitsubishi M701F4 gas-vapor combined cycle heating unit, whose structure is shown in Fig. 4. There are 20 flame canisters in the combustion chamber, and 20 thermocouples are arranged to measure the exhaust temperature at the outlet of the turbine, and they are uniformly distributed. Thermal modeling was performed on the Mitsubishi M701F4 gas-vapor combined cycle heating unit based on relevant data. In the construction of the combustor model, a sea of fault conditions from actual operations is used as a reference and the relevant properties of the flame and smoke flow are taken into account. In this study, 10921 groups of samples under the normal operating condition of the data acquisition gas turbine were analyzed, with 70% as the training set samples and 30% as the test set samples, and the sampling time interval was 1 min. The gas turbine boundary condition parameters and equipment measurable parameters involved in the gas turbine combustor model are shown in Table 1.

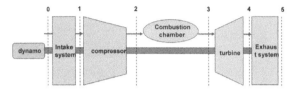

Fig. 4. Thermodynamic operation of heavy-duty gas turbines.

Table 1. Table captions should be placed above the tables.

Item	Parameters	Symbol	Unite
Boundary condition parameter	Atmospheric pressure	p_0	Pa
	Atmospheric temperature	t_0	°C
	Atmospheric relative humidity	R_{RH}	%
	IGV opening position	IGV	%
	Generator output power	N_e	kW
Equipment measurable parameters	Compressor inlet pressure	p_1	Pa
	Compressor inlet temperature	t_1	°C
	Natural gas mass flow rate	G_f	kg/s
	Compressor outlet pressure	p_2	Pa
	Compressor outlet temperature	t_2	°C
	Turbine outlet pressure	p_5	Pa
	Turbine outlet temperature	t_5	°C
	Rotational speed	n	r/min

Because the combustion chamber of a gas turbine in a power plant operates continuously in an elevated temperature and pressure environment, the internal temperature of the combustion chamber is too high to be directly measured by sensors. Currently, the exhaust gas temperature of the exit section of the turbine is commonly used to indirectly measure and judge the anomaly in the combustion chamber. Because the exhaust temperature of gas turbine outlet section is controlled by a number of parameters, such as environmental conditions, operating conditions, production and installation deviations of combustion units, and swirl factors in the combustion process, etc., the influence of swirl and other factors has been considered in the previous model construction process. The exhaust temperature of a gas turbine is measured by thermocouples uniformly arranged at the outlet of the turbine end to indirectly monitor the performance of the thermal components, and the distribution of gas turbine combustion chambers and thermocouples is shown in Fig. 5.

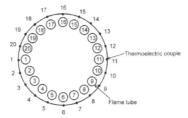

Fig. 5. Distribution diagram of flame tube and thermocouple.

The convolutional layers of the CNN-GRU model studied in this paper use a reduced convolutional kernel to reduce the amount of computation while ensuring the same perceptual range, thus enabling deeper features to be extracted. At the same time, the network parameters are reduced, more nonlinearities are introduced and overfitting is suppressed. In this paper, the 2D convolutional neural network is selected based on the features of the data, and the 2D convolution process is shown in Fig. 6.

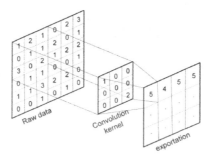

Fig. 6. 2D convolution process.

The proposed network model architecture is shown in Table 2. Referring to the well-known network model architecture and validation results in, two convolutional layers are used with the number of hidden neurons being 32 and 64, and the activation function of the convolutional layer is the ReLu function. After the convolutional layers, pooling is performed separately to extract features. After convolution, sequential augmentation is performed, a flattening operation is performed, and then GRU layers with 5 hidden units are connected. Finally, there are two fully connected layers with activation functions that are softmax functions. The model construction is based on many empirical suggestions to set parameters [7], the learning rate is set to 1e−2, the number of learning cycles is 400, and the regularization parameter is set to 1e−04. Regularization is used in this paper as a method to overcome the overwhelming problem.

3.2 Evaluation Metrics and Performance Analysis

Since the gas turbine exhaust temperature correlation data are directly measured by the temperature measurement thermocouple, they are susceptible to environmental factors

Table 2. Neural Network Model Architecture Table.

Name	Feature dimension	Hidden layer parameters
Image Input	n × 20	—
Reshape	n × 5 × 2 × 2	—
conv_1	n × 4 × 1 × 32	kernel number = 32,kernel size = 2 × 2,stride = 1,padding,lc = 1e−4
batchnorm_1	n × 4 × 1 × 32	batch normalization = 1e−05
ReLu_1	n × 4 × 1 × 32	—
maxpooling_1	n × 3 × 1 × 32	maxpooling = 2 × 1, stride = 1
conv_2	n × 2 × 1 × 64	kernel number = 64,kernel size = 2 × 1,stride = 1,padding,lc = 1e−4
batchnorm_2	n × 2 × 1 × 64	batch normalization = 1e−05
ReLu_2	n × 2 × 1 × 64	—
maxpooling_2	n × 1 × 1 × 64	maxpooling = 2 × 1,stride = 1
FLATTEN	n × 640	—
GRU	n × 5	
FC1/FC2	n × 21/n × 2	Hidden neuron number = 21
softmax	n × 1 × 1 × 2	—
classoutput	—	output = 21

or fault problems of the temperature measurement thermocouple, and the instability of the data distribution can easily interfere with model training. We choose two stabilization parameters a and b to evaluate the training of the model.

$$E_1 = 1 - \frac{\sum_{i=1}^{M} (TP_1 - TN_1)}{M} \tag{8}$$

$$E_2 = 1 - \frac{\sum_{i=1}^{N} (TP_2 - TN_2)}{N} \tag{9}$$

Using the above evaluation metrics, the accuracy changes during the training of the diagnostic model are obtained, as shown in Fig. 7. In iteration 22, the test set accuracy has reached 95.58%. Although the accuracy and loss still fluctuate during the iterations, they have stabilized within 200 iterations. The final training resulted in an accuracy of 97.66% and a loss of only 0.1975. The test and training accuracies are in reasonable agreement, indicating that the proposed model has strong generalization performance and can avoid overfitting.

To further verify the superiority of the proposed combustor fault diagnosis model, different smart analysis models were selected for comparison testing on data from the same gas turbine with the same operation time. This data is based on simulated operation data for the combustor model described above and actual operation data for the power plant.

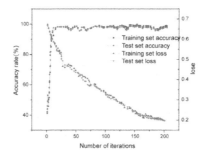

Fig. 7. Plot of accuracy change during model training.

The selected model methods include CNN-GRU network model, BP neural network model (BP), support vector machine (SVM), convolutional neural network model (CNN), genetic algorithm optimized BP (GA-BP) and Random forest algorithm (RF). The BP neural network model consists of two hidden layers with 32, 64 hidden neurons and a learning rate of 0.01. The penalty factor of the SVM was 5, the radial basis function (RBF) was used as the kernel function, and the kernel function parameter was 0.1. The CNN is a dual convolutional layer with 32 and 64 convolutional kernels, 1e−4 learning rate, 200 learning epochs, and 1e−5 regularization parameters. GA-BP has a number of hidden layer nodes of 5, an error threshold of 1e−6, 25 genetic generations, and a population size of 5. The number of decision trees in RF is 50 and the minimum number of leaves is one. A comparison of the accuracy of the AI learning methods is shown in Fig. 8.

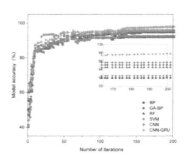

Fig. 8. Comparison of accuracy of artificial intelligence learning methods.

In Fig. 8, through the comparison of each model method, it can be seen that in the analysis of gas turbine combustion chamber smoke exhaust temperature data, compared with BP neural network model, SVM, GA-BP, CNN and RF, The CNN-GRU network model designed in this paper shows superior data analysis ability, and its data analysis accuracy is significantly higher than that of other model algorithms, which is 5.84%, 5.32%, 2.68%, 2.09% and 3.26% higher than that of BP, SVM, GA-BP, CNN and RF algorithm models, respectively. The above experiments were performed using only 20% of the simulated experimental data of the Mitsubishi M701F4 gas-vapor combined cycle heating unit. Figure 9 shows how the stability of the model changes as the data sample size is enlarged.

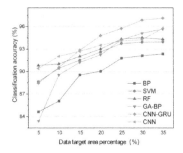

Fig. 9. Stability change diagram of expanded data sample size model.

As can be seen from Fig. 9, when the target data samples are miniscale (below 15%), the model diagnosis method of CNN-GRU does not significantly outperform the accuracy of other algorithm models. As the number of target data samples increases to 20%, the experimental approach already reflects its superiority and shows a clear advantage over other algorithmic models in terms of classification accuracy. When the sample data volume reaches 20% (or even larger), the performance of BP, SVM, GA-BP, CNN, and RF mode diagnosis methods tends to be stable with the CNN-GRU algorithm model, but there is still a large accuracy gap. Therefore, the proposed gated recurrent unit optimized convolutional neural network fault diagnosis model approach is an effective method for fault detection in heavy-duty gas turbine combustion chambers.

3.3   A Subsection Sample

In further research, data features are extracted through the model. In this paper, the Unified Manifold Approximation and Projection (UMAP) method [8] is selected to carry out visual analysis of dimensionality reduction. This approach takes a dataset with more than two features and outputs a low-dimensional image for exploring the dataset, thus showing the clusters of samples in the high-dimensional space and the relationships between sample points on the low-dimensional image. And significantly speed up, better saving data globally. In Fig. 10a), 4200 data are selected and set to 21 categories, and category T is the fault-free case. Categories F1–F20 correspond to 20 combustion cylinder faults in the combustion chamber of the Mitsubishi M701F4 gas turbine, respectively. Without visual analysis, the initial data feature distribution is irregular and features of different classes are distributed interactively, which does not intuitively reflect the clustering properties of the data. After visual clustering analysis by UMAP method, as shown in Fig. 10b), the same class features are aggregated with each other, forming 21 particle clusters. Although there is still some data clustering error between each class, the overall data is accurately clustered and the error clustering has slight impact on the overall accuracy. UMAP visual analysis has excellent overall feature

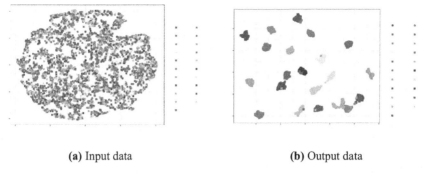

(a) Input data   **(b)** Output data

Fig. 10. Data feature visualization analysis process diagram.

extraction and clustering analysis results with extensive development and optimization space.

After the data visualization analysis in Fig. 10, the model effectively distinguishes the normal operation data of the combustion chamber from the fault data in terms of data characteristics. In order to reduce the learning difficulty and improve the analysis efficiency, some data cases are selected and analyzed by comparing with radar chart [9], as shown in Fig. 11.

Fig. 11. Radar diagram of gas turbine exhaust temperature distribution.

In Figs. 11, 10 sets of cases from the analyzed data are selected for radar map representation. Radar Fig. 1 shows the ideal exhaust temperature profile, radar Fig. 2 shows the normal operating exhaust temperature profile, and radar Figs. 3–10 show the faulty exhaust temperature profile for a single combustion cylinder. With radar maps, it is clear, intuitive and convenient to find anomalous gas turbine exhaust temperatures, analyze the cause and eliminate the fault at the first time, thus achieving the purpose of safe operation of gas turbines.

4 Conclusion

In this paper, we propose a fault diagnosis method for gas turbine combustion based on the analysis of a convolutional neural network model, which combines the advantages of multiple intelligent approaches for the gas turbine combustion fault problem. Based on the analysis of the results, the following conclusions can be drawn:

1) Different from the traditional approach of fault diagnosis of gas turbine combustor based on exhaust gas temperature through threshold setting, this experiment uses an artificial intelligence approach to build a CNN-GRU network model and solves the problem of unstable model data analysis by iteratively adding neurons and expanding the structure layer. This approach largely avoids the false alarm rate and missed alarm rate of conventional gas turbine fault diagnostics and improves the safety of the unit operation.
2) UMAP was used to visualize the data, and radar maps were plotted for comparison. The analysis results show that the UMAP algorithm is advantageous in obtaining and identifying the feature information of the data. During the efficient clustering analysis of the simulated run data, the overall clustering analysis results in positive results, although there are still some cases of erroneous clustering. Through the radar map, it is clear, intuitive and convenient to find the abnormal gas turbine exhaust temperature, analyze the cause, eliminate the fault at the first time and ensure the normal operation of the gas turbine. UMAP visualization combined with radar chart comparison analysis also provides advanced ideas for gas turbine exhaust temperature analysis.
3) Compared to other machine learning and deep learning methods, the experimental model shows a wide range of great diagnostic performance, with an average diagnostic accuracy of 97.66%. It has particularly himmense application potential and optimization value, and can achieve the objectives of improving security and reducing cost requirements in a variety of data analysis problems.

References

1. Jin, Y., Ying, Y., Li, J., et al.: A gas path circuit diagnosis method for gas turbine based on model and data hybrid drive. Thermal Power Gener. **50**(9), 66–71 (2021)
2. Pelasved, S.S., Attarian, M., Kermajani, M.: Failure analysis of gas turbine burner tips. Engineering Failure Analysis ISSN 1350–6307 (2019)
3. Ying, Y., Li, J.: An improved performance diagnostic method for industrial gas turbines with consideration of intake and exhaust system. Appl. Thermal Eng. **3**(222), 1–19 (2023)
4. Ying, Y., Li, J., Pang, J., et al.: Review of gas turbine gas-path fault diagnosis and prognosis based on thermodynamic model. Proc. CSEE **39**(3), 731–743 (2019)
5. Bai, M., Yang, X., Liu, J., et al.: Convolutional neural network-based deep transfer learning for fault detection of gas turbine combustion chambers. Appl. Energy **302** (2021)
6. Chen, G., Su, Y., Kou, H., et al.: Research on convolution gating cyclic residual network for bearing fault diagnosis. China Measure. Test 1–5 (2023)
7. Liu, J., Bai, M., Long, Z., et al.: Early fault detection of gas turbine hot components based on exhaust gas temperature profile continuous distribution estimation. Energies **13**(22), 5950 (2020)
8. Qi, X., Cheng, Z., Cui, C., et al.: Fault diagnosis method of planetary gearbox based on JS-VME-DBN and MS-UMAP. J. Aerospace Power 1–12 (2023)
9. Qiu, W.: Application of radar map on the analysis of M701F4 gas turbine BPT big deviation alarm events. Gas Turbine Technol. **31**(02), 68–72 (2018)

Research on Key Intelligent System in Unmanned Surface Vessel

Yongguo Li[1], Xiangyan Li[2(✉)], Caiyin Xu[2], and Xuan Tang[2]

[1] Shanghai Engineering Research Center of Marine Renewable Energy, Shanghai 201306, China
[2] College of Engineering Science and Technology, Shanghai Ocean University, Shanghai 201306, China
844384355@qq.com

Abstract. With the rapid development of automated control and mobile communication technologies, unmanned ship technology is a derivative product of robotics. This paper aims to study the key technologies of intelligent and autonomous design of unmanned surface vessel (USV). Starting from the analysis of the development history and current situation of USV at home and abroad, the intelligent technologies are discussed, including environmental perception technology based on multi-sensors, control technology for route generation and path planning, autonomous obstacle avoidance technology based on intelligent control, positioning technology of underactuated USV, and wireless communication technology based on modern communication system. Through the research and summary of the intelligent system of unmanned ship, it can be concluded that the unmanned ship technology, which integrates the control and automation system, communication and network system, positioning system and other intelligent systems, will develop to high intelligence, standardization, high stability and serialization in the future. Consistent with cross-domain expectations, not only the integration of multiple technologies on unmanned ship platforms, but also the cooperation and integration of multiple new and traditional technologies should be developed in the process of seeking intelligence.

Keywords: Unmanned surface vessel · Intelligentization · Perceptual recognition · Obstacle avoidance navigation · Motion control · Wireless communication

1 Introduction

The design and manufacturing of unmanned devices has contributed to the rapid development of the unmanned industry with technological ingenuity. There are mature unmanned aerial vehicles in the sky, gradually maturing unmanned vehicles and unmanned supermarkets on land, and rapidly developing unmanned ships and unmanned underwater vehicles in the ocean. As one of the countries with the longest coastline [1], China's potential in Marine resource development, maritime shipping development, Marine environmental protection, and maritime sovereignty maintenance should not be underestimated.

© ICST Institute for Computer Sciences, Social Informatics and Telecommunications Engineering 2024
Published by Springer Nature Switzerland AG 2024. All Rights Reserved
J. Li et al. (Eds.): 6GN 2023, LNICST 553, pp. 375–389, 2024.
https://doi.org/10.1007/978-3-031-53401-0_33

Various smart devices have been developed to protect this ocean. Among them, the surface intelligent unmanned vessel, referred to as the unmanned surface vessel (USV), is a small-scale surface vessel carrying different functional modules with certain intelligence and autonomous/semi-autonomous capabilities [2] in a narrow sense, which is the research object of this paper. Due to its outstanding characteristics such as miniscale size, high sensitivity, low production cost and module integration, the USV has an irreplaceable role compared to manned ships and has received attention and development from various countries.

2 Research Status of USV

2.1 Development Status of USV Abroad

First of all, the United States, which has the most mature technology of USV in the world, has developed the "Owl MK II" unmanned vehicle with side-scan sonar and camera functions as early as the 1990s when technology was limited [3]. Ocean Infinite subsequently collaborated in the 2014 search for Malaysia Airlines flight MH370, using eight onboard acoustic detection devices at the same time for the first time. The USV, developed by SailDrone, took part in an international maritime exercise on Jan 31, 2022, carrying multiple sensors to measure photosynthetic radiation, carbon dioxide and more to build a shared image of the surrounding ocean. It can be seen that the US has been leading the development of the USV, using a variety of techniques to perform different difficult operations.

Israel, as a developed country, followed suit and was early to recognize the smart ship. In 2005, Elbit developed the Stingary unmanned ship, which can complete tasks such as coastal target recognition and intelligent patrol [4]. In 2007, the "Silver Marlin" unmanned vessel was developed with automatic obstacle avoidance and autonomous navigation [5]. In 2017, the unmanned boat unit [6] was unveiled.

Worldwide, Italy has developed the two-body USV Charlie to gather atmospheric data and sample water surfaces in a planned area. France has the "InspectorMk" series USV for hydrological monitoring, surface target detection, etc. Britain has developed an unmanned trimaran called the Mayflower, which uses wind and solar power as its driving force. It can be seen that the development of USV abroad was initiated early and heavily invested.

2.2 Development Status of Domestic USV

In contrast, the world has well-established technologies for reference, despite China's late start. There have also been many achievements in recent years, mostly concentrated in universities, research institutes and businesses. As the leader of universities in the field of unmanned ships, Shanghai University has created the first domestic unmanned ship series "Jinghai" since 2013 [7]. Jilin Radio and TV University proposed an unmanned driving technology based on multi-sensor data fusion [8].

Figure 1 shows a small unmanned catamaran developed by Qingdao Institute of Marine Geology for the first time, which uses a shallow water bathymetry system, inertial positioning system and AD ad hoc network wireless transmission system to accurately measure the lake bottom topography in Shanghe's Ruyi Lake. Yunzhou Intelligent is one of the first high-tech enterprises in China to focus on the unmanned vessel sector. It has taken the lead in the application of USVs to water quality monitoring, as shown in Fig. 2, and has launched the world's first USV for environmental measurements and the first domestically developed stealth USV. At the Spring Festival Gala in 2018, the multi-boat collaborative technology was first demonstrated to the people of the whole country, with a large scale collaborative formation of 80 small USVs led by a main ship, marking the intelligent development of China's unmanned technology on the water surface [9].

Fig. 1. Unmanned bathymetry vessel in Qingdao 'ShangheRuyi' Lake

Fig. 2. "Yunzhou Intelligent" unmanned vessel for water environment monitoring

3 Key Technologies

First, the overall system composition of the USV is understood, and then methods and techniques from different modules are intensively studied. The connectivity and working pattern of each module during autonomous navigation is shown in Fig. 3. Equipped with a variety of plug-and-play modules on the basic boat can realize the diversification of the functions of the USV, which not only improves the overall performance of the boat, but also reduces the workload of later maintenance, and the problem only needs to be detected in the corresponding module.

3.1 Sensing Technology Based on Multi-sensors

Sensing the environment is equivalent to the operation of the "sensory system" of the USV. At present, the equipment used mainly includes Automatic Identification system (AIS), radar, visible light and infrared equipment, sonar, etc. For example, Zhuang

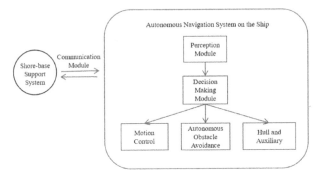

Fig. 3. Each module of the USV works with each other

Jiayuan et al. [10] from Harbin Engineering University proposed an embedded image acquisition and processing system based on target detection technology of USV in radar images. Lidar is highly accurate and can be accurately identified at short range, but weather factors can cause its performance to degrade. Visible and infrared sensor devices also have high environmental requirements, angular blind spots, and algorithms that are not yet mature.

Each device only highlights its advantages in a specific usage scenario, and a single device cannot interpret the complex and diverse environment on the water surface. To improve the applicability of USV on water surfaces, data collection and fusion techniques among various sensors are directions for future research. For example, Fig. 4 is the multi-sensor fusion environmental perception framework of the "Jinghai" series developed by Shanghai University. Based on accurate detection, recognition and tracking of obstacles, environment modeling is used to accurately understand the surrounding environment.

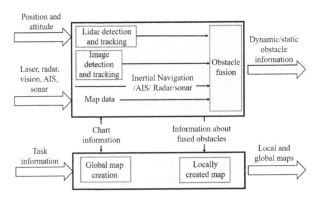

Fig. 4. "Jinghai" series multi-sensor fusion perception framework [11]

In terms of the application of specific technologies, Jungwook Han et al. [12] used the fusion of Marine radar and lidar to realize the complementarity in distance. Su Wenzhi et al. [13] focused on the fusion of AIS and radar information in unmanned driving, and proposed a weighted data fusion algorithm based on ship navigation, which made data combination possible. The team of He Sheng Wang [14] from Shanghai Jiaotong University proposed an algorithm to fuse lidar and vision information. Recently, deep learning methods have opened new avenues for data fusion. In 2017, a study [15] proposed to use deep learning to fuse different sensor data on smart ships, and proved the feasibility of using deep learning to improve environmental perception. Yu Baifeng [16] studied data fusion of vision sensors and audio sensors with appropriate deep learning algorithms to promote intelligent navigation of USV.

3.2 Planning Techniques Used for Route Generation

After perceiving the external environment and recognizing objects, it is necessary to plan the path reasonably to ensure safe navigation. The commonly used USV path planning algorithms studied in this paper can be roughly divided into three categories: graph search algorithms, intelligent algorithms, and virtual latent field methods, as shown in the following figures (Fig. 5).

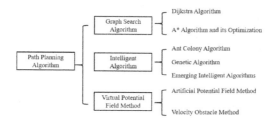

Fig. 5. Common path planning algorithms

Combined with algorithmic principles, the comprehensive comparison is summarized in Table 1 by global and local partitions. Due to the limited space, only some of the more representative results are listed in Table. In addition to well-known classical algorithms, scholars have attempted to apply emerging intelligent biological inspired algorithms to plan machine paths. Inspired by the interaction of a droplet with its environment due to gravity, the smart droplet algorithm finds the optimal path, but it inevitably suffers from premature convergence. Therefore, Xu Jiamin [22] et al. proposed a compilation optimization algorithm through nonuniformity. Bacterial foraging optimization algorithms are a new class of intelligent bio-inspired algorithms derived from observations of Escherichia coli foraging behavior in the human gut and fundamental laws of colony evolution. In solving nonlinear and multidimensional global optimization problems, it can address the common problem of slow convergence. In addition, the firefly algorithm, which simulates the firefly movement (the firefly spontaneously moves from low brightness to high brightness) to find the optimal solution through multiple iterations, is also suitable for global path planning.

Virtual potential field methods are mostly used for local path planning, where complete information is not required and the surrounding environment is obtained through sensors. Among them, the artificial potential field is analogous to the electric potential field, the target point generates attraction, and the obstacle generates repulsion. The obstacle avoidance information of the target point is obtained from the superposition of the two force fields. Due to its simple principles and fast response, it has become the most widely used local planning method. But it is inevitably prone to local minima problems. Compared to the well-established global path algorithm, the local path algorithm is still in the experimental model validation stage and its ability to handle complex water environments is weak. In the optimization of the algorithm, the advantages of each algorithm can be combined in the future, and multiple algorithms can be used to ensure the safety and accuracy of USV navigation.

Table 1. Comparison of path planning algorithms.

Item	Algorithm	Principle	Advantages	Disadvantages	Research Results
Global Path Planning	Dijkstra Algorithm	A single source search algorithm is used to calculate the shortest path from the starting point to all other nodes, which extends layer by layer from the starting point to the target point	Simple Robust High precision	Low computational efficiency, Takes up a lot of memory, Can only sail according to the predetermined route, Suitable for small areas	Zhuang Jiayuan et al. [17] effectively completed USV path planning
	A* Algorithm	Dijkstra extension algorithm with heuristic function. Each time, select the node with the minimum value of the evaluation function from the open list to expand, and add it to the close list, and repeat until the search reaches the target point	The principle is simple and efficient High extensibility The solution path is the most direct and efficient	Heavy computation Over-reliance on heuristics	Singh et al. [18] used it to search for the path hitting the wall Jin Jian 'an [19] Extended 3D A* search algorithm

(continued)

Table 1. (*continued*)

Item	Algorithm	Principle	Advantages	Disadvantages	Research Results
	Ant Colony Algorithm	The ant foraging behavior is simulated, and pheromone secretion and volatilation are used to guide the population to the optimal solution iteration	Strong robustness Ease of parallelism	The large size of the initial path or the lack of pheromone leads to the increase of the solution time and the global optimal solution cannot be obtained	Zhang Haini [20] realized automatic route generation by considering the resistance interference in the air
	Genetic Algorithm	It simulates the evolution of biological population, and each path is similar to the population individual, which is represented by the code. It iterates continuously and converges to the optimal solution	The fitness function is designed to solve the nonlinear problem with multiple constraints	Slow convergence The real-time performance is poor Easy to precocious Low stability	Kim et al. [21] realized path planning under complex sea conditions Jin Jian 'an [19] considered the improved genetic algorithm with multiple constraints
	Emerging Intelligent Algorithms	Intelligent water drop algorithm, Bacterial Foraging optimization algorithm, firefly algorithm	Positive feedback mechanism The self-organization property is good	Poor search ability Easy to converge and premature	Long et al. [23] Chen et al. [24]

(*continued*)

Table 1. (*continued*)

Item	Algorithm	Principle	Advantages	Disadvantages	Research Results
Local Path Planning	Artificial Potential Field Method	The potential field is established in the obstacle avoidance environment, the obstacle is regarded as the repulsive force field, and the target point is regarded as the gravitational field	Easy to calculate and fast to react Most widely used	A local optimal solution appears when surrounded by obstacles	Chen Chao [25] optimized the traditional function to solve the local minimum problem
	Velocity Obstacle Method	Consider the obstacle velocity	Strong obstacle avoidance safety High accuracy	The dynamic change of obstacle speed is not considered Slow solution process	Kuwata implements locally safe obstacle avoidance Kim et al. [26]

3.3 Autonomous Obstacle Avoidance Technology Based on Intelligent Control

If the sensors are compared to the human senses and the path planning is compared to the thinking, then the motion control module is injected into the flesh and blood, imitating the human thinking to control the motion of the hull through the "nerve" connection. It is the core of the surface USV system and the intuitive embodiment of its intelligence and autonomy. Domestic and international control approaches for USV in the face of complex water surface environments and emergencies can be summarized as two core issues. The first is to ensure the stability of the control system and to compensate for unmeasured and nonlinear stochastic interference caused by wind and wave currents. The other is to increase the accuracy and timeliness of USV control to address the poor performance caused by objective factors such as speed, load and water depth. Figure 6 summarizes the common control methods currently used for USVs, which are briefly described below.

Quadratic programming methods and model predictive controllers are commonly used to obtain optimal control commands in optimal control. As the name suggests, feedback linearization can linearize the system, but it is difficult to set the control rate in the nonlinear scenario of large-angle steering, which does not satisfy the accuracy requirements. Again, the use of backstepping must establish an accurate model. While this stabilizes the global situation, the multi-level backstepping control used to improve

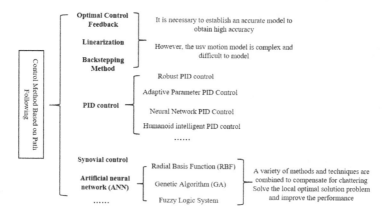

Fig. 6. Commonly used control method for USV

robustness increases the intermediate virtual quantities. In summary, the above three methods are model-based and have a strong dependence on model accuracy. However, the USV has a complex system and is difficult to model. Academics continue to search for better solutions.

Proportional-integral-derivative (PID) control method, as a classical algorithm in the motion control of USV, is favored by scholars because of its simple structure, high control accuracy and strong ductility. To remedy the shortcomings of classical PID control, such as weak adaptation, difficulty in tuning control parameters, and limited global control, researchers have explored various new control methods based on traditional PID control. For example, the advantages of artificial intelligence algorithm in the nonlinear field are used to combine the neural network PID control with fuzzy control [27]. Human-like intelligent PID control with improved accuracy and timeliness under human-like intelligent system [28]. Although the intelligence of the method is generalized to a certain extent to optimize the effect, there is still a deviation between the estimated output value and the true value when dealing with external perturbations. It is necessary to continuously adjust the PID parameters during navigation and smooth the model during training to enhance the performance.

When the training learning and classification functions of Artificial Neural Network (ANN) are applied to the control of motor system, the key is to continuously adjust and determine the connection structure and weight parameters of Ann, rely on Radial Basis Function (RBF) gradient optimization, and Genetic Algorithm (GA) is used to optimize the RBF to avoid falling into local optimum. Wang Renqiang et al. [29] used RBF and GA to design a USV motion sliding mode intelligent control method to effectively solve the input chattering problem, and verified that the intelligent control method could accurately avoid obstacles and track targets through comparative optimization process, and realized the neural network intelligent control of USV. Xue Yinku [30] integrated the logical reasoning function into the neural network by using the fusion combination method, introduced the fuzzy neural network of conjugate gradient method to improve the convergence speed, and used one technology to solve another technical problem to achieve complementarity.

In the process of sorting out the pros and cons of optimization, it has been found that in order to remedy the shortcomings of methods, scholars usually complement and modify existing algorithms, integrate multiple approaches, and strengthen machine learning for intelligent control.

3.4 Localization Techniques for Underactuated USV

When an USV performs most outdoor tasks on a surface, it is first necessary to ensure the accuracy of its own localization. High-precision positioning and navigation technologies are not only relevant to the path and how a vessel is moved, but also to the premise of outdoor applications and safety guarantees for water operations. Generally, unmanned ships have only two inputs, thrust and transship torque [31], which are less than the input dimension of the whole hull. It is typically an underactuated control system and has some difficulties in dynamic localization. The factors that affect the ship's positioning can be divided into two categories: external perturbations from the environment, and internal perturbations due to uncertainties in its own model. Figure 7 shows the procedure taken to solve the perturbation and eventually achieve accurate navigation of the USV.

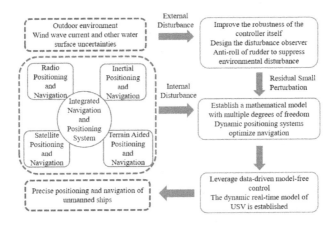

Fig. 7. Implementation of accurate positioning of USV under disturbance

The integrated navigation and positioning system of the USV combines different navigation technologies to improve localization performance by leveraging its own characteristics. Among them, inertial navigation is a common technology, which obtains the speed and position through the analysis of the motion state by the Inertial Measurement Unit (IMU) [32], but the error will accumulate with the increase of time [33]. In addition, Global Positioning System (GPS) is widely used in outdoor positioning because of its mature technology, which can provide real-time position information for USV. However, due to the small hull and easy occlusion, the localization information lags or vanishes when the GPS signal is perturbed. IMU sensors are often fused to improve localization accuracy. In order to further obtain the surrounding environment for accurate obstacle avoidance, GPS/IMU data is usually fused into simultaneous localization and mapping (SLAM). The kinematic properties of a USV with six degrees of freedom pose great difficulties for accurate modeling. For the resulting inaccurate positioning problem, the paper [31] proposed a solution to establish a mathematical model with multiple degrees of freedom. Moreover, for under-actuated UAVs, dynamic localization techniques can be used to optimize the navigation system. Dynamic positioning consists of an integrated software and hardware system. Compared with traditional anchor positioning, dynamic positioning has fast positioning speed, high accuracy and is not affected by terrain depth [34]. Taking the "Jinghai" series as an example, the Jinghai intelligent unmanned measurement vessel adopts the bow controller to stabilize the desired direction and the ship speed controller to stabilize the desired point position. The two controllers are used separately [11]. The peculiar dynamic properties of the jet pump actuator combined with the control algorithm achieve dynamic accurate tracking of the underactuated unmanned vessel.

3.5 Wireless Communication Technology Based on Modern Communication System

At this point, the main technology of the autonomous navigation system on the ship is illustrated. As a bridge between the ship-borne terminal and the shore-based terminal support system, the wireless communication technology can not only receive mission information from the shore, but also act as a timely alert in case of distress, thus playing a role in security protection. The communication function acts as the "language system" of the unmanned ship, which can realize the horizontal communication between "ship-coast-ship" and the vertical connection between "sky, water and air". Generally speaking, there are VHF data exchange systems (VDES), satellite navigation systems, radio navigation systems and so on.

Unmanned ships typically have low-powered antennas and are located close to the water surface, making communication signals vulnerable to interference from obstacles. In the effort to realize communication in different environments, many scholars have conducted studies. LI et al. [35] used WiFi technology to complete ship-to-shore communication, and the ship was successfully applied to lake water monitoring and water sample collection. XU et al. [36] used cloud platform communication based on 4G Internet of things technology to expand the working distance and realize remote real-time monitoring.

Due to the rapid increase in the amount of network data nowadays, how to effectively optimize communication resources has become another key research of USV [37]. Emerging low-power WAN technologies and deep learning-based optimal scheduling algorithms provide new research directions for smart ship communications. For example, ZigBee wireless communication technology, which has been used for the longest time and the most applications in the Internet of Things technology, was successfully used in the communication of remote control USV by LI [38]. Long Range Radio (LoRa) has become a new research direction for ship communication because of its wide communication range and low power consumption. ZHOU [39] carried out the feasibility test and analysis of LoRa technology of ship internal communication system, and verified the prominent advantages of low-power wide area network in this field. Through the research literature, intelligent technologies have naturally been applied to surface communication, and it is inevitable that there will be information accumulation. In order to reasonably optimize the scheduling of information resources, Gao ZHAI et al. [40] proposed a deep learning algorithm for information optimization of unmanned ship communication network, which greatly reduces the delay, improves the information processing efficiency, and provides stable communication network support for USV in unstable environment.

4 Discussion

1) Perception technology
   Multi-sensor fusion is undoubtedly the key technology of the future at the perception level, and it is necessary to mine data fusion and deep learning algorithms to overcome the shortcomings of sensors. The difficulty of current techniques is object recognition based on navigation videos, and there are few studies on data fusion for panoramic vision. In the future, we should make full use of the salient features of each sensor to improve the timeliness and accuracy of perception.

2) Planning technique
   Path planning is to find a collision-free safe path from the initial state point to the final goal point according to certain criteria (shortest path, least energy consumption, highest safety factor, etc.). This aspect is the focus of intelligent USV research. At present, many new intelligent bionic algorithms have emerged. Although there are still some problems in the initial stage, they have great possibilities in the field of play and outstanding innovation points, which are worth exploring and optimizing by scholars.

3) Obstacle avoidance technique
   If it is difficult to secure safe navigation by one method alone, it is necessary to combine them by taking advantage of their respective advantages. For example, in the search for better quadratic programming, the parallel computing power of dynamic ANN is used to improve speed. When the PID adaptive control improves the response, the fuzzy control improves the accuracy, and the neural network adapts to the effects of external perturbations. The introduction of reinforcement learning theory with self-learning capability into ANN provides a new research idea for intelligent control with broad promise in engineering applications.

4) Positioning technology
   Based on the current information, remote control and autonomy of USV are still intractable problems. To compensate for the nonlinearity and uncertainty of the model, many intelligent methods are used. Although the effect is clearly improved, the computational time complexity increases and the timeliness of the localization decreases. The whole motion control process still needs to be improved in terms of theoretical research and engineering applications.

5) Communication technology
   In the future, satellite communications will play a more important role in ship communication. At the same time, the security of the received and sent network information cannot be neglected. Networking communication is also a big challenge for the future.

5 Conclusion

Intelligent technology is an important driving force for USVs, and unmanned ships are the main force of "Marine power". In this paper, the overall system framework of USV is introduced and the key techniques and principles of various optimization methods are described in detail. In terms of sensor perception, the sensitivity of obstacle recognition needs to be improved and more sonar, camera, and other devices need to be explored. In terms of autonomous obstacle avoidance, when satisfying multi-task autonomous operation, the optimization direction should not only achieve mutual cooperation of the modules, but also ensure relative independence of each part. In terms of communications, land-based communication technologies have been implemented on the water, with technical difficulties arising from the uncertainty of the environment and the endurance of the ships themselves. Network security and surface multi-aircraft cooperation have also been studied and are directions for future development. Each system module interacts with each other, and the use of a single technology cannot meet the multi-functional and intelligent requirements of USVs, which require multi-disciplinary integration. In the future, the intelligence of technology, the fusion of sensors and the optimization of algorithms will be the focus of research. Using the existing cloud platform of smart technologies to build a new ship network system, we will finally achieve a breakthrough in intelligent generalization of the entire ship system.

References

1. Liu, B.Q., Meng, W.Q., Zhao, J.H., et al.: Variation of coastline resources utilization in China from 1990 to 2013. J. Natl. Resourc. **30**(12), 2033–2044 (2015)
2. Liu, C.G., Chu, X.M., Wu, Q., et al.: A review and prospect Of USV research. China Shipbuild. **2014**(4), 194–205 (2014)
3. Bertram, V.: Unmanned surface vehicles-a survey. Copenhagen, Denmark, (2008)
4. Yan, R., Pang, S., Sun, H., et al.: Development and missions of unmanned surface vehicle. J. Mar. Sci. Appl. **2010**(4), 451–457 (2010)
5. Wan, J.X.: Status and development trends of foreign military unmanned surface boats. Nat. Defense Technol. **35**(05), 91–96 (2014)
6. Chen, Y.B.: Overview of the development status and key technologies of unmanned ships. Sci. Technol. Innov. **2019**(02), 60–61 (2019)
7. Chang, J.Q., Pu, J.Q., Zhuang, Z.Y., et al.: Application analysis of unmanned vehicle in the field of marine survey. Ship Eng. **41**(01), 6–10 (2019)
8. Yan, Y.S., Ju, W.B.: Application of data fusion technology in the unmanned marine vehicle autopilot technology. Ship Sci. Technol. **38**(20), 1–3 (2016)
9. Wang, Y.: Yunzhou: "Road and air integration" unmanned system water path. Transport Construct. Manage. **476**(03), 114–117 (2020)
10. Zhuang, J.Y., Xu, Y.R., Wan, L., et al.: Target detection of an unmanned surface vehicle based on a radar image. J. Harbin Eng. Univ. **33**(2), 129–135 (2012)
11. Di, W.: On the design and key technologies of unmanned intelligent survey boats. China Maritime Safety **07**, 52–57 (2019)
12. Han, J., Kim, J., Son, N.: Persistent automatic tracking of multiple surface vessels by fusing radar and lidar. IEEE Conferences, OCEANS, vol. 2017, pp. 1–5. Aberdeen (2017)
13. Su, W.Z., Chen, P.: Application of data fusion technology in unmanned ship. Ship Sci. Technol. **39**(10), 134–136 (2017)

14. Zhang, X.L., Wang, H.S., Cheng, W.D.: Vessel detection and classification fusing radar Vision-Data. In: Seventh International Conference on Information Science and Technology. Da Nang, Vietnam, April 16–19 (2017)
15. Liu, H.X., Chen, W.H., Liu, S.Y.: Multi-frame point cloud fusion algorithm based on IMU and dynamic target detection. Appl. Res. Comput. **38**(07), 2179–2182 (2021)
16. Yu, B.F.: Overview of intelligent navigation research based on sensor and AI technology. J. New Indust. **11**(02), 65–67 (2021)
17. Zhuang, J.Y., Su, Y.M., Liao, Y.L., et al.: The local path planning of unmanned surface vessel based on the marine radar. J. Shanghai Jiaotong Univ. (Chin. Ed.) **46**(9), 1371–1375 (2012)
18. Singh, Y., Sharma, S., Sutton, R., et al.: A constrained A*approach towards optimal path planning for an unmanned surface vehicle in a maritime environment containing dynamic obstacles and ocean currents. Ocean Eng. **169**, 187–201 (2018)
19. Jin, J.A.: Research on path planning and real-time obstacle avoidance for USVs. Zhejiang University, Hangzhou (2020)
20. Zhang, H.N.: Automatic generation andpath planning of unmanned boat routes based on ant colony optimization algorithm. Ship Electron. Eng. **39**(3), 46–49, 97 (2019)
21. Kim, H., Kim, D., Shin, J.U., et al.: Angular rate-constrained path planning algorithm for unmanned surface vehicles. Ocean Eng. **84**, 37–44 (2014)
22. Xu, J.M., Ye, C.M.: A review of path planning algorithms for unmanned surface vehicles. Logist. Sci.-Tech. **38**(08), 28–31 (2015)
23. Long, Y., Su, Y.X., Shi, B.H., et al.: A multi-subpopulation bacterial foraging optimisation algorithm with deletion and immigration strategies for unmanned surface vehicle path planning. Intel. Serv. Robot. **14**(2), 303–312 (2021)
24. Chen, X.C., Zhou, M., Huang. J., et al.: Global path planning using modified firefly algorithm. In: Proceedings of International Symposium on Micro-Nanomechatronics and Human Science, pp. 1–7. IEEE Press, Piscataway, NJ (2017)
25. Chen, C., Geng, P.W., Zhang, X.C.: Path planning research on unmanned surface vessel based on improved potential field. Ship Eng. **37**(09), 72–75 (2015)
26. Kim, Y.H., Son, W.S., Park, J.B., et al.: Smooth path planning by fusion of artificial potential field method and collision cone approach. MATEC Web Conf. **75**, 05004 (2016)
27. Zhang, H.Y., Liu, T.: Miniature unmanned ship control system design and research of heading control methods. Comput. Measur. Control **25**(01), 88–90+93 (2017)
28. Hu, J.: Research on unmanned ship autopilot system based on human simulated intelligent control. Ship Sci. Technol. **39**(14), 46–48 (2017)
29. Wang, R.Q., Wang, Z.Y., Deng, H.: Genetic optimized RBF neural network based intelligent control algorithm of USV. J. Guangzhou Maritime Univ. **28**(01), 31–34 (2020)
30. Xue, Y.K.: A Study of Heading Control of Unmanned Surface Vehicle Based on Fuzzy Neural Network. Dalian Maritime University (2020)
31. Pei, Z.Y., Dai, Y.T., Li, L.G., et al.: Overview of unmanned surface vehicle motion control methods. Mar. Sci. **44**(03), 153–162 (2020)
32. Lan, X.: Research on Navigation Positioning and Path Planning of Unmanned Ships. Chongqing University of Posts and Telecommunications (2021)
33. Mu, X., He, B., Wu, S., et al.: A practical INS/GPS/DVL/PS integrated navigation algorithm and its application on Autonomous Underwater Vehicle. Appl. Ocean Res. **106**(3), 102441 (2020)
34. Yang, Z.H.: Development and Technical Verification of Unmanned Surface Vessel System with Dynamic Positioning Capability. Shanghai Jiaotong University (2020)
35. Li, F.: The Key Technology Research ofUnmanned Surface Vehicle. Shangdong University (2016)

36. Xu, H.E., Xiang, H.H., Shao, X., et al.: Design and implementation of unmanned ship cloud control system based on 4G Internet of Things technology. Software Guide **16**(06), 56–58 (2017)
37. Zhang, S.L., Liu, J., Yan, L.C., et al.: Failure diagnosis and intelligent reconstruction method based on deep learning in electric backbone communication networks. Software **39**(03), 194–198 (2018)
38. Li, X.W.: Unmanned ship remote control system based on wireless sensor network. Ship Sci. Technol. **40**(08), 10–12 (2018)
39. Zhou, J.G.: Application of Narrowband Internet of Things in Ship communication system. Changjiang Inform. Commun. **34**(02), 175–177 (2021)
40. Zhai, W.F., Xia, W.Y.: Application of deep learning in information resource optimal scheduling of unmanned ship communication network. Ship Sci. Technol. **43**(04), 163–165 (2021)

Simulation Research on Thermal Management of Hydrogen Fuel Cell for UAV

Zixuan Chang, Yi Fan[⊠], Weiting Jiang, Jingkui Zhang[⊠], and Jiakai Zhang

College of Energy and Mechanical Engineering, Shanghai University of Electric Power, Shanghai 200090, China
yifan0112@shiep.edu.cn, zk_neu@163.com

Abstract. The relatively high energy density of fuel cell UAVs has attracted more attention, and the working efficiency of fuel cell has a great impact on fuel cell UAVs. This study examines the influences of three cooling techniques—air cooling, liquid cooling, and pulsing heat pipe cooling on fuel cells operation and temperatures distribution of fuel cells. The factors' sensitivity analysis including the amount of water present, oxygen mole fraction, and current density are also studied under 5 m/s cruising speed of the fuel cell UAV. In addition, the research compares the cooling methods on pulsing heat pipe between the interior fan cooling and exterior natural convection cooling. The results show that under the same external wind speed, liquid cooling has the best cooling effect on fuel cells. Under the operating voltage of 0.35 v. 0.3 m/s flow rate has the largest heat flux. However, utilizing pulsing heat pipes can improves the temperature consistency of fuel cell. The outcomes demonstrate that the fuel cell's performance is not significantly impacted by the heat pipe casing's outer extension length. The heat pipe section can be appropriately extended from the outside of the housing during the design process to reduce the volume occupied by the fuel cell inside the UAV housing.

Keywords: Proton exchange membrane fuel cell · Thermal management · Numerical simulation

1 Introduction

Nowadays, businesses and researchers endeavour to develop new energy systems and technologies that can address the related environmental issues, and one of the most exciting technologies is fuel cell. The aviation sector produces significant amounts of carbon emissions and air pollution. The unmanned aerial vehicles (UAVs), which has the benefit of long working hours, requires efficient and clean energy propulsion systems. UAVs can operate with new hybrid engines that use fuel cells [1], Compared with lithium batteries, fuel cells have higher energy density, low noise, no pollution, and long endurance characteristics [2].

J. Li et al. (Eds.): 6GN 2023, LNICST 553, pp. 390–408, 2024.
https://doi.org/10.1007/978-3-031-53401-0_34

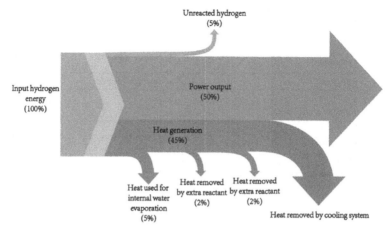

Fig. 1. Energy flow diagram of the whole PEMFC obtained by experiment [3].

Nomenclature		σ	surface tension coefficient,N/m
A	area, m²	ε	porosity
C	molar concentration,mol m^{-3}	η	surface over potential, V
c_f	the fixed charge concentration,mol m^{-3}	κ	Electrical conductivity, $S \cdot m^{-1}$
C_p	specific heat capacity, $Jkg^{-1}K^{-1}$	λ	water content
D	diffusion coefficient, $m^2 \cdot s^{-1}$	μ	fluid viscosity, $kg \cdot m^{-1}s^{-1}$
d	Heat pipe diameter, m	ζ	specific active surface area, m^{-1}
F	Faraday constant, $96487 Cmol^{-1}$	ρ	density of the fluid, $kg \cdot m^{-3}$

FR	filling ratio,%	δ	the evaporation coeffcient, $kg \cdot m^{-2} s^{-1}$
g	acceleration of gravity, $m \cdot s^{-2}$	ϕ	electric potential,V
H	height,m	Subscript and superscripts	
Δh_{fg}	the heat of evaporation, $J \cdot kg^{-1}$	a	anode
i	current density, $A \cdot m^{-2}$	act	activation
j	exchange current density, $A \cdot m^{-3}$	av	average value
K	permeability, m^2	bp	bipolar plate
k	thermal conductivity, $W \cdot m^{-1} K^{-1}$	c	cathode;cross-sectional area
L	length,m	cell	Fuel cell
M	molar mass, $kg \cdot mol^{-1}$	ch	channel
n_f	charge number of the sulfonic acid ion	con	condenser
p	pressure,Pa	ct	catalyst layer
Δp	pressure difference,Pa	d	diffusion layer
Q	heat load,w	eff	effective value
q	heat flux, $W \cdot m^{-2}$	eva	evaporator
R	electrical resistance, Ω ; thermal resistance, $K \cdot W^{-1}$	ex	experiment
RH	relative humidity,%	H_2	hydrogen
S	source term	H_2O	water
t	time,s	i	species: H_2, O_2, H_2O
U	index of uniform temperature,K	l	liquid phase
u	velocity vector, $m \cdot s^{-1}$	m	mass;membrane phase
v	velocity, $m \cdot s^{-1}$	O_2	oxygen
V	voltage,V	oc	open circuit voltage,V
W	width,m	ohm	ohmic polarization

Greek symbols		ref	reference value
α	Charge transfer coefficient	s	solid phase;surface
γ	concentration exponent	sim	Simulation

The fuel cell in Fig. 1 has about 50% efficiency in converting energy into electricity, which means that during operation, more than approximately fifty percent the energy generated will be released as heat, including entropic heat of reaction (about 30% of total heat), irreversible heat of electrochemical reaction (about 60% of total heat), Joule heat of ohm resistance (about 10% of total heat), and latent heat of water phase transition are the primary heat sources in fuel cells. Effective thermal management technology may lengthen the range and service life of the UAV by boosting the heat dissipation of the fuel cell thus enhancing the stack's internal temperature distribution's uniformity [4].

The selection of cooling methods is particularly critical, and the common thermal management methods are air cooling, liquid cooling, heat pipe and so on [5]. Air cooling is a typical thermal management technique used in fuel cells and is commonly used to portable equipment [6], The impact of utilizing metal foam at the cathode side on the air cooling effect of fuel cells was experimentally examined by Wan et al. [7]. The study found that expanding the metal foam's breadth improved the thermal control of air-cooled fuel cells and that the air was better mixed as a result of the metal foam move field's influence on temperature distribution. The metal bipolar plate air-cooled battery stack was created by Chang et al. [8], and the experimental findings indicate that raising the flow rate may significantly enhance the battery's internal temperature uniformity under high current density. A single battery's voltage consistency is influenced by hydrogen pressure, nonetheless, the fuel cell's total output energy to the stack is not significantly affected. Numerous scholars also investigated the water cooling method. The effects of three distinct coolant inlet temperatures—50 °C, 60 °C, and 70 °C—on the thermal properties and power exportation of fuel cell stacks were experimentally studied by Liu et al. [9]. They discovered that raising the coolant's temperature might result in a more even distribution of temperatures across the stack, thus enhancing the PEMC stack's power efficiency. Mahmoodi et al. [10] investigated the interaction between a fuel cell and a metal oxide (HM) tank, optimized the condenser's structure using the finite element technique, and investigated the effects of the quantity of copper wires in the core layer and heat pipes and heat sinks. To carry out thermal control of PEM fuel cells, Silva et al. [11] suggested connecting a number of capillary pumps with continuous conduction heat pipes back, and found this technology is capable of keeping PEM fuel cells' operating temperature within the required temperature range.

Some scholars also combined PEMFC and UAV. Horde et al. [12] numerically simulated PEMFC and analyzed its impact on fuel cells under different altitudes and air stoichiometric factors representing aircraft cruise, and found that the decline of environmental pressure would lead to the decline of fuel cell performance. The safe operating range of the electric pile was compiled by Zeng et al. [13] after researching the effects of cooling water flow at various flight heights on the electric pile. The key to ensuring the smooth functioning and stability of the UAV battery system is thermal control of the UAV fuel cell. Applying the proper cooling techniques and thermal management techniques may significantly improve the size and cost.

The fuel cell UAV needs to maintain a lightweight and compact design while ensuring the high-density energy conditions of the fuel cell to ensure its carrying capacity and flight performance. In order to boost the fuel cell's energy density, it is currently usual practice to boost the hydrogen storage tank's ability to hold hydrogen. For instance, Sung et al. [14] developed a new type of cylinder to improve the hydrogen storage density. To increase the energy density of fuel cells, Mohame et al. [15] proposed an energy management method based on frequency separation rules to save fuel consumption, so as to increase the driving range without changing the volume. There are few relevant studies considering reducing the occupied volume of fuel cells. How to reasonably place the system in the limited space inside UAV to minimize its occupied volume is also an optimization idea of the UAV. The impact of decreasing the fuel cells' occupied volume in a compelled convection environment on their efficiency is examined in this work.

Pulsating heat pipes can maintain a relatively stable working state under microgravity or weightlessness conditions, and can better adapt to the complex and changeable working conditions of UAS. It is not placed in the forced air convection cooling environment at the normal cruise speed of the UAV to consider the impact on its working condition. In this study, three cooling techniques—heat pipe, water, and air are studied to investigate the operation influences of variation of temperature, water content in the film, oxygen mole fraction, and current density when they are used in a wind field that simulates the UAV's cruise. In addition, the characteristics of various pulsing heat pipes extension lengths were studied. The effect of the fuel cell's working efficiency under the condition that part of the heat pipe's extension spares the space occupied by the fuel cell inside the shell was also discussed.

2   Geometric Modeling

A three-dimensional model of a fuel cell served as the foundation for this simulation. Reactant gas flows in a laminar and two-phase manner. The model is applied the following presumptions:

1. PEMFC functions in non-isothermal steady state settings
2. Laminar flow is thought to be the flow state and the gas reactant in the battery is treated as an incompressible ideal gas
3. Assume that the catalyst layer and gas diffusion layer are both homogeneous, isotropic porous media
4. Assume that the liquid phase speed in the reaction gas channel is equal to the gas phase
5. The material's properties are unaffected by the battery's compression circumstances
6. This model takes multi-component diffusion into account
7. Ignore the contact resistance

2.1 Model Description

The PEMFC's construction and cooling flow channel are shown in the following cross-section, which makes the simple calculation that the PEMFC's outer wall thickness is 0. The UAV is subsequently scaled down to a 0.12 m * 0.12m * 0.12 m space, a fan and convection hole are placed on the opposite wall, a wind field with a fixed velocity of wind is used on the outside of the fuel cell to simulate the UAV's operating speed, and the air is forced to cool the exterior of the fuel cell to represent the actual operating conditions inside the UAV (Figs. 2 and 3).

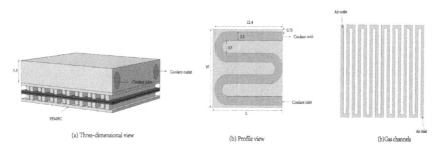

(a) Three-dimensional view   (b) Profile view   (b)Gas channels

Fig. 2. Fuel cell cooling channel.

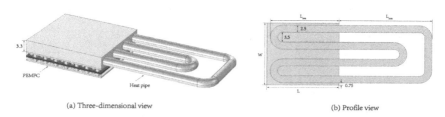

(a) Three-dimensional view   (b) Profile view

Fig. 3. Fuel cell heat pipe.

Figure 4 depicts the position relationship between the fuel cell and the fan and vent in the assumed inner space of the UAV case, with the fan assumed to be circular and the vent assumed to be square. Figure d5 depicts the distance between the fuel cell, which uses a heat pipe for heat management after cutting the inner space of the UAV case, and the cutting part and the outer wall of the case.

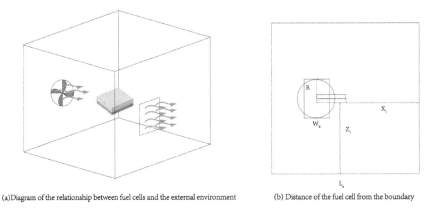

(a)Diagram of the relationship between fuel cells and the external environment   (b) Distance of the fuel cell from the boundary

Fig. 4. Fuel cell inside the drone housing.

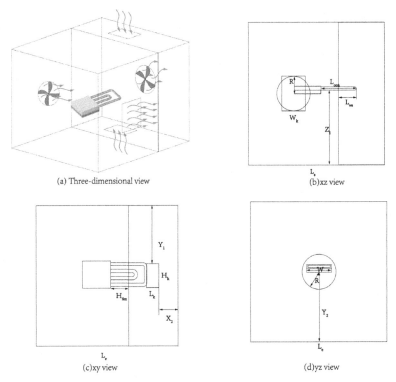

(a) Three-dimensional view   (b)xz view

(c)xy view   (d)yz view

Fig. 5. Heat pipe cooling cutting.

2.2 Governing Equation

In this example, the transport of reactants and water is examined. The model takes into account electrochemical currents in GDL, porous electrodes, and polymer films as well as mass and acceleration transfer processes in flow channels, GDL, and porous electrodes.

This example uses the "Darcy's Law" interface to solve for the pressure and the resulting velocity vector.

The battery's operating temperature is 70 °C. For reference equilibrium potentials at higher temperatures, the reference state of each reaction is calculated according to the following formula and based on the standard free energy of the generated entropy and the reaction entropy

$$E_{eq,ref}(T) = \frac{(\Delta H - T\Delta S)}{nF} \tag{1}$$

where n is the number of electrons involved in the electrode reaction, T represents the temperature,

According to the following formula, an expression of type Butler-Volmer can be used to define the electrode dynamics of the cathode

$$i_{loc,O_2} = i_{0,ref,O_2}\left(\left(\frac{p_{H_2O}}{p_{ref}}\right)^2 \exp\left(\frac{\alpha_{a,O_2}F\eta_{ref,O_2}}{RT}\right) - \frac{p_{O_2}}{p_{ref}}\exp\left(-\frac{\alpha_{c,O_2}F\eta_{ref,O_2}}{RT}\right)\right) \tag{2}$$

where η_{ref} is the overpotential relative to the reference state, defined as follows

$$\eta_{ref,O_2} = \phi_s - \phi_l - E_{eq,ref,O_2} \tag{3}$$

According to the following mass action law, it can be assumed that this model has ideal kinetic properties:

$$\alpha_{a,O_2} + \alpha_{c,O_2} = n \tag{4}$$

For the anode, the dynamics are assumed to be very fast, so that a linear Butler-Volmer expression can be used at the anode boundary.

$$i_{loc,H_2} = i_{0,ref,H_2}\left(\frac{p_{H_2}}{p_{ref}}\exp\left(\frac{\alpha_{a,H_2}F\eta_{ref,H_2}}{RT}\right) - \exp\left(-\frac{\alpha_{c,H_2}F\eta_{ref,H_2}}{RT}\right)\right) \tag{5}$$

Global variables of heat and energy balance: Symbols represent different integral domains: Ω geometric domains, $\partial\Omega_{ext}$ external boundaries, $\partial\Omega_{int}$ internal boundaries. The total heat source Q_{Int} includes all domain sources, inner boundaries, edge sources and point sources, as well as radiation sources at inner boundaries:

$$Q_{Int} = \int_{\Omega} Q d\omega + \int_{\partial\Omega_{int}} Q_b d\omega + \int_{\partial\Omega_{int}} Q_r d\omega \tag{6}$$

The total internal energy is E_{i0} defined as

$$E_{i0} = E_i + \frac{\boldsymbol{u} \cdot \boldsymbol{u}}{2} \tag{7}$$

For a fixed model with incompressible convection, the thermal equilibrium becomes:

$$\text{ntfluxInt} = QInt \tag{8}$$

Heat transfer interface: Using temperature gradient and effective thermal conductivity to evaluate thermal conductivity flux variable dflux:

$$\text{dflux} = -k_{\text{eff}}\nabla T \tag{9}$$

Heat transfer to turbulent fluids

$$k_{\text{eff}} = k + k_T \tag{10}$$

where k_T is the thermal conductivity of turbulence.
Turbulent heat flux

$$\textit{turbflux} = -k_T\nabla T \tag{11}$$

Turbulent kinetic energy k transport equation

$$\rho\frac{\partial k}{\partial t} + \rho\boldsymbol{u}\cdot\nabla k = \nabla\cdot\left(\left(\mu + \frac{\mu_T}{\sigma_k}\right)\nabla k\right) + P_k - \rho\varepsilon \tag{12}$$

The capillary resistance in the tube's walls rises as the inner diameter decreases, which has a strong influence on the pulsing heat pipe's ability to function. The following formula can be used to determine the pulsing heat pipe's inner diameter:

$$0.7\sqrt{\frac{\sigma_1}{(\rho_1 - \rho_v)g}} \le d \le 1.8\sqrt{\frac{(\rho_1}{(\rho_v - \rho_v)}g} \tag{13}$$

For ease of calculation, the pulsating heat pipe is regarded as a solid with high thermal conductivity.

$$R = (T_e - T_c)/Q_{\text{PHP}} \tag{14}$$

k_{eff} is calculated as the formula:

$$k_{\text{eff}} = \frac{l_{\text{eff}}}{RA_c} \tag{15}$$

Among them, The formula can be used to determine the optimum length of the pulsing heat pipe:

$$l_{\text{eff}} = \frac{\int_{l_{eva}}(\int_0^x q_e dx)dx + l_{adia}q_{c,\max} + \int_{l_{con}}(\int_{l_e+l_{adia}}^x q_e dx)dx}{q_{c,\max}} \tag{16}$$

Assuming that the heating and heat removal in both the evaporation and condensing sections are uniform, the effective length is:

$$l_{\text{eff}} = l_{\text{eva}}/2 + l_{\text{adia}} + l_{\text{con}}/2 \tag{17}$$

2.3 Model Verification

Based on the model employed by Nuttapol and Limjeerajarus [16], this paper's physical model of the PEMFC is being examined. Figure 6 displays the I-V polarization curve pairings that were discovered during model validation. The findings of the numerical simulation in this paper's polarization curve are well congruent with those found in the literature [16]. Assuming that it works in a steady state, the pulsating heat pipe's effective thermal conductivity is fixed. The pulsing heat pipe is treated as a solid with a thermal conductivity of 4056.35 W/(m K) in the study that follows.

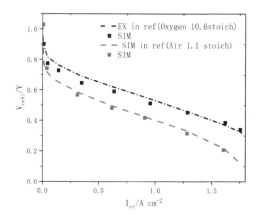

Fig. 6. PEMFC polarization curve verification.

3 Results and Discussion

In this section, the effects of three cooling modes of fuel cell, namely air cooling, water cooling and heat pipe, on fuel cell temperature, oxygen mole fraction, water content in film and current density under different operating voltages were analyzed under the circumstance that the operation of the UAV was simulated by an external wind field.

3.1 Applied Wind Field

Temperature Distribution Under Different Cooling Methods
Figure 7(a) compares the average and maximum temperatures of air-cooled water cooling and heat pipe under three working conditions with an additional 5m/s wind field. According to the figure, the lower the voltage is, the higher the fuel cell temperature will be. Because a more intense degree of electrochemical reaction is required to meet

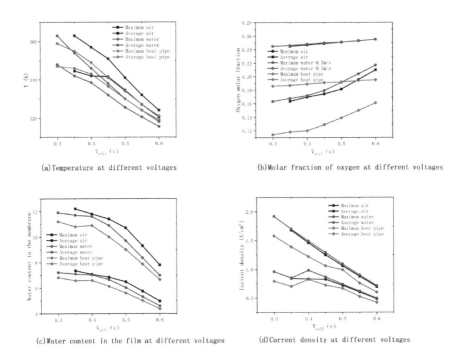

Fig. 7. Performance of fuel cells with different cooling modes at different operating voltages.

the output of the battery at a lower operating voltage, the reaction gas inlet selected in this work has a steady flow rate, this will cause the fuel cell's own heat generation to be uneven, hence widening the temperature gap at the CCL/M. Compared with the temperature distribution without external wind field in the original model, the external wind field effectively reduces the temperature difference between different cooling methods, and the coolant temperature is the same as the external temperature, which is more convenient for application in actual operation. Water cooling at 0.3 m/s may lower the fuel cell's maximum temperature to 354 k under an operational voltage of 0.35 v, which is 9 k lower than that of air cooling under the same working condition. In the low voltage working range (0.3 v–0.45 v), such as 0.35 v, the air-cooled method has a 37 k variation in temperature, the water-cooled 24 k, and the heat pipe is only 18 k. The temperature distribution map of the CCL/M interface during the water cooling-assisted thermal control of the fuel cell is shown in Fig. 8. As can be observed, the temperature distribution of the fuel cell is relatively uneven in the low voltage working section, and local hot spots are prone to appear. In contrast, heat pipes provide better protection against localized fuel cell hot spots. The fuel cell's operational life can be successfully extended by it, particularly in low voltage, high temperature operating areas. The use of heat pipe cooling may also successfully minimize the interior volume of the occupied UAV housing, taking into account the volume and load issues of the fuel cell UAV.

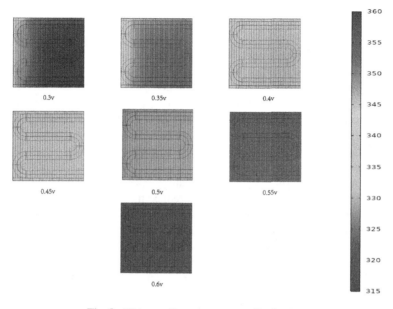

Fig. 8. Water cooling temperature distribution.

Oxygen Molar Fraction Under Different Cooling Methods

In Fig. 7(b), the average and maximum oxygen molar fraction of air cooled water cooled and heat pipe under the conditions of 5 m/s wind field are compared. At the operating voltage of 0.3 v, the difference between the maximum and average oxygen molar fraction of water-cooled water is 0.08. At the operating voltage of 0.55 v, the difference is 0.04, and the lower the operating voltage, the more intense the electrochemical reaction, the more fuel is consumed, and the less oxygen remains at the interface. Compared with air cooling and water cooling, heat pipe has better temperature equalization ability, the reaction effect in low pressure region is the best, and the oxygen remaining is the least. The oxygen molar fraction rises as the voltage rises, and while the temperature progressively falls as the voltage rises, the activity of catalyst is inhibited, the electrochemical reaction rate decreases, and the amount of residual oxygen increases. It can be seen that the change range of maximum oxygen molar fraction is relatively stable, while the change range of average oxygen molar fraction gradually increases after 0.45 v. This indicates that there is always an inactive region, but when the voltage is greater than 0.45 v, the overall reaction rate slows down, resulting in an increase in residual oxygen at the interface. The mass circulation rate at the inlet of the fuel cell model is set to a constant value. If the reactants are not supplemented in time along the flow path, the oxygen molar fraction at the interface will become smaller. Since heat exchange among the fuel cell and the general environment is dominant in the case of external wind field, the cooling effect of air in the cooling flow path is negligible. For the variables of oxygen mole fraction, water content in the film and current density, the circulation rate of the cooling fluid in the cooling conduit has almost no influence. The subsequent variable analysis does not introduce the influence of air cooling and water cooling flow rate separately, but only introduces the three cooling methods.

Water Content in the Film Under Different Cooling Methods

Figure 7(c) compares the average and maximum water content in the membrane under three conditions of air-cooled water cooling and heat pipe under a 5 m/s wind field. An appropriate quantity of water in the membrane can encourage proton transport throughout the fuel cell activation phase and avoid flooding. The possibility of flooded areas in the fuel cell increases when the membrane has an excessive amount of water in it. The flooding phenomena causes the fuel cell's pores to get blocked, which slows down the reaction rate. Under high voltage operation, the water content is mainly affected by the battery operating temperature range. Lower battery operating temperature will reduce the chemical reaction rate of the battery and reduce the water content of the film. In the region with low voltage (0.3 v–0.5 v), the battery will increase the water content of the film due to the high activity of the catalyst. At 0.4 v, the average water content in the film is 7.08 for water cooling and 7.02 for heat pipe and 6.59 for heat pipe. The proton conductivity of the PEM will decrease when the internal water content is too low, decreasing the output voltage of the system, decreasing the efficiency and durability.

The Average Current Density Under Different Cooling Methods

In Fig. 7(d), the average current density and the maximum current density under three working conditions of cooling methods under an additional 5 m/s wind field are compared. Voltage is less than 0.4 v, the average current density becomes more uniform. The amalgamated impacts related to temperature and homogeneity of temperature have caused this change in current density. The consistency of temperature, water content in the membrane interface, and current density must all be taken into account in order to enhance battery performance. For example, the average current density of the fuel cell at 0.4 v is 0.73 A/cm^2, that of the water cooling is 0.98 A/cm^2, and that of the heat pipe is 0.82 A/cm^2. This is because the oxygen transfer rate leads to the current density change. The catalyst's activity is hindered when the working temperature exceeds 353.15 k, and the exchange membrane's substantial water loss results. Which will lead to the damage of the reaction components of the battery, reduce the conductivity of the membrane, and local hot spots will also cause the same problem.

3.2   Heat Pipe Cutting

According to the above analysis, although at the same external air flow speed, water cooling can maximize the cooling of fuel cells to achieve a lower operating temperature, but considering the temperature uniformity and working volume, heat pipe is a more suitable choice for drones. Next, we will discuss the addition of fans inside the fuel cell housing, so that the air flow rate inside the housing is 9.8 m/s. The heat pipe part extends from the outside of the housing to the outside air for forced air convection heat transfer of 5 m/s (the fixed cruise speed of the fuel cell UAV). The following contents discuss the effects of the extension length of 5mm and 15mm on the fuel cell temperature, oxygen mole fraction, water content in the film and current density.

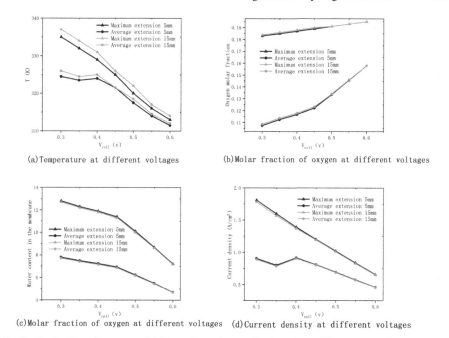

(a) Temperature at different voltages

(b) Molar fraction of oxygen at different voltages

(c) Molar fraction of oxygen at different voltages   (d) Current density at different voltages

Fig. 9. Fuel cell performance of different heat pipe cutting lengths at different operating voltages.

The Influence on Temperature After Reducing the Volume Occupied by the Housing

Figure 9(a) examines the fuel cell's temperature variations at various heat pipe extension lengths in a 5 m/s wind field. First of all, the space of the high-flow area (9.8 m/s) caused by space cutting becomes smaller, and the forced convection heat transfer effect is strengthened. It can be seen that the change of the overhanging length basically only brings about a temperature change of 2 K. At the operating voltage of 0.45 v of the fuel cell, the average temperature can even be balanced, which effectively proves that not only can it simplify cooling structure, but also add heat pipe extension appropriately to reduce the volume of the general design of the fuel cell inside the drone housing, thereby reducing the overall volume of the drone housing part.

The Effect of Reducing the Volume Inside the Housing on the Oxygen Molar Fraction

Figure 9(b) demonstrates the changing trend of the oxygen mole fraction on the membrane interface for fuel cells with overhang lengths of 5mm and 15mm. The difference in overhang length may have a small impact on the oxygen molar percentage at the CCL/M interface when the fuel cell is running at low operating voltage (0.3v–0.5v). Due to the low overall fuel cell temperature at high operating voltage, the electrochemical reaction rate is modest, maintaining the same oxygen molar percentage. It can be concluded that the extension length of the heat pipe has minimal effect on the oxygen molar fraction of the CCL/M interface of the fuel cell when taking into account the volume of the system in the housing and the actual working state.

The Effect of Reducing the Volume Inside the Housing on the Water Content in the Film

Figure 9(c) demonstrates the water content of the membrane's fluctuation trend between the 5mm and 15mm overhang lengths for UAV fuel cells. It can be seen that neither the membrane's peak water content nor its average water content shows a discernible decelerating tendency when the membrane's water content is below 0.45 v. The membrane's water content diminishes and the chemical reaction proceeds slowly. Although the temperature uniformity at low voltage is lower than that at high voltage, the electrochemical reaction will be uneven, but the overall temperature of the battery at this time is more suitable for chemical reaction, so the water content in the membrane at this time is relatively high.

The Effect of Reducing the Volume Inside the Housing on the Current Density

Figure 9(d) depicts the connection between average current density and voltage. At low operating voltage, the average current density will first increase and then decrease with the increase of voltage. The highest voltage and average voltage only differ by a modest amount, 0.3 A/cm^2, after 0.5 v. The current density will fall when the temperature drops because the fuel cell's response rate will slow down. The influence of overhang length on current density is negligible. This demonstrates that the heat pipe section may be adequately relocated out to lower the surface area of the PEMFC in the inside of the UAV housing and will not have a significant influence on the PEMFC's efficiency when the fuel cell UAV adopts it for the fuel cell's heat control.

4   Conclusion

In this paper, the fuel cell characteristics, i.e., temperature, oxygen mole fraction, water content in the film and current density, are discussed under the original forced cooling model and the external wind field forced cooling model. The cruising speed of the fuel cell UAV during operation is set 5 m/s.

It is found that the air flow rate in the cooling channel has no effect on the fuel cell because of the dominant effect of forced convection heat transfer.

The flow velocity of coolant water still has effect on the fuel cell's temperature when it is used to cool serpentine fuel cells. As several factors, like working temperature, reactant flow velocity, and chemical reaction activity, can pose impacts on the oxygen molar percentage, the water content in the film, and the current density, the influence of the cooling water flow rate is comparatively minor.

It is assumed that the heat pipe is a metal with fixed thermal conductivity when it is used to cool the system. The gap among the highest temperature and the median temperature at the temperature reading interface is less at low operating voltage, despite the fact that the overall temperature and cooling impact on the fuel cell are not as good as those of cooling water. This demonstrates that when using a heat pipe to cool the fuel cell, the cooling equality-temperature is superior than using water. The system's local hot spot will shorten its useful life, and because water cooling requires external devices to cycle, the fuel cell's volume will definitely rise. The UAV just has a little weight. Additionally, it works better for cooling fuel cell drones.

Next, based on the fuel cell that uses heat pipe, the heat pipe part of the fuel cell extends out of the UAV shell for forced convection heat transfer under the original working wind speed. At the same time, the two variables of fan at the UAV shell and heat pipe extension length are discussed. Additionally, comparisons and discussions were made about the fuel cell temperature of operation, oxygen mole fraction, water content in the membrane, and current density. According to the simulation's findings, Cutting in the original space reduces the space for forced convection and improved the heat transfer impact. Additionally, a partial extension of the heat pipe may have the potential to diminish the volume of the system inside the UAV housing. The 5 mm extension and 15 mm extension only have a temperature difference of 2 K in temperature, and the difference in oxygen mole fraction, water content in the film and current density is very small, which proves that on the basis of the original structure, it can be considered to reduce the space occupied by the heat pipe in the interior of the UAV to save the internal space of the fuel cell UAV shell for other equipment structures.

Appendix

Table 1. Size of physical model.

Parameter	Symbol	Numerical value	Parameter	Symbol	Numerical value
Membrane height(m)	H_m	5×10^{-5}	Fan diameter(mm)	R	205
Catalyst layer height(m)	H_{cl}	1.5×10^{-5}	Wind speed(m/s)	s	9.8
Gas diffusion layer height(m)	H_{dl}	1.9×10^{-4}	UAV shell side length(m)	L_e	0.12
Gas channel height(m)	H_{ch}	8×10^{-4}	Fuel cell x axial distance from boundary(m)	X_1	0.0568
Gas channel width(m)	W_{ch}	8×10^{-4}	Axial distance of the fuel cell from the boundary z(m)	Z_1	0.066
MEA length(m)	L	2.32×10^{-2}	Vent width(m)	W_k	0.002
MEA width(m)	W	2.2×10^{-2}	Heat pipe cutting length(m)	H_{fen}	0.015 0.005
Cell active area(m^2)	A_m	5.104×10^{-4}	Cut part vent width(m)	L_k	0.01
PHP evaporation segment length	L_{eva}	2.24×10^{-2}	Cut part vent height(m)	H_k	0.02

(*continued*)

Table 1. (*continued*)

Parameter	Symbol	Numerical value	Parameter	Symbol	Numerical value
PHP condensing segment length	L_{con}	3.23×10^{-2}	Distance between the Y-axis direction of the cut part air vent and the border(m)	Y_1	0.05
Fan speed(rpm)	n	2900	Cut part of the vent x axis direction distance(m)	X_2	0.021
Fan volume(m^3/min)	Q	18.38	Distance between the cutting part of the fan and the boundary(m)	Y_2	0.06

Table 2. Electrochemical parameters and material characteristics of PEMFC.

Parameter	Symbol	Numerical value	Parameter	Symbol	Numerical value
Anode reference exchange current density (A/m^2)	$i_{a,ref}$	7.17	Diffusion layer porosity	ε_{dl}	0.6
Cathode reference exchange current density (A/m^2)	$i_{c,ref}$	7.17×10^{-5}	Catalyst layer porosity	ε_{cl}	0.4
Anodic charge transfer coefficient	α_a	1	Permeability of the catalyst layer (M)	K_{cl}	10^{12}
Cathode charge transfer coefficient	α_c	1	Membrane equivalent (kg/mol)	M	1.1
H_2 reference concentration (mol/m^3)	$c_{H_2,ref}$	0.88	Thermal conductivity of the collector plate [W/(m K)]	k_{bp}	100
O_2 reference concentration (mol/m^3)	$c_{O_2,ref}$	0.88	Thermal conductivity of diffusion layer [W/(m K)]	K_{dl}	10

(continued)

Table 2. (*continued*)

Parameter	Symbol	Numerical value	Parameter	Symbol	Numerical value
Anode concentration index	γ_a	1	Thermal conductivity of catalyst layer [W/(m K)]	k_{cl}	10
Cathode concentration index	γ_c	1	Thermal conductivity of proton exchange membrane [W/(m K)]	k_m	0.16
Faraday constant (C/mol)	F	96487	Diffusion layer conductivity (S/cm)	κ_{dl}	2.8
H_2 diffusion coefficient (m^2/s)	D_{H_2}	3×10^{-5}	Conductivity of catalyst layer (S/cm)	κ_{cl}	50
O_2 diffusion coefficient (m^2/s)	D_{O_2}	3×10^{-5}	Conductivity of the collector plate (S/cm)	κ_{bp}	1×10^4
H_2O diffusion coefficient at the cathode (m^2/s)	D_{c,H_2O}	3×10^{-5}	Effective specific surface area (1/m)	ζ	1.127×10^{-7}
H_2O diffusion coefficient at the cathode (m^2/s)	D_{a,H_2O}	3×10^{-5}	Open-circuit voltage (v)	V_{oc}	1.05

Table 3. Experimental parameters of heat pipe.

Q/W	FR/%	T_{eva-ex}/K	R_{ex}/(K/W)
18	40	320.70	0.845
18	60	332.00	1.32

References

1. Bahari, M., et al.: Performance evaluation and multi-objective optimization of a novel UAV propulsion system based on PEM fuel cell. Fuel **311**, 122554 (2022)
2. Çalışır, D., et al.: Benchmarking environmental impacts of power groups used in a designed UAV: Hybrid hydrogen fuel cell system versus lithium-polymer battery drive system. Energy **262**, 125543 (2023)
3. Shabani, B., Andrews, J.: An experimental investigation of a PEM fuel cell to supply both heat and power in a solar-hydrogen RAPS system. Int. J. Hydrogen Energy **36**(9), 5442–5452 (2011)

4. Yh, A., et al.: Thermal management of polymer electrolyte membrane fuel cells: A critical review of heat transfer mechanisms, cooling approaches, and advanced cooling techniques analysis

5. Fan, Z., Song, K., Zhang, T.: A review on performance optimization of air-cooled proton exchange membrane fuel cell. Automobile Technol. **4**, 1–8 (2020)

6. Chen, Q., et al.: Thermal management of polymer electrolyte membrane fuel cells: A review of cooling methods, material properties, and durability. Appl. Energy **286**(1), 116496 (2021)

7. Wan, Z., et al.: Thermal management improvement of air-cooled proton exchange membrane fuel cell by using metal foam flow field. Appl. Energy **333**, 120642 (2023)

8. Chang, H., et al.: Experimental study on the thermal management of an open-cathode air-cooled proton exchange membrane fuel cell stack with ultra-thin metal bipolar plates. Energy **263**, 125724 (2023)

9. Liu, Q., et al.: Experimental study of the thermal and power performances of a proton exchange membrane fuel cell stack affected by the coolant temperature. Appl. Therm. Eng. **225**, 120211 (2023)

10. Mahmoodi, F., Rahimi, R.: Experimental and numerical investigating a new configured thermal coupling between metal hydride tank and PEM fuel cell using heat pipes. Appl. Thermal Eng. **178**, 115490 (2020). https://doi.org/10.1016/j.applthermaleng.2020.115490

11. Silva, A.P., et al.: A combined capillary cooling system for fuel cells. Appl. Therm. Eng. **41**, 104–110 (2012)

12. Hordé, T., Achard, P., Metkemeijer, R.: PEMFC application for aviation: Experimental and numerical study of sensitivity to altitude. Int. J. Hydrogen Energy **37**(14), 10818–10829 (2012)

13. Zeng, Z., et al.: A modeling study on water and thermal management and cold startup of unmanned aerial vehicle fuel cell system. eTransportation **15**, 100222 (2023)

14. Cho, S.M., et al.: Lightweight hydrogen storage cylinder for fuel cell propulsion systems to be applied in drones. Int. J. Press. Vessels Pip. **194**, 104428 (2021)

15. Boukoberine, M.N., et al.: Hybrid fuel cell powered drones energy management strategy improvement and hydrogen saving using real flight test data. Energy Convers. Manage. **236**, 113987 (2021)

16. Nuttapol, et al.: Effect of different flow field designs and number of channels on performance of a small PEFC. Int. J. Hydrogen Energy (2015)

Performance Analysis on a Coupled System of Gas Turbine and Air Cycle Driven by the Waste Heat of Flue Gas

Jingkui Zhang[1,2], Wen Yan[1], Zhongzhu Qiu[1(✉)], Yi Fan[1], Puyan Zheng[1], and Jiakai Zhang[1(✉)]

[1] College of Energy and Mechanical Engineering, Shanghai University of Electric Power, Shanghai 200090, China
quizhongzhu@shiep.edu.cn, 18867153037@163.com
[2] Shanghai Non-Carbon Energy Conversion and Utilization Institute, Shanghai 200240, China

Abstract. In this work, a novel gas-air combined cycle of power generation system is proposed to enhance the generation capability of gas turbines, which uses the Ericsson cycle to transform gas turbine waste energy into a more useful form of energy. In this paper, the temperature-entropy diagram of the combined system is given, and the gas-air combined cycle system is built and simulated. After validation of the simulation, the effects of air flow, ambient temperature, pressure ratio and gas turbine load on the power generation capability and energy utilization efficiency are investigated. The results show that the air circle greatly improves the overall power generation efficiency with the same gas supply. The maximum power generation efficiency of the gas-air combined cycle is 31.95% at the rated load of the gas turbine with an optimized air flow of 14 kg/s, in which the power generation efficiency of the air cycle is 4.47%. The increase of air flow rate can improve the net output power, while the power generation efficiency of air circle and combined circle all increases first and then decreases. As the increase of pressure ratio of compressors, the power generation efficiency increases first and then decreases, while the energy utilization efficiency increases significantly with the decreasing of outlet temperature of smoke from the high-pressure heater and low-pressure heater. The ambient temperature effects on the power generation of the air circle and the combined circle are almost linear. The generation efficiency of the air circle and combined circle are all increased with the increase of gas turbine load. Therefore, under the selection of suitable working conditions, the gas-air combined cycle power generation system can effectively improve the total power generation efficiency and has great potential in waste heat utilization.

Keywords: Ericsson cycle · gas-air combined circle · Waste heat utilization · Power generation efficiency

J. Li et al. (Eds.): 6GN 2023, LNICST 553, pp. 409–427, 2024.
https://doi.org/10.1007/978-3-031-53401-0_35

1   Introduction

Climate change due to large emissions of carbon dioxide has been broadly recognized as one of the significant problems facing society, mainly due to activities such as burning fossil fuels [1]. In the Fourth Assessment Report (AR4) of the United Nations Intergovernmental Panel on Climate Change (IPCC) in 2007, it was stated that rising concentrations of greenhouse gases due to human emissions over the past 50 years may be responsible for the warming, and this is even more evident in the Fifth Assessment Report (AR5) released in 2013, where the concentration of carbon dioxide has increased from about 280 ppm in pre-industrial times to 391 ppm in 2011 [2]. To cope with this global problem, on the one hand, a strong focus should be placed on sustainable energy generation technologies such as solar and wind power [3–6]. On the contrary, technologies to raise energy use efficiency and improve the energy use structure should be developed [7–9]. However, since renewable energy technologies have problems such as intermittency and integrating their generation into the grid, which is a great challenge for the reliability and security of the grid, monitoring and scheduling of loads, power sources, storage systems and the grid is needed to cope with this risk [10], where the construction and maintenance of some fossil energy plants are essential. Thermal power generation can lead to a drastic decrease in the share of power generation due to climate issues and it has problems such as high pollution and long start-up times, which are detrimental to the future of integrated energy use. Compared with coal power, gas-fired power generation will take a huge role in the coming energy layout due to its advantages of high power density, fast start-up, and high quality of power generation [11], so it is crucial to develop the efficiency of the gas-fired power station for power generation and primary energy efficiency. So far, researchers have improved the efficiency of gas-fired power stations by improving the thermal efficiency of gas turbines, hybrid co-generation, and the low-temperature thermal energy gradient effect [12–15].

Since there is currently a large number of surplus heat resources in the gas-fired power generation industry, whose exhaust temperatures can be as high as 400–600 °C [16], various technologies need to be developed and introduced to improve the utilization of gas turbine exhaust waste heat and reduce primary energy consumption to improve the quality of surplus heat energy utilization. Currently, the mainstream surplus heat utilization technologies include the organic Rankine cycle, Kalina cycle, as well as Ericsson cycle. The organic Rankine cycle (ORC) technology is considered feasible engineering [17], which uses low boiling point organic matter as the cycle workpiece to turn low-grade heat energy into electrical energy [18]. A large amount of ORC waste heat recovery plants have been put into operation so far [19]. Henrik Öhman evaluated the ORC low-temperature waste heat recovery system in a Swedish pulp mill to demonstrate the advantages of ORC technology [20]. Jian Song et al. analyzed the waste heat recovery for marine diesel engines using ORC engineering and determined the optimal operating conditions [21]. Zhang Xiang Wu et al. offered a novel dual-function system of inverse Carnot cycle-organic Rankine cycle to recover low-grade surplus heat for power generation or high-grade heat energy [22]. The Kalina cycle also has an important position in low-temperature thermal resource power generation, which mainly uses ammonia water as the working medium and is fit for hot gas heat sources such as gas turbine exhaust, and can provide 32% more power in industrial residual heat applications in comparison to the traditional Rankine cycle [23]. The Ericsson cycle is rarely mentioned compared to the previous two thermal cycles. Heat engines based on the Ericsson cycle principle have great performance at low-powered levels, such as micro-cogeneration using solar energy, biomass, etc., or coupled with natural gas combustion systems to form cogeneration [24–26].

As for large gas-fired combined cycle power systems, the gas-steam combined circulation power system has become the most used cycle power plant with many benefits such as low environmental emissions, high efficiency, and short construction cycle [27]. It is mainly a cycle structure consisting generator together with a waste heat boiler, a gas turbine engine and a steam turbine [28]. The high-temperature flue gas from the work done by the turbine is recovered and converted to high-temperature steam by the waste heat boiler, eventually, the steam is injected into the steam turbine for power generating, which is why the performance of a combined gas-steam circulation power station is superior to that of a traditional power generation system using a gas turbine or a steam turbine. Of course, many scholars have studied gas-steam combined cycle power generation systems, Felipe R et al. [29] analyzed the impact of ambient temperature on a power station consisting of two Siemens AG 501F gas turbines coupled to three pressure levels of heat recovery steam generators (HRSG) and reheated by supplementary combustion and steam turbines and found that ambient temperature has a substantial influence on the power output of such Meeta et al. [30] studied the improvement of combined cycle efficiency by using different types of HRSG layouts and performed a thermomechanical analysis of the perfect performance of the HRSG. Dan-Teodor et al. [31] proposed an exhaust heat utilization scheme by adding an organic Rankine circulation unit downstream of a gas-steam combined circulation power station, and the simulation analysis displayed that the efficiency of the plant was increased by about 1.1%.

As the current gas-steam combined cycle power plants are MW level, they can take up the pressure of power supply and peaking tasks in large cities, while for remote areas with low power demand, gas-steam combined circulation power stations are less suitable due to the large size of the building and the imbalance between supply and demand, so some scholars proposed micro combined cycle power generation systems, aiming to provide power and other forms of supply for areas with low population density. It is designed to provide power and other forms of energy supply to meet the normal load requirements of customers, and can also serve as a peaking function to relieve the pressure on the local power grid during times of high power consumption. Small and microturbines currently offer many potential advantages over other small-scale power generation technologies, especially for distributed generation [32]. Schneider et al. [33] reviewed the commercial development and studied activities of biomass-fueled CHP systems in Europe, focusing on small-scale CHP in combination with Stirling engines, intending to provide attractive small-scale CHP solutions. Invernizzi et al. [34] investigated the improvement of micro gas turbine performance by adding a bottom organic Rankine cycle and found that the combined configuration with a 100 kW micro gas turbine could raise the net power by about 1/3 and the power production efficiency by up to 40%.

The gas-air combined cycle power generation system mentioned in this essay consists of a kW-class gas turbine and an air cycle power generation structure. The innovation of this combined system lies in the effective use of micro gas turbine waste heat on the one hand, and the use of air as the circulating mass on the other hand, to achieve the goal of improving the power production efficiency of the gas turbine. The main principle is based on the ideal Ericsson cycle, using two-stage compression with middle cooling and two-stage expansion with intermediary heating to replace the isothermal compression and isothermal expansion processes in the ideal cycle, to achieve an approximate Ericsson cycle effect. After that, the effects of air flow, ambient temperature, pressure ratio and combustion engine load on the power production performance of the combined system are investigated by modelling the gas-air combined cycle system using commercial software, aiming to offer theoretical support for the practical engineering application of the new waste heat utilization system.

2 System Description

2.1 Ideal Thermodynamic Model

The ideal Ericsson cycle power system contains a compressor, a regenerator, as well as an expander with air serving as the working fluid. Figure 1 illustrates the system configuration in an open operating environment, where the intake working fluid comprises ambient air at atmospheric pressure and temperature, as well as the heat source is waste heat from gas turbines, steam turbines, and other sources. The system comprises four working processes: air compression in the compressor and heat absorption in the exchanger. Subsequent transfer of high-temperature flue gas heat to the air. Finally, expander-driven expansion of the air to generate useful work.

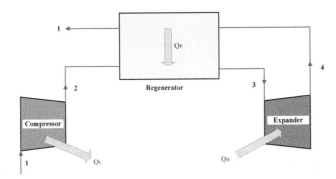

Fig. 1. System composition and working process diagram of Ericsson cycle.

Subsequently, the thermodynamic model of the Ericsson circulation is presented. It includes four main processes: two isobaric and two isothermal processes. To better describe this phenomenon, its temperature-entropy diagram is shown in Fig. 2. The model is on the basis of the thermodynamic transformation of the working mass at a stable state so that air is deemed an ideal gas in this cycle. As can be observed in Fig. 1, 1-2 is the isothermal compression process, where the mass releases heat to the low-temperature heat source (Q_L). 2-3 is the isobaric heating process, where the mass absorbs heat from the regenerator (Q_P). 3-4 is the isothermal expansion process, where the mass soaks up heat from the high-temperature heat source (Q_H). as well as 4-1 is the isobaric exothermic process, where the mass releases heat to the regenerator (Q_P).

Based on the earlier discussion, 1-2 and 3-4 are external transformation processes of the cycle that must complete energy conversion in a finite time, and hence, they are defined as external irreversible processes. Assuming an ideal regenerator in the circulation, the heat released by the heat source during the 4-1 process equals the heat absorbed by the cold source during the 2-3 process. Thus, the area b-3-2-a of the temperature-entropy diagram under the 2-3 path is equal to the area d-4-1-c under the 4-1 path. Commercial regenerators are generally more than 90% efficient, making this assumption reasonable [15]. Subsequently, for the ideal state, the heat transfer relation equation can be formulated as follows:

$$|Q_{41}| = |Q_{23}| \tag{1}$$

During energy transfer in the regenerator, process 2-3 provides heat to the working liquid, and the only source of heat from the external environment occurs along path 4-1. Assuming the working mass to be an ideal gas, we can use the definition of work to derive the energy conservation equation for the input energy during 3-4. Therefore, it can obtain:

$$Q_H = Q_{34} = mRT_H \ln \varphi_p \tag{2}$$

where φ_p is the pressure ratio, p_2/p_1.

In the same way, the equation for the transfer of energy from 1-2 to the outside is,

$$Q_L = -Q_{12} = mRT_L \ln \varphi_p \tag{3}$$

For the whole thermal system cycle 1-2-3-4-1, its energy conservation equation is,

$$W_{net} = Q_{net} = Q_H - Q_L \tag{4}$$

The substitution of Eqs. (2) and (3) into Eq. (4) results in,

$$W_{net} = mR \ln \varphi_p (T_H - T_L) \tag{5}$$

2.2 Approximate Thermodynamic Model

The proposed exhaust heat utilization power production system in this paper aims to mimic the ideal Ericsson cycle, which is challenging to implement in practical conditions. To approximate the Ericsson cycle, the system incorporates double-stage compression with intercooling in the compressor and double-stage expansion with intermediate heating in the expander. The purpose of using double-stage compression with intercooling is to approach the ideal constant temperature compression process. The more compression stages are employed, the closer the system gets to the ideal state. However, due to cost constraints and other limitations, the number of compression stages in practical applications typically does not exceed four. Figure 3 in the paper depicts the diagram of the waste heat utilization power generation system with an approximate Ericsson cycle. This

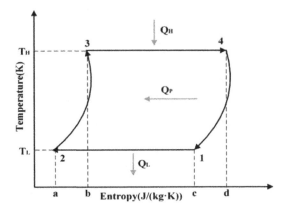

Fig. 2. Temperature-entropy diagram of Ericsson cycle.

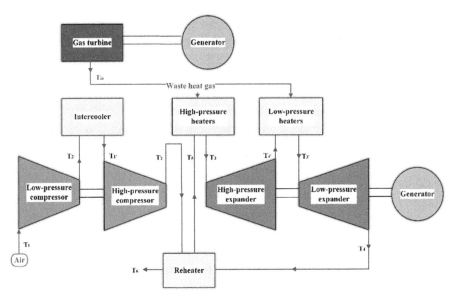

Fig. 3. Diagrammatic diagram of Gas-air combined cycle system.

diagram illustrates the arrangement and components of the system, showcasing how it operates and utilizes waste heat to generate power.

In the study of a gas-air united cycle system in steady-state operation, the following suppositions are created in the essay for the modelling and simulation of the combined system operating under different operating conditions:

a. The working gas used in the air circulation system is air, which is presumed to be an ideal gas and to conform to the ideal gas formula of state.
b. Assuming that the specific heat capacity of the flue gas and fuel properties remain steady in this study.

c. Assuming that the isentropic efficiency of the pressurizer and expander remains constant and the heat exchanger efficiency remains constant.

d. Neglecting the flow pressure losses and heat dissipation losses of the working mass in the air system in the ducts and heat exchangers.

e. Considering the output work of the expander is lost at the generator, the generator generation efficiency in the model is 0.98.

Figure 4 depicts the temperature-entropy drawing of the gas-air combined circle to better understand the gas-air co-generation system. Process 1-2' involves the air entering the low-pressure compressor (LC), which compresses the air and increases its temperature to T_2. In process 2'-1', the air enters the intercooler (IC), which cools its temperature to T_1'. The 1'-2' process entails the air entering the high-pressure compressor (HC), where it is compressed, leading to an increase in temperature to T_2. Next, in processes 2-5, the air enters the recuperator for heating, causing the temperature to rise to T_5. During processes 5-3, air and high-temperature flue gas exchange heat in the high-pressure heat exchanger (HH), increasing temperature to T_3. In processes 3-4', high-pressure and high-temperature air undergo work in the high-pressure expander (HE), resulting in a decrease in temperature to T_4. Process 4'-3' involves air and high-temperature flue gas exchanging heat in the low-pressure heat exchanger (LH), increasing temperature to T_3. During process 3'-4, high-pressure and high-temperature air undergo work in the low-pressure expander (LE), resulting in a temperature decrease to T_3. Finally, process 3'-4 leads to high-pressure and high-temperature air working in LE, causing a temperature drop to T_4. Process 4-6 involves the exchange between the high-temperature outlet air of LE and HC outlet air in the reheater (HR), reducing temperature to T_6.

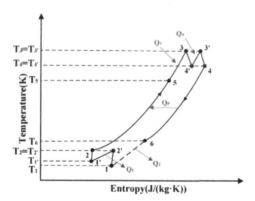

Fig. 4. Temperature-entropy diagram of Gas-air combined cycle system.

In the ideal Ericsson cycle, the heat released during processes 4-1 from the high-temperature source is supposed to be fully absorbed by the low-temperature source in processes 2-3. However, in the approximate Ericsson cycle proposed in this paper, this heat transfer is not perfect due to irreversible heat losses in the heat exchanger, low heat transfer efficiency, and other factors. As a result, some of the heat (Q_2) is discharged into the environment instead of being fully absorbed by the cold source. During the intercooling process (2'-1'), ideally, the intake temperature of the LC should be the same as the intake temperature of the HC. However, due to cooler efficiency and other factors, the intake temperature of HC is slightly higher than that of LC, as shown in Fig. 4 of the paper. Moreover, in the proposed system, when the air temperature increases after primary compression, cold water can be used in the heat exchanger to cool the air. This exchange of heat generates hot water, which can be utilized by customers through piping to improve energy use efficiency. This feature enhances the overall energy utilization of the system.

In the system secondary compression with intermediate cooling work process, assuming that the air is deemed an ideal gas, according to the definition of work can be two compressor power consumption, the formula is calculated by,

$$w_C = w_{LC} + w_{HC} = \frac{n}{n-1} R_g T_1 \left[1 - \left(\frac{p_2}{p_1} \right)^{\frac{n-1}{n}} \right] + \frac{n}{n-1} R_g T_{1'} \left[1 - \left(\frac{p_3}{p_2} \right)^{\frac{n-1}{n}} \right]$$

(6)

where w_{LC} is the power required by LC and w_{HC} is the power required by HC. n is the multivariate index. R_g is the gas constant. T_1 is LC intake temperature and $T_{1'}$ is HC intake temperature. p_1 is LC intake pressure, p_2 is LC outlet pressure and HC intake pressure, and p_3 is HC outlet pressure.

3-4 is an approximation of the expansion work process in the Ericsson cycle, and the external output power equation is calculated by,

$$w_p = w_{LE} + w_{HE} = \frac{1}{n-1} R_g T_3 \left[1 - \left(\frac{p_4}{p_3} \right)^{\frac{n-1}{n}} \right] + \frac{1}{n-1} R_g T_{3'} \left[1 - \left(\frac{p_1}{p_4} \right)^{\frac{n-1}{n}} \right]$$

(7)

where w_{LE} is the output work of LE, and w_{HE} is the output work of HE. p_3 is the intake pressure of HE, p_4 is the outlet pressure of HE, and p_1 is the outlet pressure of LE.

In the cooler, the heat absorbed by the process of 2'-1' is calculated by,

$$Q_{IC} = \zeta_{IC} \cdot c_{p,c} \cdot q_{m,c} (T_{2'} - T_{1'})$$

(8)

where ζ_{IC} is the cooler efficiency. $q_{m,c}$ is the chilling water flow rate. $c_{p,c}$ is the specific heat capacity of water. $T_{2'}$ is the LC outlet temperature and T_1 is the intake temperature of HC.

The heat input to the system from the exhaust flue gas is calculated by,

$$Q = Q_{HH} + Q_{LH} = \zeta_{HH} \cdot c_{p,g} \cdot q_{m,g} (T_{in} - T_{out})$$

(9)

where ζ_{HH} is the heater efficiency. $c_{p,g}$ is the exhaust flue gas's constant pressure-specific heat capacity. $q_{m,g}$ is the flue gas flow rate. T_{in} is the flue gas intake temperature and T_{out} is the flue gas outlet temperature.

The heat exchange in the HR is calculated by,

$$Q_r = \zeta_{HR} \cdot c_{p,h} \cdot q_{m,h} (T_4 - T_6)$$

(10)

where $q_{m,h}$ is the hot side air flow. $c_{p,h}$ hot side air constant pressure specific heat capacity, It is essential to determine the flue gas composition to reckon the constant pressure specific heat capacity. T_4 is the hot side intake temperature, and T_6 is the hot side exit temperature.

Heat dissipated by the system in the environment can be calculated by,

$$Q_{HS} = c_{p,h} \cdot q_{m,h}(T_6 - T_1) \tag{11}$$

The formula for exhaust flue gas heat is written as,

$$Q_g = c_{p,g} \cdot q_{m,g} \cdot T_{in} \tag{12}$$

The air circulation efficiency can be calculated by,

$$\eta_{Air} = \frac{w_P - w_C}{Q} \tag{13}$$

The power generation efficiency of a gas-air cycle system can be calculated by

$$\eta_{Gas-air} = \frac{(w_P - w_C) + w_{GTE}}{Q_{LNG}} \tag{14}$$

where w_{GTE} is the gas turbine generation. Q_{LNG} is the calorific value of natural gas.

The equation for the primary energy utilization of a gas-air cycle system is

$$\eta_{Per} = \frac{w_P + w_{GTE}}{Q_{LNG}} \tag{15}$$

The flue gas energy efficiency of the gas-air circulation is calculated by,

$$\eta_{Fer} = \frac{Q}{Q_g} \tag{16}$$

2.3 Model Verification

After completing the modelling using commercial software, a power cycle system from the literature [36] was selected for comparison in this essay to verify the validity of the model. This power cycle system is similar to the part of the air cycle system used in this paper for the combined cycle system, and both use ideal air as the circulating mass. Using the same parameters as in the literature (shown in Table 1), a similar part of the circulation system was simulated for separate operations and the model simulation outcomes were compared to the calculated results from the literature [36], and the outcomes are shown in Table 2. As can be viewed from Table 2, the simulation conclusions in this essay are within ± 4% relative error to the literature values, indicating that the relevant modules in the commercial software used in this paper are reliable and the simulation results of the constructed aerodynamic cycle part are valid, which provides a reliable basis for the combined mechanism of the air circulation system and the gas turbine.

Table 1. Input parameters for the validation model

Parameters	Value
Air flow	5.807 kg·s^{-1}
The intake temperature of LC	300 K
The intake pressure of LC	100 kPa
The intake temperature of HC	300 K
The intake pressure of HC	300 kPa
The outlet temperature of HH	1400 K
The outlet temperature of LH	1400 K
The intake pressure of HE	1000 kPa
The intake pressure of LE	300 kPa
Compressor efficiency	0.8
Expander efficiency	0.8

Table 2. Comparison of the simulation results in this paper with those in reference

Parameters	Literature [36]	Results of this article	Relative Error
The power consumption of LC	806.650	803.985	−0.330%
The power consumption of HC	897.240	894.664	−0.287%
The output power of HE	1948.829	1924.606	−1.243%
The output power of LE	1801.331	1779.698	−1.201%
The heat exchange of HR	3486.523	3371.028	−3.313%
The heat exchange of HH	2672.962	2641.123	−1.191%
The heat exchange of LH	1948.829	1927.761	−1.081%
Backpressure ratio	45.435%	45.856%	0.927%
Efficiency	44.274%	43.898%	−0.849%

3   Analysis of the Thermodynamic Characteristics of Gas-Air Combined Cycle Power Generation Systems

Firstly, in this paper, commercial software is selected to simulate the gas-air combined cycle power generation system, and the effects of air flow, pressure ratio, ambient temperature and gas turbine load on the combined cycle power generation system are investigated. The combined cycle system model under different operating conditions is analyzed and calculated using the established thermodynamic model. The design parameters of some components in the gas-air combined cycle power generation structure are provided in Table 3.

Table 3. Parameters of gas-air co-generation system

Parameters	Value
Air intake pressure	101.3 kPa
Cooling water temperature	20 °C
Cooling water flow	10000 kg/h
Compressor efficiency	0.78
Expander efficiency	0.8
Intercooler efficiency	0.85
Heater efficiency	0.8

3.1 Influence of Air Flow

Air flow plays a crucial role in determining the thermodynamic parameters of a gas-air cycle system. In Fig. 5, it is observed the impact of air flow rate on a gas-air combined circulation power generation system. The range of air flow rate is set from 3 kg/s to 15 kg/s. The air circulation pressure ratio is set at 1.6, the ambient temperature is 20 °C, and the gas turbine is operating at full load.

From Fig. 5(a), we can see that the power generating efficiency of the air cycle system remains constant from 3 to 8 kg/s as the flow rate increases. However, the efficiency gradually decreases after reaching 8 kg/s. In Fig. 5©, the gas generation rate remains constant, but the generating efficiency of the gas-air cycle system initially increases and then decreases. It reaches its maximum value of 31.95% when the air flow rate reaches 14 kg/s. Additionally, the proportion of air cycle generation to total generation is highest at 4.47% when the air flow rate is 14 kg/s. This is due to the gas turbine power generation maintaining a constant value of 3082.17 kW. As the air flow rate increases, the air cycle output and power consumption depicted in Fig. 5(b) show a linear increase. At 14 kg/s, the net output power of the air cycle system reaches its peak value of 500.735 kW.

Figure 5(d) demonstrates that with an increase in air flow rate, under a specific temperature and flow of exhaust flue gas, the flue gas outlet temperature of the heater decreases rapidly while the heat exchange volume increases significantly. The increase in net output work of the air circulation system is smaller than the increase in heat exchange volume in the heat exchanger, leading to a decrease in power generation efficiency after 8 kg/s. Consequently, the contribution of air circulation to the total power generation rate is significant compared to the 27.48% power generation efficiency of the gas turbine without air circulation.

Furthermore, Fig. 5(e) illustrates that as the air flow rate rises, both primary energy utilization and flue gas utilization of the system increases rapidly. This is because the increase in flow rate results in a substantial rise in air circulation output work, given a specific amount of gas turbine power generation. Under the condition of a specific amount of exhaust flue gas heat, the flue gas outlet temperature decreases significantly, leading to an increase in heat exchange in the heater and subsequently higher flue gas utilization.

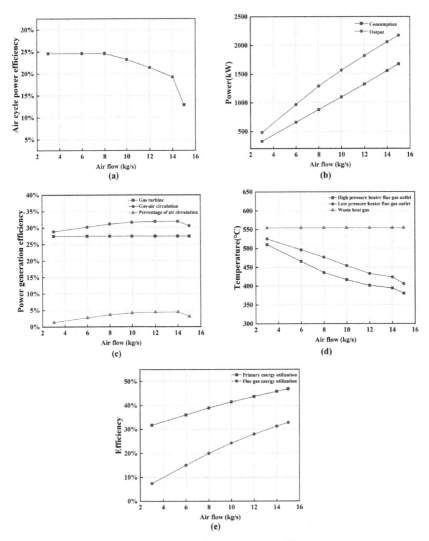

Fig. 5. Variation of thermodynamic parameters at different air flow rates

3.2 Influence of Pressure Ratio

According to Eq. (6), the pressure ratio primarily affects the power consumed by the compressor in the air cycle. Setting a mass flow rate of 7 kg/s, an ambient temperature of 20 °C, and the gas turbine operating at full load, it can be observed from Fig. 6(a) that the air cycle power generation efficiency initially increases and then decreases with an increasing pressure ratio. The maximum efficiency of 24.61% is achieved at a compressor pressure ratio of 1.6.

Figure 6(b) shows that as the compressor pressure ratio gradually increases, the power consumption, output power, and heat exchange of the air cycle also gradually increase. However, after a pressure ratio of 1.6, the increase in net output power is smaller compared to the rise in heat exchange, resulting in a decrease in air cycle power generation efficiency.

In Fig. 6(c), the power generation efficiency of the gas-air cycle also exhibits a similar trend of initially increasing and then decreasing. This is because the power generation of the gas turbine remains constant in the cycle, while the net output work of the air cycle shows an initial increase followed by a decrease. At a pressure ratio of 2.2, the net output work reaches its maximum value of 489.83 kW, resulting in a peak power

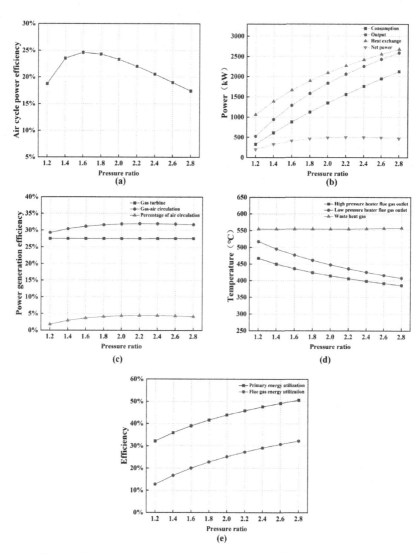

Fig. 6. Variation of thermodynamic parameters at different pressure ratio

generation rate of 31.94% for the gas-air cycle. The contribution of the air cycle to the total power generation is 4.45%.

Figure 6(d) demonstrates that, under certain flue gas flow rate and temperature conditions, the outlet temperature of the heater gradually decreases. This is due to the increased heat exchange between the air and flue gas as the compressor pressure ratio increases, resulting in a decrease in the flue gas outlet temperature. This trend is also supported by the energy utilization rate shown in Fig. 6(e), indicating that an increase in pressure ratio has a positive effect on the utilization of flue gas waste heat. The utilization rate of primary energy in the system in Fig. 6(e) also shows a substantial increase. When the gas turbine's power generation remains constant, an increase in pressure ratio leads to higher heat exchange in the heater, resulting in increased output power from the air cycle and consequently a considerable increase in the utilization rate of primary energy.

3.3 Influence of Ambient Temperature

Figure 7 presents the impact of ambient temperature on the combined gas-air circulation system. Initially, the air flow rate in the air cycle is set to 7 kg/s, the pressure ratio is 1.6, and the gas turbine operates at full load.

From Fig. 7(a), it can be seen that the air cycle power generation rate decreases with increasing ambient temperature. Similarly, the gas turbine power generation rate and the total gas-air cycle power generation efficiency also exhibit a slight decreasing trend in Fig. 7(c). According to Eq. (6), the increase in ambient temperature (T_1) leads to an increase in the work consumed by the air cycle compressor. Figure 7(b) reveals that the work consumed by the air cycle increases at a faster rate than the work output, resulting in a decrease in the net output work of the air cycle system. The heat absorbed by the air cycle, shown in Fig. 7(d), remains constant. Consequently, the air cycle power generation rate decreases.

The gas turbine power generation rate declines because the temperature rise causes a decrease in the heat from natural gas, and this decline is faster than the decrease in the gas turbine power generation, as depicted in Fig. 7(d). As a result, the total cycle power generation efficiency also decreases. This demonstrates that the impact of ambient temperature on the gas-air cycle power generation efficiency is significant. However, the proportion of power generation contributed by the air cycle gradually increases and reaches a maximum of 3.89% at 35 °C.

In Fig. 7(e), the flue gas temperature shows a tendency to increase with rising ambient temperature, and there is a slight increase in the heater outlet temperature. The flue gas energy utilization rate in Fig. 7(f) also exhibits a small increase. This is because the residual heat from the gas turbine decreases significantly (as shown in Fig. 7(d)), while the heat absorption of the air circulation system remains constant. This indicates that the ambient temperature has a minimal impact on the heat exchange rate. The slight increase in primary energy utilization is mainly due to the faster decrease in natural gas heat compared to the total cycle power generation, resulting from the increase in ambient temperature.

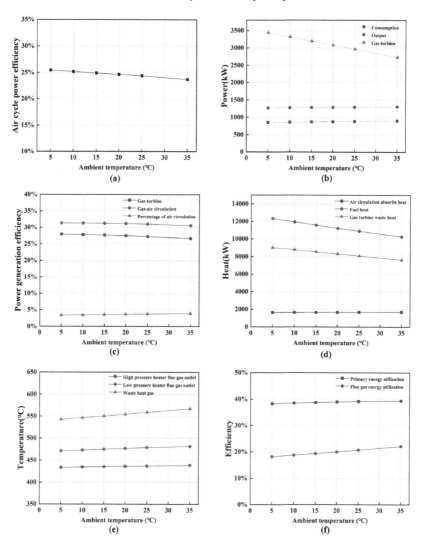

Fig. 7. Variation of thermodynamic parameters at different ambient temperature

3.4 Influence of Gas Turbine Load

Figure 8 illustrates the impact of gas turbine load on the gas-air circulation system. To better represent this trend, the gas turbine load is set at a rated power of 3000 kW, starting from 50% of the rated power. The flow rate in the air circulation is set at 7 kg/s, the pressure ratio is 1.6, and the ambient temperature is 20 °C.

From Fig. 8(a), it can be observed that the air cycle grows linearly with the increase in turbine load. This is because, under the conditions of a constant pressure ratio, air flow rate, and ambient temperature, an increase in gas load leads to higher turbine exhaust flue gas temperature and residual heat (as shown in Figs. 8(d) and 8(e)). This, in turn, increases the heat exchange between the air and flue gas in the heater, increasing the air cycle output work. However, the power consumption of the air cycle compressor remains constant at 879.3 kW, as depicted in Fig. 8(b). As a result, the net output work of the air cycle rises. Additionally, Fig. 8(d) shows a slight increase in the heat exchange of the air cycle in the heater, even though the air flow rate remains constant.

In Fig. 8(c), as the turbine load increases, the power generation rate of both the gas turbine and the gas-air cycle system also increases. The contribution of the air cycle to the total power generation reaches a maximum of 3.44% at full load, indicating its significant role in the gas-air cycle power generation system. Furthermore, the utilization rate of primary energy in the gas-air cycle system increases with the increase in gas turbine load,

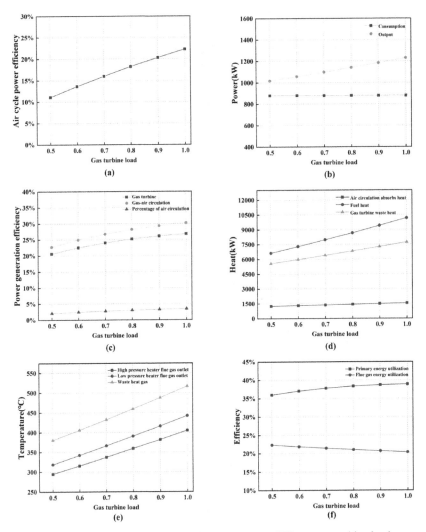

Fig. 8. Variation of thermodynamic parameters at different gas turbine load

as shown in Fig. 8(f). However, the flue gas energy utilization rate exhibits a decreasing trend. This is mainly because the increase in turbine load leads to higher output work of the air cycle and power generation of the gas turbine, increasing the system's primary energy utilization rate.

Figure 8(d) demonstrates that the residual heat of the gas turbine engine increases with the increase in gas turbine load. However, the heat absorption of the air cycle only shows a slight increase. The exit temperature of the two heaters, shown in Fig. 8(e), indicates that although the flue gas intake temperature increases significantly, the difference between the intake and outlet temperatures remains almost unchanged. This suggests that there is no significant increase in the heat absorption of the air circulation, and if it is assumed that the specific heat capacity of flue gas and flow rate is constant, according to Eq. (16), the waste heat of the gas turbine gradually increases, resulting in a decreasing trend in the rate of change of its flue gas energy utilization rate.

4   Conclusion

In this paper, a gas-air co-generation system based on an approximate Ericsson cycle is proposed to utilize the waste heat from a gas turbine and to improve the total power generation efficiency. The main conclusions from the thermodynamic analysis of this combined system are as follows:

1. The impact of air flow rate on the gas-air combined cycle power generation system is very obvious. Compared with the gas turbine power generation efficiency (27.48%), the whole power generation efficiency of the gas-air combined cycle system peaks at 31.95% with a suitable air flow, and the contribution of the air cycle to the total cycle is the largest at 4.47%. The increase of air flow rate can obviously improve the net output power, while the power generation efficiency of air circle and combined circle all increase first and then decrease.
2. Choosing the right pressure ratio is crucial for achieving optimal compressor power consumption and air cycle power generation efficiency. With the optimized pressure ratio 1.6, the maximum air cycle power generation efficiency is 24.61%, and the gas-air circulation power generation efficiency is 31.14%, with the 3.65% contribution of the air cycle. As the increase of pressure ratio of compressors, the power generation efficiency increases first and then decreases, while the energy utilization efficiency increases significantly with the decreasing of outlet temperature of smoke from the high-pressure heater and low-pressure heater.
3. The impact of ambient temperature on both the gas turbine and air cycle system is relatively large. In a lower temperature environment, the gas-air cycle power generation efficiency is higher, When the ambient temperature increases from 5 °C to 35 °C, the total power generation efficiency of the combined cycle system decreases from 31.32% to 30.55%, and the air cycle power generation share changes from 25.44% to 23.70%.
4. A higher gas turbine load leads to higher net output and power generation efficiency for the gas-air circulation system, but the flue gas utilization rate tends to decrease. Increasing the air flow rate can help maintain high flue gas waste heat utilization rates.

5. The proposed combined system can effectively utilize the waste heat utilization and power generation of gas turbines, which provides a new way to improve the total efficiency of the gas turbine It is worth mentioning that in the future, energy-saving methods such as cooling, heating and power tri-generation can also be considered to take full advantage of waste heat resources to adapt to different users' needs.

References

1. Huisingh, D., Zhang, Z., Moore, J.C., Qiao, Q., Li, Q.: Recent advances in carbon emissions reduction: policies, technologies, monitoring, assessment and modelling. J. Clean. Prod. **103**(1), 1–12 (2015)
2. Yue, T.-X., Zhao, M.-W., Zhang, X.-Y.: A high-accuracy method for filling voids on remotely sensed XCO2 surfaces and its verification. J. Clean. Product. **103**, 819–827 (2015). https://doi.org/10.1016/j.jclepro.2014.08.080
3. Johansson, V., et al.: Value of wind power – Implications from specific power. Energy **126**(1), 352–360 (2017)
4. Vargas, S.A., Esteves, G.R.T., Maçaira, P.M., Bastos, B.Q., Cyrino Oliveira, F.L., Souza, R.C.: Wind power generation: A review and a research agenda. J. Clean. Product. **218**(1), 850–870 (2019)
5. Singh, G.K.: Solar power generation by PV (photovoltaic) technology: A review. Energy **53**(1), 1–13 (2013)
6. Hayat, M.B., Ali, D., Monyake, K.C., Alagha, L., Ahmed, N.: Solar energy—A look into power generation, challenges, and a solar-powered future. Int. J. Energy Res. **43**(3), 1049–1067 (2019)
7. Remeli, M.F., Kiatbodin, L., Singh, B., Verojporn, K., Date, A., Akbarzadeh, A.: Power generation from waste heat using heat pipe and thermoelectric generator. Energy Procedia **75**(1), 645–650 (2015)
8. Kanoglu, M., Dincer, I., Rosen, M.A.: Understanding energy and exergy efficiencies for improved energy management in power plants. Energy Policy **35**(7), 3967–3978 (2007)
9. Cormos, C.-C.: Hydrogen and power co-generation based on coal and biomass/solid wastes co-gasification with carbon capture and storage. Int. J. Hydrogen Energy **37**(7), 5637–5648 (2012)
10. Toledo, O.M., Oliveira Filho, D., Diniz, A.S.A.C.: Distributed photovoltaic generation and energy storage systems: A review. Renew. Sustain. Energy Reviews **14**(1), 506–511 (2010)
11. Wang, G., et al.: Key problems of gas-fired power plants participating in peak load regulation. In: H., Shiyan (eds.) International Conference on Energy Internet 2022, ICEI, pp. 147–152. Springer, Norway (2022)
12. Badran, O.O.: Gas-turbine performance improvements. Appl. Energy **64**(1), 263–273 (1999)
13. De Sa, A., Al Zubaidy, S.: Gas turbine performance at varying ambient temperature. Appl. Therm. Eng. **31**(14), 2735–2739 (2011)
14. Liu, H., Qin, J., Ji, Z., Guo, F., Dong, P.: Study on the performance comparison of three configurations of aviation fuel cell gas turbine hybrid power generation system. J. Power Sources **501**(1), 230007 (2021)
15. Kakaras, E., Doukelis, A., Leithner, R., Aronis, N.: Combined cycle power plant with integrated low temperature heat (LOTHECO). Appl. Therm. Eng. **24**(11), 1677–1686 (2004)
16. Samarasinghe, T., Abeykoon, C., Turan, A.: Modelling of heat transfer and fluid flow in the hot section of gas turbines used in power generation: A comprehensive survey. Int. J. Energy Res. **43**(5), 1647–1669 (2019)

17. Imran, M., Haglind, F., Asim, M., Zeb Alvi, J.: Recent research trends in organic Rankine cycle technology: A bibliometric approach. Renew. Sustain. Energy Rev. **81**(1), 552–562 (2018)
18. Tchanche, B.F., Lambrinos, G., Frangoudakis, A., Papadakis, G.: Low-grade heat conversion into power using organic Rankine cycles – A review of various applications. Renew. Sustain. Energy Rev. **15**(8), 3963–3979 (2011)
19. Lecompte, S., Huisseune, H., van den Broek, M., Vanslambrouck, B., De Paepe, M.: Review of organic Rankine cycle (ORC) architectures for waste heat recovery. Renew. Sustain. Energy Rev. **47**(1), 448–461 (2015)
20. Öhman, H.: Implementation and evaluation of a low-temperature waste heat recovery power cycle using NH3 in an Organic Rankine Cycle. Energy **48**(1), 227–232 (2012)
21. Song, J., Song, Y., Gu, C.: Thermodynamic analysis and performance optimization of an Organic Rankine Cycle (ORC) waste heat recovery system for marine diesel engines. Energy **82**, 976–985 (2015). https://doi.org/10.1016/j.energy.2015.01.108
22. Wu, Z., Sha, L., Zhao, M., Wang, X., Ma, H., Zhang, Y.: Performance analyses and optimization of a reverse Carnot cycle-organic Rankine cycle dual-function system. Energy Convers. Manage. **212**(1), 112787 (2020)
23. Zhang, X., He, M., Zhang, Y.: A review of research on the Kalina cycle. Renew. Sustain. Energy Rev. **16**(7), 5309–5318 (2012)
24. Creyx, M., Delacourt, E., Lippert, M., Morin, C., Desmet, B.: Modélisation des performances d'un moteur Ericsson à cycle de Joule ouvert. Revista Termotehnica **1**(1), 64–70 (2014)
25. Bădescu, V.: Optimum operation of a solar converter in combination with a Stirling or Ericsson heat engine. Energy **17**(6), 601–607 (1992)
26. Bonnet, S., Alaphilippe, M., Stouffs, P.: Energy, exergy and cost analysis of a micro-cogeneration system based on an Ericsson engine. Int. J. Therm. Sci. **44**(12), 1161–1168 (2005)
27. Shin, J.Y., Jeon, Y.J., Maeng, D.J., Kim, J.S., Ro, S.T.: Analysis of the dynamic characteristics of a combined-cycle power plant. Energy **27**(12), 1085–1098 (2002)
28. Poullikkas, A.: An overview of current and future sustainable gas turbine technologies. Renew. Sustain. Energy Rev. **9**(5), 409–443 (2005)
29. Arrieta, F.R.P., Lora, E.E.S.: Influence of ambient temperature on combined-cycle power-plant performance. Appl. Energy **80**(3), 261–272 (2005)
30. Sharma, M., Singh, O.: Thermodynamic study of multi-pressure HRSG in gas/steam combined cycle power plant. Journal of the Institution of Engineers (India): Series C **100**(2), 361–369 (2019)
31. Bălănescu, D.-T., Homutescu, V.-M.: Performance analysis of a gas turbine combined cycle power plant with waste heat recovery in Organic Rankine Cycle. Procedia Manufac. **32**(1), 520–528 (2019)
32. Pilavachi, P.A.: Mini- and micro-gas turbines for combined heat and power. Appl. Therm. Eng. **22**(18), 2003–2014 (2002)
33. Schneider, T., Müller, D., Karl, J.: A review of thermochemical biomass conversion combined with Stirling engines for the small-scale cogeneration of heat and power. Renew. Sustain. Energy Rev. **134**(1), 110288 (2020)
34. Invernizzi, C., Iora, P., Silva, P.: Bottoming micro-Rankine cycles for micro-gas turbines. Appl. Therm. Eng. **27**(1), 100–110 (2007)
35. Blank, D.A., Wu, C.: Power limit of an endoreversible Ericsson cycle with regeneration. Energy Convers. Manage. **37**(1), 59–66 (1996)
36. Moran, M., Shapiro, H.: Ericsson and stirling cycles. In: Fundamentals of Engineering Thermodynamics, pp. 550–553. Wiley, America (2010)

Author Index

Printed in the United States
by Baker & Taylor Publisher Services